5th International Conference on Advances in Optoelectronics and Micro/Nano-optics (AOM 2015)

AA001065

Journal of Physics: Conference Series Volume 680

Hangzhou, China
28 – 31 October 2015

Editors:

Min Qiu   **Xiaocong Yuan**
Michael Fiddy

ISBN: 978-1-5108-2090-6
ISSN: 1742-6588

Printed from e-media with permission by:

Curran Associates, Inc.
57 Morehouse Lane
Red Hook, NY 12571

Some format issues inherent in the e-media version may also appear in this print version.

Copyright© (2015) by the Institute of Physics
All rights reserved. The material featured in this book is subject to
IOP copyright protection, unless otherwise indicated.

Printed by Curran Associates, Inc. (2016)

For permission requests, please contact the Institute of Physics
at the address below.

Institute of Physics
Dirac House, Temple Back
Bristol BS1 6BE UK

Phone: 44 1 17 929 7481
Fax: 44 1 17 920 0979

techtracking@iop.org

Additional copies of this publication are available from:

Curran Associates, Inc.
57 Morehouse Lane
Red Hook, NY 12571 USA
Phone: 845-758-0400
Fax: 845-758-2634
Email: curran@proceedings.com
Web: www.proceedings.com

Table of contents

Volume 680

5th International Conference on Advances in Optoelectronics and Micro/Nano-optics (AOM 2015)
28–31 October 2015, Hangzhou, China

Accepted papers received: 29 December 2015
Published online: 3 February 2016

Preface
011001

5th International Conference on Advances in Optoelectronics and Micro/Nano-optics (AOM 2015) OPEN ACCESS Min Qiu, Michael Fiddy and Xiaocong Yuan

011002

Peer review statement OPEN ACCESS

Papers

012001

A Finger Vein Identification Method Based on Template Matching OPEN ACCESS Hui Zou, Bing Zhang, Zhigang Tao and Xiaoping Wang   pg. 1

012002

Analysis of Relationship between Wavelength Selectivity and Angular Selectivity of Rugate coating OPEN ACCESS Zheng Guangwei and Wang yang   pg. 7

012003

Gold/Silicon nanowire arrays modified by Gold nanosphere as the surface-enhanced Raman spectroscopy substrate OPEN ACCESS Long Zhou, Mingyu Li, Longhua Tang and Jian-Jun He   pg. 13

012004

Silicon-on-Insulator Nanowire Based Optical Waveguide Biosensors OPEN ACCESS Mingyu Li, Yong Liu, Yangqing Chen and Jian-Jun He   pg. 17

012005

Microlens Array Diffuser with Randomly Distributed Structure Parameters OPEN ACCESS Tianyi Guo, Chao Yu, Haifeng Li, Chen Su, Yinxu Bian and Xu Liu pg. 24

012008

Reflective Characteristics of Spatially-Bounded Laser Beam by Rugate coating OPEN ACCESS Zheng Guangwei pg. 30

012009

A Novel Atomic Force Microscope with Multi-Mode Scanner OPEN ACCESS Chun Qin, Haijun Zhang, Rui Xu, Xu Han and Shuying Wang pg. 38

012010

Photoacoustic sensor of Temperature with Linear-shaped light source OPEN ACCESS Yuanyuan Peng, Shulian Wu, Dongqing Peng, Zhifang Li and Hui Li pg. 42

012011

Design of high resolution panoramic endoscope imaging system based on freeform surface OPEN ACCESS Qun Liu, Jian Bai and Yujie Luo pg. 49

012012

Full-aperture long focal-length measurement based on divergent beam OPEN ACCESS Jia Luo, Jian Bai and Kaiwei Wang pg. 56

012013

Three-Photon Luminescence of Gold Nanorods Excited by 1040 nm Femtosecond Laser for High Contrast Tissue and In Vivo Imaging OPEN ACCESS Shaowei Wang, Xinyuan Zhao, Hequn Zhang, Fuhong Cai and Jun Qian pg. 63

012014

Classification and recognition of texture collagen obtaining by multiphoton microscope with neural network analysis OPEN ACCESS Shulian Wu, Yuanyuan Peng, Liangjun Hu, Xiaoman Zhang and Hui Li pg. 68

012015

The optical design of highly efficient cold shield in IR detector based on ASAP OPEN ACCESS Xianjing Zhang, Jian Bai and Shuang Yin   pg. 77

012016

Plasmonic Fano resonances in compositional heterogenous Al- Au nanorod dimers OPEN ACCESS Botao Wu, Yingxian Xue, Qiang Ma, Chengjie Ding, Youying Rong, Yan Liu, Lingxiao Chen, E Wu and Heping Zeng   pg. 84

012017

High Accuracy Tiny Crack Detection in Metal by Low Frequency Electromagnetic Technique OPEN ACCESS Weimin Lou, Changyu Shen, Fengying Shentu, Guanghai Li, Yu Chang and Xinyuan Lu   pg. 90

012018

An optical liquid level sensor based on core-offset fusion splicing method using polarization-maintaining fiber OPEN ACCESS Weimin Lou, Debao Chen, Changyu Shen, Yanfang Lu, Huanan Liu and Jian Wei   pg. 93

012019

Fast Perturbation Monte Carlo simulation for heterogeneous medium and its utilization in functional near-infrared spectroscopy OPEN ACCESS Y M Song, J W Li and F H Cai   pg. 97

012020

Enhanced Second Harmonic Generation in AU/AI2O3/AU absorber OPEN ACCESS Fenglun Huang, Songang Bai, Qiang Li, Yurui Qu and Qiu Min   pg. 102

012021

The ordering alignment of gold nanorods in liquid crystals and its applications to polarization-sensitive SERS OPEN ACCESS Y L Wang, L Y Chen, Q K Liu, F H Cai and J Qian   pg. 107

012022

Study of humidity on the structure and optical properties of cesium iodide thin film OPEN ACCESS Bo Liang, Shuang Liu, Lina Guo, Dejun Chen, Yong Liu, Zhiyong Zhong and Liufeng Xiong   pg. 117

012023

Polarization-dependent refractive index fiber-optic sensor based on the core-offset with a taper OPEN ACCESS Youqing Wang, Changyu Shen, Weimin Lou and Fengying Shentu   pg. 122

012024

An Incorporate Ultrasonic Coupling Device for Long-Focal- Zone Photoacoustic Imaging System OPEN ACCESS Dong-qing Peng, Yuan-yuan Peng, Shu-lian Wu and Hui Li   pg. 127

012025

A New Method for Axial Decay Function Calibration of Evanescent Field in Multi-Angle Total Internal Reflection Fluorescence Microscopy OPEN ACCESS Jian Wu, Peng Xiu, Luhong Jin, Di Nan, Cuifang Kuang, Xiaoxiang Zheng, Yingke Xu and Xu Liu   pg. 131

012026

Design of visible/infrared double-band spectral imager OPEN ACCESS Tang Tianjin, Zhang Zhuo and Wang Bao-hua   pg. 136

012027

Sacrificial solder based nanowelding of ZnO nanowires OPEN ACCESS Guoping Liu, Qiang Li and Min Qiu   pg. 142

012028

Laser assisted welding of gold nanowires OPEN ACCESS Lina Zhou, Gongping Liu, Si Luo, Qiang Li and Min Qiu   pg. 147

012029

Optical microfiber-based photonic crystal cavity OPEN ACCESS Yi-zhi Sun, Yang Yu, Hui-lan Liu, Zhi-yuan Li and Wei Ding   pg. 150

012030

Identification of high explosive RDX using terahertz imaging and spectral fingerprints OPEN ACCESS Jia Liu, Wen-Hui Fan, Xu Chen and Jun Xie   pg. 158

012031

Micro-nano Structurized Gold Chip for SPR Imaging Sensor OPEN ACCESS Bing Zhang, Kai Pang, Chunfei Shi, Yi Sun, Wei Dong and Xiaoping Wang pg. 167

012032

Optical microfiber knot resonator (MKR) and its slow-light performance OPEN ACCESS Liyong Ren, Yiping Xu, Chengju Ma, Yingli Wang, Xudong Kong, Jian Liang, Haijuan Ju, Kaili Ren and Xiao Lin pg. 171

012033

Fluorescence enhancement with metamaterial mirrors OPEN ACCESS Jian Qin, Wei Wang, Si Luo, Xingxing Chen, Min Qiu and Qiang Li pg. 184

012034

Measuring acetone using microstructured optical fiber and Raman spectroscopy OPEN ACCESS Fenghong Chu and Jianping Wu pg. 187

012035

Direct phase extraction of self-mixing displacement measurement using Hilbert transform OPEN ACCESS Yufeng Tao, Ming Wang, Dongmei Guo and Jiahuan Zhang pg. 191

012036

Convex Aspherical Surface Testing Using Catadioptric Partial Compensating System OPEN ACCESS Jingxian Wang, Qun Hao, Yao Hu, Shaopu Wang, Tengfei Li, Yuhan Tian and Lin Li pg. 201

012037

Implement of Digital Moire technique on DSP for alignment of partial compensation interferometer OPEN ACCESS Yuhan Tian, QunHao, YaoHu, Shaopu Wang, Tengfei Li and Jingxian Wang pg. 208

012038

System of Thermal Micro/Nano Printing and its Application in Metallic Glass OPEN ACCESS Y. Xu, X.L. Hu, L.B. Sun, L.S. Wang, S.Q. Ding, J. Liu, J.Z. Jiang and D.X. Zhang pg. 215

012039

Ultra-broad band absorber made by tungsten and aluminium OPEN ACCESS Wei Wang, Ding Zhao, Qiang Li and Min Qiu pg. 222

PREFACE

The 5th International Conference on Advances in Optoelectronics and Micro/Nano-Optics (AOM 2015) was held in Hangzhou, China during 28-31, October 2015. It was the fifth event of a series of OSA topical meetings that began in Tianjin in 2009, and consequently took place in Guangzhou (2010), Hong Kong (2013), and Xi'an (2014). Following the successes of the previous conferences, the aim of this year's conference was to present the most recent advances in the fields of micro/nano-optics and optoelectronics, especially the new ideas and concepts in modeling, fabrication and testing of photonic materials, structures, devices and systems, and their applications.

The conference consisted of 4 plenary lectures (by Yuri Kivshar, Shanhui Fan, Susumu Noda, and Peter Nordlander, respectively), 9 keynote lectures, and 110 invited talks, in which a wide range of topics were covered and the most recent significant results were presented. To encourage the participation of young researchers, oral and poster presentations were also included. Some selected works presented in the conference are published in this edition of Journal of Physics: Conference Series.

The AOM 2015 was organized by the State Ley Laboratory of Modern Optical Instrumentation of Zhejiang University, and the Joint International Research Laboratory of Photonics of Zhejiang University, in collaboration with the Optical Society of Zhejiang Province. With 287 international participants from 17 countries, the conference was a great success. We wish to thank all the participants, and the support of the sponsors. We also gratefully acknowledge the support of K. C. Wong Education Foundation.

Finally, we would like to take this opportunity to thank the technical program committee, the local organizing committee, the reviewers, as well as the chairs of the conference sessions.

Min Qiu, *Zhejiang University, China*
Michael Fiddy, *University of North Carolina at Charlotte, USA*
Xiaocong Yuan, *Shenzhen University, China*

Content from this work may be used under the terms of the Creative Commons Attribution 3.0 licence. Any further distribution of this work must maintain attribution to the author(s) and the title of the work, journal citation and DOI.
Published under licence by IOP Publishing Ltd

Technical Program Committee

Michael Fiddy	University of North Carolina at Charlotte, USA
Qihuang Gong	Beijing University, China
Min Gu	Swinburne University of Technology, Australia
Joseph W. Haus	University of Dayton, USA
Aaron Ho	Chinese University of Hong Kong, China
Gong-Ru Lin	National Taiwan University, Taiwan
Chao Lu	Hong Kong Polytechnic University, China
Stefan A. Maier	Imperial College, UK
Ting Mei	Northwestern Polytechnical University, China
Colin Sheppard	Italian Institute of Technology, Italy
Mike Somekh	The Hong Kong Polytechnic University/University of Nottingham
Charles Surya	Hong Kong Polytechnic University, China
Kartarina Svanberg	Lund University, Sweden
Sune Svanberg	Lund University, Sweden
Jinghua Teng	IMRE, Singapore
Limin Tong	Zhejiang University
Dinping Tsai	Research Center for Applied Sciences, Academia Sinica, Taiwan
Jingjun Xu	Nankai University, China
Baoli Yao	Xi'an Institute of Optics and Precision Mechanics, CAS, China
Xiaocong Yuan	Shenzhen University, China
Andreas Zumbusch	University of Konstanz, Germany

Organizing Committee

Co-Chairs

Min Qiu	Zhejiang University, China
Michael Fiddy	University of North Carolina at Charlotte, USA
Xiaocong Yuan	Shenzhen University, China

Local Organizing committee:

Xu Liu, Liming Tong, Ke Si, Jun Lu, Liying Chen, Yanfang Lu

Planary Speakers

Yuri Kivshar	The Australian National University, Australia
Shanhuo Fan	Stanford University, USA
Susumu Noda	Kyoto University, Japan
Peter Nordlander	Rice University, USA

Session Keynote Speakers

Javier García de Abajo	The Institute of Photonic Sciences, Spain
Toshihiko Baba	Yokohama National University, Japan
Che Ting Chan	Hong Kong University of Science and Technology, Hong Kong
Harald Giessen	University of Stuttgart, Germany
Byoungho Lee	Seoul National University, Korea

Yi Luo	Tsinghua University, China
Stefan A. Maier	Imperial College, UK
Colin Sheppard	Italian Institute of Technology, Italy
Din Ping Tsai	National Taiwan University, Taiwan
Anatoly V. Zayats	King's College London, UK

Peer review statement

All papers published in this volume of *Journal of Physics: Conference Series* have been peer reviewed through processes administered by the proceedings Editors. Reviews were conducted by expert referees to the professional and scientific standards expected of a proceedings journal published by IOP Publishing.

Content from this work may be used under the terms of the Creative Commons Attribution 3.0 licence. Any further distribution of this work must maintain attribution to the author(s) and the title of the work, journal citation and DOI.

Published under licence by IOP Publishing Ltd

A Finger Vein Identification Method Based on Template Matching

Hui Zou, Bing Zhang, Zhigang Tao and Xiaoping Wang *

State Key Laboratory of Modern Optical Instrumentation, College of Optical Science and Engineering, Zhejiang University, Hangzhou, China

E-mail: xpwang@zju.edu.cn

Abstract. New methods for extracting vein features from finger vein image and generating templates for matching are proposed. In the algorithm for generating templates, we proposed a parameter-templates quality factor (TQF) - to measure the quality of generated templates. So that we can use fewer finger vein samples to generate templates that meet the quality requirement of identification. The recognition accuracy of using proposed methods of finger vein feature extraction and template generation strategy for identification is 97.14%.

1. Introduction

Biometric identification is a technology that uses human bodies' physiological characteristics to distinguish status of individuals. It is known as the most secure and convenient identification method. Fingerprint identification is the most widely used biometric technology currently. Like fingerprint, finger vein patterns of individuals are also unique and therefore can be used in biometric identification. Compared with fingerprint identification, finger vein identification has several advantages :(1) The characteristics of finger vein will not be affected by wounds or stains on skin; (2) According to medical statistics, about 5% of people's fingerprints can hardly be collected due to physiological defects, while finger vein characteristics are more universal; (3) From a security standpoint, one's finger vein characteristics can hardly be acquired by others because they lie inside of human body. With these unique advantages, finger vein identification technology is becoming a hot spot in current research. In this study, New methods to extract the features of the finger vein and synthesis templates are adopted. In the algorithm for generating templates, we proposed a parameter-templates quality factor (TQF) - to measure the quality of generated templates. The experimental results show that high recognition accuracy is achieved with proposed finger vein features extraction methods and template generation strategy for identification.

2. Materials and Methods

2.1. Image acquisition

A finger vein acquisition device is used to obtain finger vein image. Finger vein is located beneath the skin. To obtain the finger vein image, near infrared light which can penetrate fingers is selected as the light source. When near-infrared light transmits through fingers, because of the strong absorption effect on the near-infrared light of hemoglobin in the finger vein dark stripes are formed in the transmission light, which are the patterns of the finger vein.

The images acquired are 256 grey levels per pixel and 861×303 pixels in size.

Content from this work may be used under the terms of the Creative Commons Attribution 3.0 licence. Any further distribution of this work must maintain attribution to the author(s) and the title of the work, journal citation and DOI.

Published under licence by IOP Publishing Ltd

2.2. Finger vein feature extraction

Original images cannot be used for identification directly. Only after a series of image processing procedures to remove redundant information, can we extract the finger vein features for identification.

Step 1 Edge detection. We need to separate fingers from background information. The gray gradient of the image has a mutation at finger edges. By using a Sobel operator to detect finger edges, we can separate finger from background information. To reduce computation cost, all images will be normalized to 45 × 121 pixels in size.

Step 2 Vein segmentation. We need to separate fingers veins from other biological tissues in fingers. The gray value of finger veins is always less than that of other biological tissues near it, which means pixels on finger veins have local minimum gray values. Considering this phenomenon, this paper presents a "multi-directional finding method" for finger veins. For a finger vein image I, as shown in Fig.1,

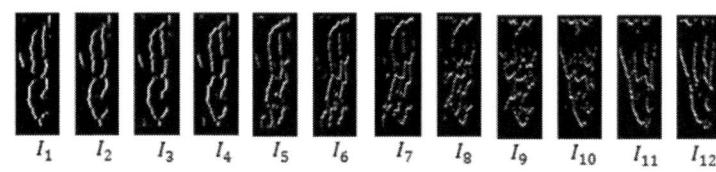

Fig 1. Finger vein image.　　　Fig 2. Finger vein pixels found in different direction.

Starting with 0 degree (the horizontal direction), according to counter-clockwise order, we search for finger vein pixels which have local minimum gray values every 15 degrees. Results of each search are stored in matrix I_i (i=1,2,3, •••, 12) , as shown in Fig.2.

$$I_i(x,y) = \begin{cases} 1, & I(x,y)\, is\, finger\, vein\, pixel \\ 0, & otherwise \end{cases} \tag{1}$$

Searching will be conducted for 12 times in the range of $[0, \pi)$. Then superimposing these 12 matrixes, as shown in Fig.3, we can have I_{sum}, namely:

$$I_{sum}(x,y) = \sum_{i=0}^{11} I_i(x,y) \tag{2}$$

Considering the existence of noise, let

$$I'_{sum}(x,y) = \begin{cases} 1, & I_{sum}(x,y) > 1 \\ 0, & otherwise \end{cases} \tag{3}$$

Then processing graphics operation on I'_{sum}, we can get binarized finger vein lines, as shown in Fig.3.

Fig 3. Binarized finger vein lines.

Step 3 Thinning finger vein lines. To reduce computation cost, extracted finger vein lines will be thinned to one pixel wide. Below is the introduction of the thinning algorithm. For pixels in a binarized finger vein image, the gray value of a pixel is 1 if this pixel lies in finger vein lines, otherwise the gray value is 0. For a pixel P_0, we define its 8-neighborhood as shown in Fig.4.

P_1	P_2	P_3
P_8	P_0	P_4
P_7	P_6	P_5

Fig 4. 8-neighborhood of a pixel.

Where R_i is the gray value of P_i.

Then we can traverse every nonzero pixel in a binarized finger vein image. A pixel will be deleted (setting its gray value to 0) if the following conditions are satisfied:

$$\begin{cases} 1 < n_{number} < 6 \\ t_{number} = 2 \end{cases} \tag{4}$$

Where n_{number} is the number of nonzero neighbors in P_0' 8-neighborhood, that is,

$$n_{number} = \sum_{i=0}^{8} R_i \tag{5}$$

And t_{number} is the number of 0-1 transitions in P_0' 8-neighborhood, that is,

$$t_{number} = \sum_{i=0}^{8} |R_{i+1} - R_i| \quad (\text{Let } R_9 = R_1) \tag{6}$$

Repeating the above processes until all pixels that need to be get rid of are set to 0. The thinned finger vein features are as shown in Fig.5.

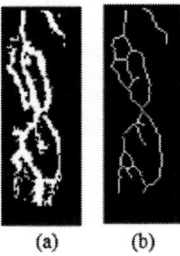

(a)   (b)

Fig 5. (a) Finger vein lines; (b) finger vein features.

2.3. Generating templates

Due to the inevitable discrepancies of illumination condition, finger position and side movement in each sampling, there are some delicate differences in each vein samples of the same finger. To avoid the impact of discrepancies on identification accuracy, we sample a finger for a several times, thus sampling results will contain the information of above discrepancies. Then we abstract finger vein features from these sampling results. Abstracted vein features of one finger will be used to synthesize a feature template. By using this template in the matching process, we can eliminate the influence of above discrepancies on identification accuracy in sampling. Specific steps are: Firstly, sampling procedure will be conduct m times of one finger. And we abstract m vein features from the sampling results, vein features are denoted by

$$C_i(i = 1, 2, \cdots, m) \tag{7}$$

Then superimpose these vein features, we can have C_T, namely:

$$C_T(x, y) = \sum_{i=0}^{m} C_i(x, y) \tag{8}$$

A higher value of a pixel in C_T indicates a greater probability of vein feature appears in this position. Some pixels in C_T have relatively small values, which indicates the probabilities of vein feature appears in these positions are not stable. If these pixels involved in matching process, they may affect the accuracy. So these pixels must be eliminated (setting its gray value to 0). Considering this, let

$$C_T'(x, y) = \begin{cases} 1, & C_T(x, y) > 1 \\ 0, & \text{otherwise} \end{cases} \tag{9}$$

C_T' is the synthesized finger vein feature template.

For example, let m=6, which means the sampling procedure will be conduct 6 times of one finger. Next, we abstract 6 vein features from the sampling results. Then we superimpose these 6 vein features and get C_T, as shown in Fig.6. After processing C_T according to equation (9), we can get C_T', which is the synthesized finger vein feature template of the given finger.

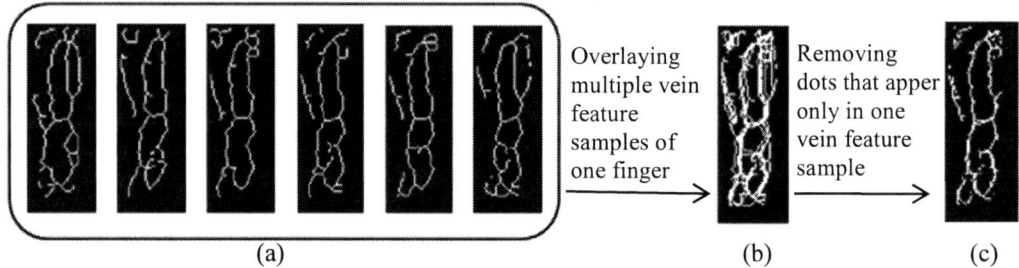

Fig 6. (a) Vein feature samples of one finger; (b) composition of vein feature samples; (c) template.

The sampling number m is a critical parameter. If the sampling number too small, the quality of generated template will be poor because of containing inadequate discrepancies. However, if the sampling number too large, the sampling time will increase significantly while the recognition accuracy will not improve much. In order to determine the appropriate value of m, we proposed an indicator-Template quality factor (TQF) - to measure the quality of generated template. TQF is defined as

$$\text{TQF} = \frac{n_1}{n_0} \cdot 100 \tag{10}$$

Where n_1 is the number of nonzero pixels in C'_T, n_2 is the number of nonzero pixels in C_T.

TQF indicates the randomness and certainty of finger vein's occurrence at a certain position in the template. If TQF is too small, the generated template will contain too much randomness while lack of certainty, therefore cannot be used for matching. A relatively larger TQF indicates the generated template contains adequate certainty. In practice, TQF increased with the increment of samples involved in generating the template. However, too large a value of TQF will lower the recognition accuracy because the template has little tolerance, and will obviously increase the sampling time. Using a different number of samples to generate templates and comparing the recognition accuracy of those templates, we found that the template quality can achieve the recognition accuracy when the TQF value is 35. Strategies for generating templates are: Firstly, we use 3 samples to generate a template, if the TQF no less than 35, the template is available. Otherwise, we add one more sample and generate a new template. Repeating above steps until the TQF of generated template is no less than 35.

2.4. Matching strategy

We use modified Hausdorff distance (MHD) to match samples and templates. MHD is defined as in equation (11)

$$H(A, B) = \frac{1}{N_A} \sum_{a \in A} \min_{b \in B} ||a - b|| \tag{11}$$

Where point set A contains all nonzero pixels of a template in sequential order, point set B contains all nonzero pixels of a feature sample in sequential order. a and b are elements from points A and B. $||a - b||$ represents the Euclidean distance between point sets A and B. $H(A, B)$ indicates the similarity between point sets A and B, the less $H(A, B)$ is, the more similar point sets A and B are.

For a new finger vein feature sample needs to be identified, we calculate the MHD between it and every saved template. Then we compare calculated MHD values and find out the minimum MHD value. Template that corresponding to the minimum MHD value will be identified as the matching template of input vein feature sample. The advantage of this method is that it can overcome the deflection and inclination of fingers in each sampling. To ensure the accuracy of recognition, a finger to be recognized needs to be sampled 3 times. Thus we can extract 3 finger vein feature samples for this finger. Each sample will be processed with the method previously mentioned to identify its matching template. Recognition will not success unless more than one results out of three are the same. Footnotes

Footnotes should be avoided whenever possible. If required they should be used only for brief notes that do not fit conveniently into the text.

3. Experiment result

To test the performance of the proposed method, 35 subjects provide their finger vein images. Each finger was sampled 10 times across different sessions. All finger vein images are from the finger vein acquisition device we designed. Each pixel of images has 256 grey levels. Images are scaled into 45×121 pixels.

All sampled images are processed into finger vein features. Each finger has 10 features, we randomly select 7 of them to generate a template, and the remaining 3 feature are used for testing. The recognition accuracy of our matching strategy is 97.14%.

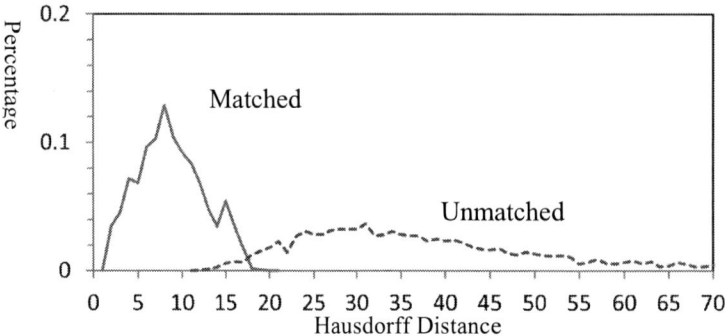

Fig 7. Distribution of Hausdorff distances in each test.

The distribution of Hausdorff distances in each test is shown in Fig.7. X axis indicates the Hausdorff distance, and the Y axis indicates the percentage of the corresponding distance. As shown in Fig.8, if samples match templates, the Hausdorff distance values are mainly distributed in the region of 5 to 15, if samples mismatch templates, the Hausdorff distance values are mainly distributed in the region of 20 to 60. These two distribution curves barely intersect with each other, indicating that this method can effectively distinguish between different finger vein samples.

4. Conclusion

In this study, a new finger vein identification method based on template matching is proposed. Firstly, we use a finger vein acquisition device to obtain finger vein image. Secondly, a series of image processing procedures are carried out to extract the features of finger vein. New methods for extracting finger vein features and generating templates for matching are proposed. Template quality factor (TQF) is proposed as an indicator to measure the quality of generated template. The similarity between sample and each feature template is determined by the modified Hausdorff distance (MHD). Experimental results show that this matching method has high accuracy.

Though the database used in this study is relatively small, and it is inadequate to confirm that this method has high recognition accuracy in the process of mass population (in terms of million users). The experimental results do indicate that finger vein recognition, as a new recognition technology, is feasible and of great long term potential. With continuous research, finger vein identification technology will be developed more mature, and will be able to be applied to many situations of social production and the lives of people.

References
[1] WANG Jun et al 2010 A novel genetic programming based morphological image analysis algorithm *New York* 979-980
[2] Yu C B et al 2009 Finger-vein image recognition combining modified hausdorff distance with minutiae feature matching *Interdisciplinary Sciences: Computational Life Sciences* **4** 280-289
[3] Yang G et al 2012 Finger vein recognition based on a personalized best bit map *Sensors* **12** 1738-1757
[4] Qin H et al 2013 Finger-vein verification based on multi-features fusion *Sensors (Basel, Switzerland)* **13** 15048-15067
[5] Wang K et al 2011 A finger vein recognition method using improved oriented filter and modified hausdorff distance *Journal of Computer-Aided Design and Computer Graphics* **3** 385-391
[6] Liu T et al 2015 Finger-vein recognition with modified binary tree model *Neural Computing and Applications* **26** 969-977

Analysis of Relationship between Wavelength Selectivity and Angular Selectivity of Rugate coating

Zheng Guangwei[1], Wang yang[2]

[1] Information and Navigation College, Air Force Engineering University, 710077, Xi'an, Shannxi, China

[2] China Satellite Maritime Tracking and Control Department, Jiangyin, Jiangsu, China

Email: zgw198196@126.com

Abstract. Based on Bragg law，Airy's formulae, and second-order Taylor series expansion, the relationships between wavelength selectivity and angular selectivity of the ordinary and phase-shifted Rugate coatings are investigated, respectively. And their expressions of the wavelength selectivity bandwidth and the angular selectivity bandwidth of these two types of Rugate gratings are put forward. The results show that when the incidence angle is far away from 0 rad, the bandwidth of the wavelength selectivity is proportional to that of the angular selectivity. And when the incidence angle approaches or even equals 0 rad (frequently-used cases), the bandwidth of the wavelength selectivity is squarely proportional to that of the angular selectivity. The results are instructive for the design and application of Rugate coatings.

1. Introduction

Rugate coating is a kind of thin film with the refractive index changing continuously along its thickness[1]. Compared with the traditional graded coatings, Rugate coating has a lot of merits, such as no or few harmonious reflective bands, low stress between different layers, high laser damage threshold, and so on[2-4]. With the development of the modern fabrication methods, the characteristics of Rugate coatings are analyzed extensively[5-8]. Nowadays, the most widely-used spatial filter is pinhole filter. Due to its focusing characteristics, pinhole filter cannot be adjusted easily to the application for the high power laser beam. So the non-focusing method has attracted attention. Due to the fine wave vector selectivity of the grating, the non-spatial filtering for laser beam based on gratings has been put forward since 2000s[9-12]. The fine wave vector selectivity denotes the fine wavelength selectivity at some angular domain and the fine angular selectivity at some wavelength domain. So Rugate coatings may be a potential candidate for the pinhole filter, especially in the high power laser field, with its fine angular selectivity[13]. However, there is little attention paid to the wavelength selectivity for the performance degradation of Rugate coatings' application upon the spatial filter of laser beam, especially upon the laser pulse. In order to make the Rugate coating suitable for the spatial filter of the short pulse laser, its relationship between the angular selectivity and wavelength selectivity is analyzed in order to search for structure of Rugate coatings with the fine angular selectivity and weak wavelength selectivity. From the refractive index's distribution, Rugate coating can be categorized into four types, such as ordinary, chirped, apodized, and phase-shifted types. In this paper, we just analyze the ordinary Rugate coating and phase-shifted Rugate coating. The other two types can be investigated with the similar method.

2. The definition of angular and wavelength selectivity bandwidths

For each Rugate grating, there is a wave vector at which the transmittance or reflectance efficiency is the highest. So we define the wave vector as the central wave vector k_0. And there must be the incidence angle θ_0 and wavelength λ_0 in corresponding to the central wave vector. The angular selectivity bandwidth $\Delta\theta$ is defined as the two times incidence angle when the transmittance or reflectance efficiency η descends to one half of its maximum η_{max}. And $\Delta\theta$ can be expressed as

$$\begin{cases} \eta(\lambda_0, \theta_0) = \eta_{max} \\ \eta(\lambda_0, \theta_0 + \Delta\theta/2) = \eta_{max}/2 \end{cases} \tag{1}$$

And with the similar definition, the wavelength selectivity bandwidth $\Delta\lambda$ can be expressed as

$$\begin{cases} \eta(\lambda_0, \theta_0) = \eta_{max} \\ \eta(\lambda_0 + \Delta\lambda/2, \theta_0) = \eta_{max}/2 \end{cases} \tag{2}$$

where the parameters has the same definition as those in equation (1).

3. The relationship between angular and wavelength selectivity of ordinary Rugate coating

The reflection of laser beam by ordinary Rugate coating is shown in figure 1.

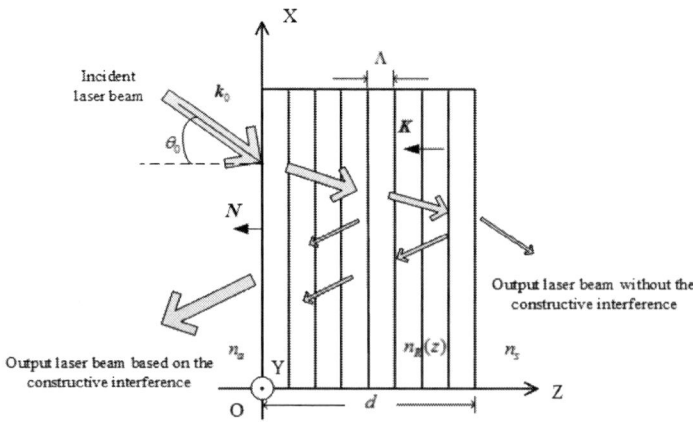

Fig1. Reflection of laser beam by ordinary Rugate coating

In figure 1, N denotes the normal to the front surface. n_a and n_S denote the refractive indexes of the ambient media and substrate, respectively. d is the thickness. K denotes the coating's vector, and is equal to $2\pi/\Lambda$, where Λ is the period of the coating. k_0 denotes the central wave vector of the incident laser. And θ_0 is the incidence angle. The refractive index of ordinary Rugate coating is sinusoidally modulated. It can be expressed as

$$n_R(z) = n_0 + n_1 \sin(Kz) \tag{3}$$

where n_0 and n_1 denote the average and modulated refractive index, respectively. From equation (3), we can conclude that the structure of the ordinary Rugate coating is the same as that of reflecting volume phase grating[14].

Due to the fact that the structure of the ordinary Rugate coating is the same as that of the reflecting volume phase grating, the reflecting characteristics of ordinary Rugate coating can be analyzed by Bragg law. From Snell's theorem and Bragg law, the central wave vector must satisfy the following equations, which can be expressed as:

$$n_a \sin\theta_0 = n_0 \sin\theta_0{'} \tag{4}$$

$$2n_0 \Lambda \cos\theta_0{'} = \lambda_0 \tag{5}$$

where $\theta_0{'}$ denotes the angle of the incidence angle θ_0 in the Rugate coating.

We make the second order Taylor series expansion for λ with the variable θ at the point $(\lambda_0, \theta_0{'})$. And then λ can be expressed as:

$$\Delta\lambda = -2n_0 \Lambda \sin\theta_0{'}\Delta\theta{'} - n_0 \Lambda \cos\theta_0{'}\Delta\theta{'}^2 + o[\Delta\theta{'}^2] \tag{6}$$

where $\Delta\theta$ denotes $\theta{'} - \theta_0{'}$, while $\Delta\lambda$ denotes $\lambda - \lambda_0$. $o[\Delta\theta{'}^2]$ denotes the higher-order infinitesimal of $\Delta\theta{'}^2$.

According to the angle θ_0, there are three situations classified for the relationship between wavelength and angular selectivity.

3.1. $\boldsymbol{\theta_0 = 0}$

When θ_0 is equal to 0, the equation (6) can be expressed as:

$$\Delta\lambda = -n_0 \Lambda \Delta\theta{'}^2 \tag{7}$$

where the higher-order infinitesimal of $\Delta\theta{'}^2$ is omitted.

We make the derivative for the equation (4), which is expressed as:

$$\Delta\theta{'} = [n_a \cos\theta_0/(n_0 \cos\theta_0{'})]\Delta\theta \tag{8}$$

Due to $\theta_0 = 0$, $\Delta\theta{'} = (n_a/n_0)\Delta\theta$.

So the relationship between wavelength selectivity bandwidth and angular selectivity bandwidth can be expressed as:

$$\Delta\lambda = -(n_a^2 \Lambda/n_0)\Delta\theta^2 \tag{9}$$

The result shows that the wavelength selectivity bandwidth is squarely proportional to the angular one, when the central vector is normally incident into the ordinary Rugate coating.

3.2. $\boldsymbol{\theta_0 \in (0, 0.1rad)}$

According to equations (4) and (8), the absolute quotient of the first two terms of the right side of equation (6) can be expressed as:

$$|2n_0 \Lambda \sin\theta_0{'}\Delta\theta{'}/(n_0 \Lambda \cos\theta_0{'}\Delta\theta{'}^2)| = |2\tan\theta_0{'}/\Delta\theta{'}| = |2\tan\theta_0/\Delta\theta| \tag{10}$$

Usually, $\Delta\theta$ is less than $0.1rad$. So when $\theta_0 \in (0, 0.1rad)$, the quantity of (10) is less than 1. The second term of the right side of (6) cannot be omitted. And with the omission of the higher order infinitesimal of $\Delta\theta{'}^2$ and equation (4), (5), and (8), $\Delta\lambda$ can be expressed as:

$$\Delta\lambda = \{(4n_0^2 \Lambda^2 - \lambda_0^2)[4(n_a^2 - n_0^2)\Lambda^2 + \lambda_0^2]\}^{\frac{1}{2}}\frac{1}{\lambda_0}\Delta\theta$$

$$+[4(n_a^2 - n_0^2)\Lambda^2 + \lambda_0^2]\frac{1}{2\lambda_0}\Delta\theta^2 \qquad (\frac{\lambda_0}{2n_0} < \Lambda < \frac{\lambda_0}{2(n_0^2 - n_a^2)^{\frac{1}{2}}}) \tag{11}$$

3.3. $\theta_0 \in [0.1, \pi/2rad)$

When $\theta_0 \in [0.1, \pi/2rad)$, the quantity of (10) is much larger than 1. So the second term of the right side of (10) and the higher order infinitesimal of $\Delta\theta'^2$ can be omitted, $\Delta\lambda$ can be expressed as:

$$\Delta\lambda = \{(4n_0^2\Lambda^2 - \lambda_0^2)[4(n_a^2 - n_0^2)\Lambda^2 + \lambda_0^2]\}^{\frac{1}{2}}\frac{1}{\lambda_0}\Delta\theta \quad (\frac{\lambda_0}{2n_0} < \Lambda < \frac{\lambda_0}{2(n_0^2 - n_a^2)^{\frac{1}{2}}}) \quad (12)$$

The result shows that the wavelength selectivity bandwidth is proportional to the angular one at this situation.

4. The relationship between angular and wavelength selectivity of phase-shifted Rugate coating

The transmission of a laser beam by phase-shifted Rugate coating is shown in figure 2.

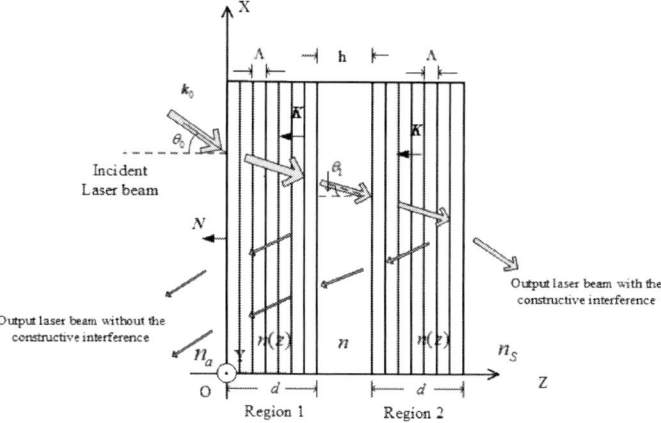

Fig2. Transmission of a laser beam by phase-shifted Rugate coating

The parameters in figure 2 are defined similarly as that in figure 1. Region 1 and region 2 are identical to each other. And their average refractive index is the same as n.

From the point that the two highly reflective coatings are placed parallel, the structure of the phase-shifted Rugate coating is similar to the Fabry-Perot interferometer. And their optical characteristics are similar, except that the Fabry-Perot interferometer has transmission harmonics, while Rugate coating has not. We just analyze the relationship between the wavelength bandwidth and angular bandwidth of the phased-shifted Rugate coating. So by Airy's formulae and without loss, the intensity of the transmitted laser beam can be expressed as[15]

$$I^T = (1 + F\sin^2\frac{\delta}{2})^{-1} \quad (13)$$

where

$$F = 4R/(1-R)^2 \quad (14)$$

$$\delta = 4\pi nh\cos\theta_1/\lambda \quad (15)$$

R denotes the reflectivity of region 1.

When I^T reduces to the half of its maximum, δ changes to δ', which can be expressed as

$$\delta' = 2\arcsin(F^{-0.5}) \quad (16)$$

From equation (15), the relationship between λ and θ_1 is expressed as:

$$2\pi nh\cos\theta_1/\arcsin(F^{-0.5}) = \lambda \quad (17)$$

Comparing equation (17) and (5), the relationships between λ and θ_0 of these two types of Rugate coatings are similar to each other. So through the similar method, we can deduce the relationship between the wavelength selectivity bandwidth and angular selectivity bandwidth of phase-shifted Rugate coating. The results are as follows.

When $\theta_0 = 0$, their relationship can be expressed as

$$\Delta\lambda = -\{2n_a^2\pi h/[n_0\arcsin(F^{-0.5})]\}\Delta\theta^2 \tag{18}$$

When $\theta_0 \in (0,0.1\mathrm{rad})$, their relationship can be expressed as

$$\Delta\lambda = \left\{\left(4n_0^2[\frac{\pi h}{\arcsin(F^{-0.5})}]^2 - \lambda_0^2\right)\left[4(n_a^2 - n_0^2)[\frac{\pi h}{\arcsin(F^{-0.5})}]^2 + \lambda_0^2\right]\right\}^{\frac{1}{2}}\frac{1}{\lambda_0}\Delta\theta$$

$$+[4(n_a^2 - n_0^2)[\frac{\pi h}{\arcsin(F^{-0.5})}]^2 + \lambda_0^2]\frac{1}{2\lambda_0}\Delta\theta^2$$

$$(\frac{\lambda_0}{2n_0} < \frac{\pi h}{\arcsin(F^{-0.5})} < \frac{\lambda_0}{2(n_0^2-n_a^2)^{\frac{1}{2}}}) \tag{19}$$

When $\theta_0 \in [0.1, \pi/2\mathrm{rad})$, their relationship can be expressed as

$$\Delta\lambda = \left\{\left(4n_0^2[\frac{\pi h}{\arcsin(F^{-0.5})}]^2 - \lambda_0^2\right)\left[4(n_a^2 - n_0^2)[\frac{\pi h}{\arcsin(F^{-0.5})}]^2 + \lambda_0^2\right]\right\}^{\frac{1}{2}}\frac{1}{\lambda_0}\Delta\theta$$

$$(\frac{\lambda_0}{2n_0} < \frac{\pi h}{\arcsin(F^{-0.5})} < \frac{\lambda_0}{2(n_0^2-n_a^2)^{\frac{1}{2}}}) \tag{20}$$

5. Conclusion

According to equation (9), (11), (18) and (19), we can conclude that for ordinary or phase-shifted Rugate coatings, when $\theta_0 \in [0,0.1\mathrm{rad})$, the bandwidth of the wavelength selectivity is almost squarely proportional to that of the angular selectivity. It means that when the incidence angle approaches or even equals 0 rad, both ordinary and phase-shifted Rugate coatings have fine wavelength selectivity and weak angular selectivity. So they can not be used as the spatial filter for the laser pulse at this situation. When the incidence angle is far away from 0 rad, the bandwidth of the wavelength selectivity is proportional to that of the angular selectivity. In order to analyze the potential substitute for pin-hole spatial filter by Rugate coating, we should design Rugate coating with fine angular selectivity and weak wavelength selectivity. With the similar method to design Rugate coating in the optical spectrum domain, we can use needle algorithm to synthesize the well-performed coating structure in the angular spectrum domain.

Acknowledgement

The authors acknowledge the support of National Natural Science Foundation of China (No.61205002).

References

[1] Bertrand G. Bovard 1993 Rugate filter theory: an overview *Appl. Opt* **32** 5427

[2] W. H. Southwell and Randolph L. Hall 1989 Rugate filter sidelobe suppression using quintic and rugated quintic matching layers *Appl. Opt* **28** 2949

[3] W. H. Southwell 1988 Spectral response calculations of rugate filters using coupled-wave

theory *Appl. Opt* **28** 2949

[4] W. H. Southwell 1989 Using apodization functions to reduce sidelobes in rugate filters *Appl. Opt.* **28** 5091

[5] Andy C. van Popta et al 2004 Gradient-index narrow-bandpass filter fabricated with glancing-angle deposition *Optics Letters* **29** 2545

[6] Stephan Fahr et al 2008 Rugate filter for light-trapping in solar cells *Optics Express* **16** 9332

[7] P. V. Usik et al 2009 Spatial and spatial-frequency filtering using one-dimensional graded-index lattices with defects *Optics Communications* **282** 4490

[8] Julien Lumeau et al 2008 Phase-shifted volume Bragg gratings in photo-thermo-refractive glass *Proc. Of SPIE* **6890** 68900A-1

[9] Ivan Moreno and J. Jesus Araiza 2004 Thin-film optical filters for spatial frequencys *Proc. of SPIE* **5524** 409

[10] Ivan Moreno et al 2005 Thin-film spatial filters *Optics Letters* **30** 914

[11] Zhang Ying et al 2015 An improved transmitting multi-layer thin-film filter *Chinese Physics B* **24** 054212-1

[12] Zhang Ying et al 2015 Analysis of the spatial filter of a dielectric multilayer film reflective cutoff filter-combination device *Chinese Physics B* **24** 104216-1

[13] Luo Zhaoming et al 2010 Low-pass rugate spatial filters for beam smoothing *Optics communications* **283** 2665

[14] H. Kogelnik 1969 Coupled wave theory for thick hologram gratings *The Bell Syst. Technol. J* **48** 2909

[15] Max Born and Emil Wolf 1999 Principles of Optics(7th edition) Cambridge University Press 325

AOM2015   IOP Publishing
Journal of Physics: Conference Series **680** (2016) 012003   doi:10.1088/1742-6596/680/1/012003

Gold/Silicon nanowire arrays modified by Gold nanosphere as the surface-enhanced Raman spectroscopy substrate

Long Zhou, Mingyu Li, Longhua Tang and Jian-Jun He

State Key Laboratory of Modern Optical Instrumentation, College of Optical Science and Engineering,Zhejiang University, Hangzhou, China 310027.

Email:limy@zju.edu.cn

Abstract. Fabrication of gold coated silicon nanowires (AuSiNW) substrate is introduced in detail and A hybrid substrate is designed for surface-enhanced Raman spectroscopy (SERS). The SERS behaviors are discussed and compared by the detection of 4, 4'-Bipyridine. Gold nanoparicles is modified on the surface of AuSiNW to form the "hot gap". Molecules captured in these "hot gap" can generate huge Raman signal. Double-fold enhancement of SERS signal has been achieved comparing with AuSiNW. The as-fabricated hybrid substrate exhibit high SERS sensitivity, long-term stability, and consistent reproducibility, highly potential for realizing a rapid, cost-effective, and label-free SERS-based biosensor.

1. Introduction

Surface-enhanced Raman spectroscopy (SERS) has gained prominence for sensitive and selective molecular identification, widely used in clinical diagnosis, food safety inspection, and environment monitoring [1-4]. Since the first demonstration of the SERS on the silver electrode in 1974 [5], many efforts have been made to fabricate a SERS substrate with ultra sensitivity, good stability, well homogeneity and perfect affinity.[6] Recently, the fast evolution of nanotechnology has reinvigorated this field. By virtue of the dramatically enhanced electromagnetic field in the proximity of nanostructured metal surfaces, ultrasensitive detection has been reported using substrates of grapheme oxide/silver nanoparticles/silicon pyramid [7], gold nanoparticles (AuNPs) [8], and Gold nanostars [9]. Up to now, electromagnetic mechanism (EM) is widely accepted as the mainly mechanisms of SERS[10]. However, challenges remain in the issues of reproducibility, throughput and stringent requirements of equipment [11].

The surface plasmons can be excited between two particles by the incident light and further contribute to the EM enhancement. A metal film over nanosphere (MFON) was proposed to improve the stability of nanostructure, where Raman signal around the hot spots can be enhanced about 10^8 fold due to the coupling of localized surface plasmon resonances [12]. A recent study of AgNPs SERS substrate shows that the molecules in the hot spots have $\sim10^7$ fold enhancement while those in not "hot" area only possess $\sim10^5$ fold enhancement [13]. Nevertheless, the active "hot-spot" areas in SERS sensors are on the scale of nanometers, making it difficult to directly detect analyte molecules diluted to femto- or atto-molar concentrations. Efforts have been made to concentrate molecules in the desired locations of hot-spots [14]. Many methods including photolithography, reactive ion etching and metal-assisted

---

Content from this work may be used under the terms of the Creative Commons Attribution 3.0 licence. Any further distribution of this work must maintain attribution to the author(s) and the title of the work, journal citation and DOI.
Published under licence by IOP Publishing Ltd

13

wet etching have been used to fabricate the silicon nanowire arrays [15]. These methods requires complex steps, dangerous reagent or expensive facility.

In this letter, silicon nanowire arrays were fabricated by polymer sphere (PS) template with short assay time and high throughput. Enhancement of SERS signal of 4, 4'-Bipyridine was measured after modifying AuNP on the gold coated silicon nanowires (AuSiNW) surface. The improved SERS signal intensity using the hybrid substrate (AuNP/AuSiNW) compared to the AuSiNWs substrate was demonstrated. This result can be well extended to open an avenue for developing various SERS-based biosensors.

2. Experiment

A self-assemble at the water-air interface technique was used to produce monolayers of PS on silicon wafer. [16] After achieving the monolayer of PS on the silicon substrate, inductively coupled plasma (ICP) etching was performed to obtain the silicon nanowires (SiNW). The PSs with reduced diameter were used as mask for etching silicon nanowires with 360 nm in diameter and 650 nm in length by STS Multiplex ICP system. For the silicon etching, SF6 and C4F8 mixed gases were applied for 6 min with flow rates of 10 sccm and 4.9 sccm, respectively. Coil power and platen power were set as 400 W and 20 W, respectively. 2 min oxygen etching was performed afterwards to remove the remaining PS. The SiNWs were coated with a 3 nm ~ 5 nm Titanium film and a 20 nm thick gold film by sputtering evaporation to form AuSiNWs. Then AuSiNWs were immersed in a mixture solution of ethanol and 4, 4'-Bipyridine with 1 mmol/L for 1 h. After cleaning by pure ethanol to lift off the spare 4, 4'-Bipyridine (4.4'-bipy) molecule, the substrate was immersed in the AuNP solution for 1 h, so that the AuNP can be adsorbed on the gold film with the assistance of 4.4'-bipy.

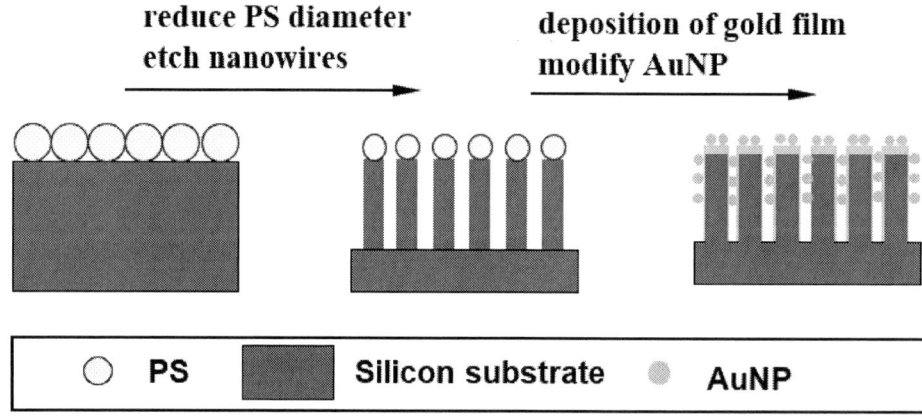

Figure.1. Schematic depiction of the fabrication process.

A schematic illustrating the process of making AuNP/AuSiNWs was shown in Figure.1. Tilt-view scanning electron microscope (SEM) image of PS was shown in Figure.2 (a). The particle size of PS is 500nm in standard. From Figure.2 (a), we can observe the PS is highly ordered and compact in a considerable area. Due to the self-assemble at the water-air interface technique, we can rapidly and conveniently get such a monolayer of PS. To fabricate expected SiNW, PS mask should be disposed of its size to obtain given duty cycle. With the help of PS mask and ICP technology, we can realize a controlled diameter and length SiNW arrays. Figure.2 (b) shows us the surface topography of AuNP/AuSiNW, we can clearly see those AuNP attached on the surface of AuSiNW with the help of 4, 4'-Bipyridine molecule. In Figure.2 (b), the diameter of the used AuNP is about 60 nm.

AOM2015   IOP Publishing
Journal of Physics: Conference Series **680** (2016) 012003   doi:10.1088/1742-6596/680/1/012003

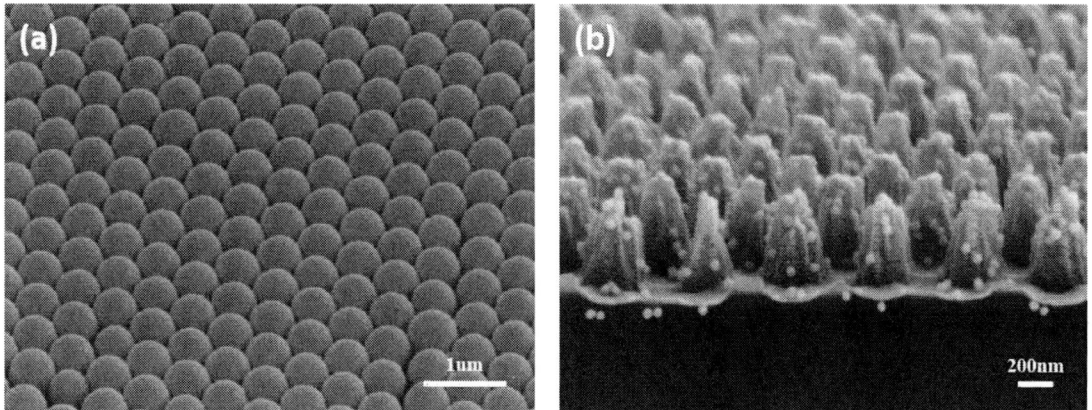

Figure.2. (a) Tilt-view SEM image of PS with 500 nm in diameter of PS; (b) Tilt-SEM image of AuSiNWs attached with AuNP.

3. Results and discussion

4, 4'-Bipyridine was used as signal molecules and was adhered to the gold film with concentration of 1mmol/L. LabRAM HR Evolution was used to detect the SERS signal. Excitation wavelength was set to be 633 nm. SERS spectra of the AuSiNW substrate and the AuNP/AuSiNW substrate were plotted in Figure. 3.

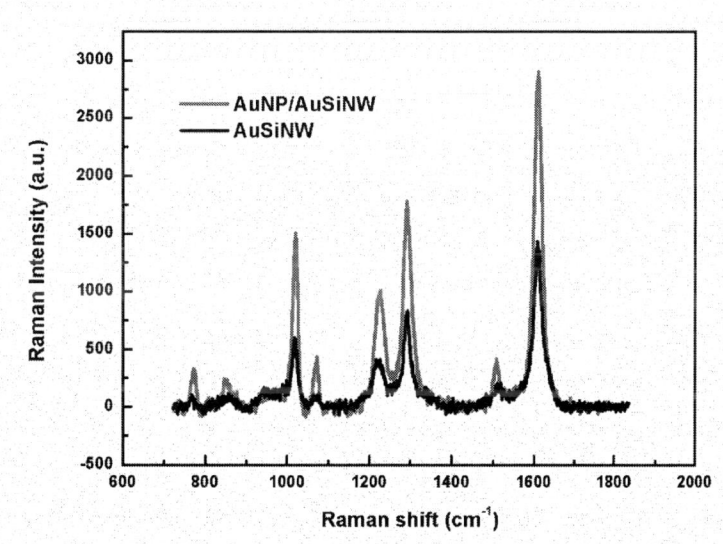

Figure. 3. The SERS spectra from AuNP /AuSiNW (red) and AuSiNW (black)

Figure.3 shows the 4, 4'-Bipyridine molecule has three mainly Raman characteristic lines, which locates at Raman shift of 1017 cm^{-1} , 1295 cm^{-1} and 1610 cm^{-1}. In view of the maximum signal intensity, we choose the signal at 1610 cm^{-1} as the intercomparable points. From Figure.3, the maximum Raman intensities of the AuSiNW substrate and AuNP/AuSiNW substrate are about 1350 and 3000, which means Raman signal on AuNP/AuSiNW substrate is beyond double times magnitude larger than that of

15

the AuSiNW substrate. As electromagnetic mechanism (EM) is widely accepted as the mainly mechanisms of SERS, we could explain these phenomena with electromagnetic field theory. In AuSiNW substrate, the wavelength of incident light is larger than the diameter of SiNW, most resonant energy is confined on the surface and resonantly coupled to the propagating surface plasmons of the Au layer. When AuNP is attached to the surface of Au layer, a so-called "hot gap" is generated between the AuNP and the AuSiNW. The electric field at the "hot gap" can be greatly enhanced. When detected molecule is captured by this "hot gap", more violent Raman signal can be obtained, which means lower detection concentration can be accepted. In general, we can confirm Raman signal on AuNP/AuSiNW substrate is larger than that on AuSiNW substrate. When AuNP/AuSiNW substrate is used for biomolecular detection or molecule sensor, lower detection limits can be obtained than that of the AuSiNW substrate.

4. Conclusion
In conclusion, AuNP/AuSiNWs for SERS enhancement were fabricated by rapidly assembling monolayer PS. The Raman intensity of the molecular captured in the gap between the AuNP and AuSiNW can be enhanced dramatically, almost double times larger than that of the AuSiNW substrate. The AuNP/AuSiNWs substrate can be utilized with a low detection limit and has the potential for realizing a rapid, low cost and label-free SERS based biosensor. From the result mentioned, we can prove electric field between the AuNP and AuSiNW is agminate and the Raman signal is enhanced. For the further study, target molecule and probe molecule will be introduced. Target molecule is modified on the AuNP and probe molecule is fixed on the AuSiNW, respectively.

5. Acknowledgments

This work was supported by National High Technology Research and Development Program of China (No.2014AA06A504), Science and Technology Department of Zhejiang Province(No.2014C31030), Fundamental Research Funds for Central Universities (No. 2014QNA5018), the National Natural Science Foundation of China (No. 61535010), Zhejiang Provincial Natural Science Foundation of China (No.LY16F050001).

References
[1]   Tan C L, Lee S K, and Lee Y T 2015 Opt. Express **23** 6254-6263
[2]   Wilson R 2010 Anal. Chem. **82** 2119-2123
[3]   Netzer N L 2011 Chem. Commun. **47** 9606-9608
[4]   Zhou H 2014 Anal. Chem. **86**, 1525-1533
[5]   Zhang C 2015 Opt.Express **23** 24811-24821
[6]   Campion A and Kambhampati P 1998 Chem.Soc.Rev. **27** 241–250
[7]   Zhang C, Jiang S Z and Huo Y Y 2015 Opt. Express **23** 24811-24821
[8]   Li T, Guo L and Wang Z 2008 Anal. Sci. **24** 907-910
[9]   Lu W, Singh A K and Khan S A 2010 J.Am.Chem.Soc. **132** 18103-18114
[10]   Kong L, Lee C and Earhart M 2015 Opt.Express **23** 6793–6802
[11]   Guo K, Xiao R and Zhang X 2015 Molecules **20** 6299-6309
[12]   Dick L A 2002 J.Phys.Chem.B. **106** 853-860
[13]   Fang Y, Seong N and Dlott D D 2008 Science **321** 388-392
[14]   Seniutinas G 2015 Opt.Express **23** 6763-6772
[15]   Huang J 2013 Nano.Let. **13** 5039-5045
[16]   Ngo H T, Wang H N and Fales A M 2013 Anal. Chem. **85** 6378-6383

Silicon-on-Insulator Nanowire Based Optical Waveguide Biosensors

Mingyu Li, Yong Liu, Yangqing Chen and Jian-Jun He

State Key Laboratory of Modern Optical Instrumentation, Zhejiang University, Hangzhou, China, 310027

jjhe@zju.edu.cn

Abstract. Optical waveguide biosensors based on silicon-on-insulator (SOI) nanowire have been developed for label free molecular detection. This paper reviews our work on the design, fabrication and measurement of SOI nanowire based high-sensitivity biosensors employing Vernier effect. Biosensing experiments using cascaded double-ring sensor and Mach-Zehnder-ring sensor integrated with microfluidic channels are demonstrated

1. Introduction

Optical waveguide sensors based on silicon-on-insulator (SOI) nanowire have received great attention due to their potential applications in many fields including bacteria and virus detection, medical diagnostics, food quality control, environment monitoring, drug development and so on. Various types of SOI sensors have been developed, including Mach-Zehnder interferometer [1], microdisks [2] and microring resonators [3]. Among various types of optical waveguide sensors, micoring resonators have been regarded as a promising solution for biological recognition and chemical analysis.

Optical waveguide biosesnsor is the device which can give the measurable signal when the target molecule is bond to another biomolecule on the surface of waveguide. Intensity interrogation and wavelength interrogation are two typical sensing methods for SOI nanowire based optical waveguide sensor. Both of these methods require a high resolution spectrometer or a narrow line-width tunable laser to achieve a high sensitivity. These instruments are very expensive and cannot be integrated on chip. To solve these problems, cascaded double microring resonators and a Mach-Zehnder interferometer (MZI) cascaded with a microring resonator employing Vernier effect were proposed [4,5]. Label-free detection capability of these two types of SOI sensors have also been demonstrated.

2. Cascaded double-ring sensor

The double-ring sensor consists of a sensing ring and a reference ring cascaded by a bus waveguide as shown in Fig.1. The whole chip was covered by Su8 upper cladding layer except that the sensing ring is exposed to the analyte sample by removing the upper cladding layer in the sensing window. The two rings have slightly different perimeter length and thus different free spectral ranges (FSRs) to produce a Vernier effect. Compared to a single microring resonator, the peak wavelength shift in the envelope function of the transmission of the cascaded double ring resonators is magnified by a Vernier amplification factor

$$F = \frac{\Delta\lambda_{FSRr}}{|\Delta\lambda_{FSRr}-\Delta\lambda_{FSRs}|} \tag{1}$$

where $\Delta\lambda_{FSRr}$ and $\Delta\lambda_{FSRs}$ are the FSR of the reference ring and sensing ring, respectively.

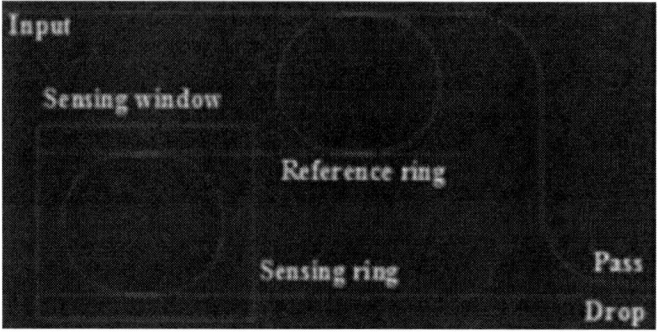

Fig. 1 Optical microscope image of the cascaded double ring sensor based on SOI nanowire.

The sensor is designed on SOI substrate with a 220 nm Si layer on a 2 μm SiO_2 layer. Since SOI is a high refractive index contrast material, the waveguide should be designed to be very small to keep the single mode of the waveguide. In our design, the widths of all the ridge waveguides are designed to be 1 μm with a shallow etched ridge height of 50 nm in order to maintain the single mode. We choose directional coupler to couple light into and out of the micro-ring resonators with the minimal distance between the bus waveguide and ring to be 1 μm, so that the optical sensor can be fabricated by contact photolithography.

The FSRs of reference ring and sensing ring are 0.7487nm and 0.6774nm, corresponding to F=10. Figure 2 shows the measured transmission spectra when sensing ring is exposed to solution of NaCl with different concentration of 2%,4%,6% and 8%. When the sample was changed, the sensing window was rinsed by the next new measuring solution. Figure 3 shows the measured central wavelength shift as a function of refractive index change of solution sample. The wavelength shift sensitivity is about 220nm/RIU. The detection limit using central peak wavelength shift detection is $3.3*10^{-3}$ RIU.

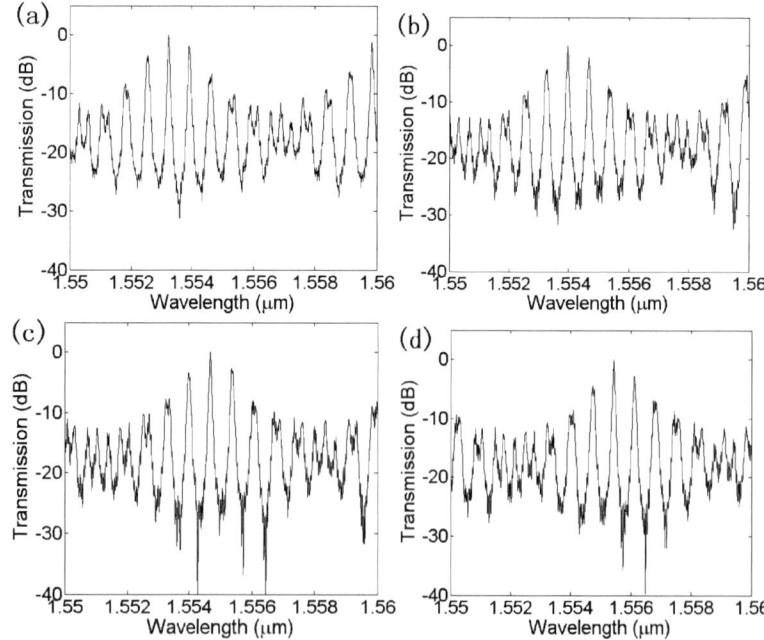

Fig.2 Measured transmission spectra of the cascaded double ring with different concentrations of NaCl solutions of 2%(a), 4%(b),6%(c) and 8%(d).

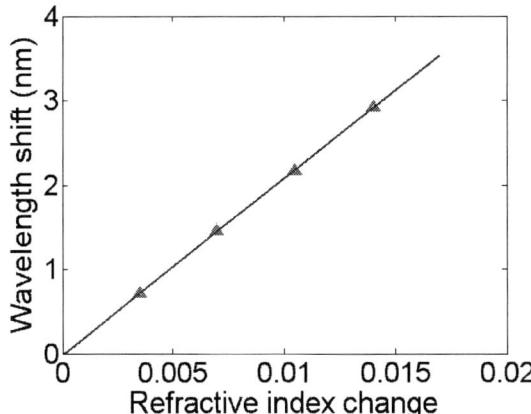

Fig. 3 Measured resonance wavelength shift of the cascaded double ring versus refractive index change of NaCl solutions

For the wavelength interrogation, the peak wavelength shift in the envelope function of the transmission of the cascaded double ring resonators is measured using an optical spectrum analyzer. Since the resonant wavelength of the reference ring is fixed, if one simply looks for the resonant peak of the maximum intensity, the minimal detectable wavelength shift $\delta\lambda_{min}$ is $\Delta\lambda_{FSRr}$, corresponding to a wavelength shift of the sensing ring of $\delta\lambda_s=|\Delta\lambda_{FSRr}-\Delta\lambda_{FSRs}|$. In order to have a lower detection limit, we made a thoroughly analysis of the transmission spectrum of the cascaded double ring, finding that the intensity ratio of the two central peaks varies with the refractive index change of the analyte. Experimentally, we proved that the detection limit can be dramatically improved by combining the detection of the intensity ratio between two central peaks [4]. In order to demonstrate this method, we used the solutions of NaCl with smaller concentration variations. Figure 4 shows the measured transmission spectra of the sensor when sensing ring is exposed to solutions of NaCl with different concentrations of 0.4%, 0.8%,1.2% and 1.6%, respectively.

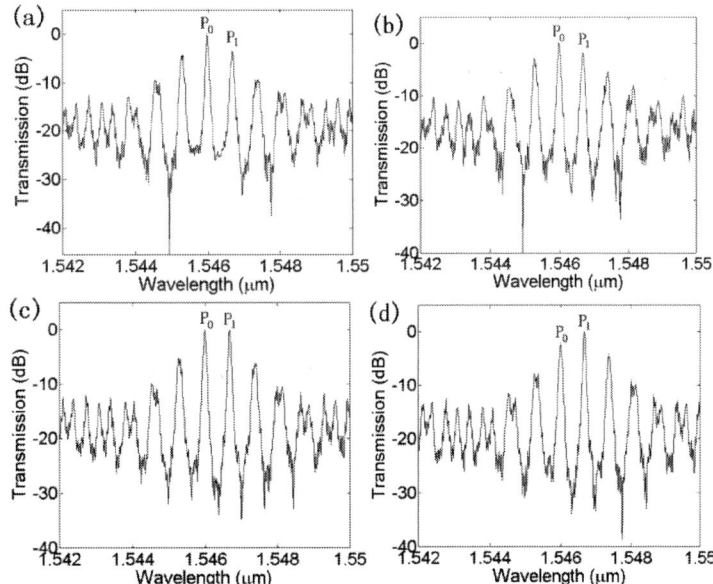

Fig. 4 Measured transmission spectra of the cascaded double ring with different concentrations of NaCl solutions of 0.4%(a), 0.8%(b),1.2%(c) and 1.6%(d)

The measured intensity ratio between the two central peaks, as a function of refractive index change of solution is shown in figure 5. The sensitivity is 2500 dB/RIU. The detection limit is $4*10^{-6}$ RIU.

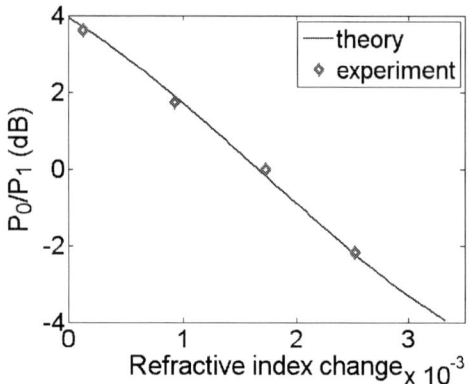

Fig.5 Measured intensities ratio P0/P1 of the two central peaks versus the refractive index change of NaCl solutions.

To reduce the cost of measurement system, the sensor can be operated in intensity interrogation using low cost broadband source without requiring wavelength information [6]. When a broadband light source such as an LED is used, the output power will change in proportion to the overlap integral of the LED spectrum and envelope function of the transmission spectrum. Assume the central wavelength of LED coincide with the peak of wavelength of the transmission envelop function initially. When n changes, the peak wavelength of the transmission shifts and they have a relative displacement. Consequently, the total output spectrum changes. Therefore, we can detect the variation of the sample refractive index by measuring the change of the total power. In this experiment, the difference of the FSRs for the two rings is 0.67%, which corresponds to an amplification factor M=150

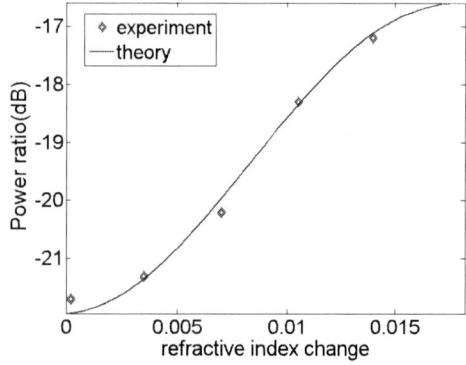

Fig.6 Power ratio between signal port and reference port vs. Refractive index change

Figure 6 shows the normalized output power versus the sample refractive index change of salt solution. By fitting the measured data with theoretical calculation, the sensitivity of the sensor reaches about 450 dB/RIU without requiring the spectral measurement. However, the detection limit of intensity interrogation is larger than that of the wavelength interrogation.

By using the standard silicon surface functionalization procedure, the anti-human IgG is immobilized on the surface of the sensing area of the sensor chip. The figure 7 shows the calibration curves for hIgG determination. From the transmission responses of the cascaded double ring exposed to solutions with different concentrations of hIgG, the concentration of 7.1–125 µg/mL was measured in the linear detection range [7]. The detection limit down to 7.1µg/ml with 0.47nm wavelength resolution. As a contrast, we also tested the chip with solution of casein and found no wavelength shift, which proves the specific recognition of the anti-hIgG functionalized surface for hIgG. The cascaded double ring sensors can be implemented in an array for potential simultaneous detection of multiple species.

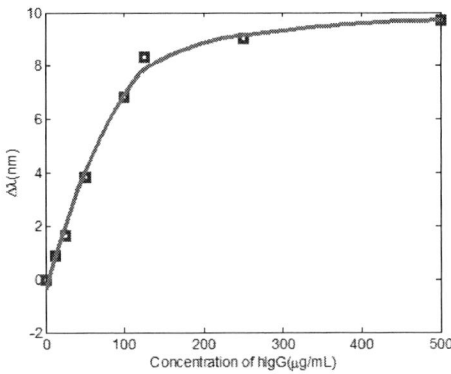

Fig.7 The wavelength shift of the transmission curve Calibration curves for hIgG determination

3. Cascaded MZI-ring sensor

Recently, an ultrahigh-sensitivity SOI nanowire optical waveguide biosensor based on cascaded MZI and ring resonator with Vernier effect using wavelength interrogation is proposed and experimentally demonstrated [5]. Fig. 8 shows the optical microscope image of the sensor. The sensor consists of a MZI cascaded with a ring resonator, in which, the MZI is used for sensing while the ring resonator acts as a reference. Using the cascaded MZI-ring, the wavelength shift of the MZI is translated into a much larger shift of the total transmission envelope function due to the Vernier effect. It's very similar to the cascaded double ring sensor. For a cascaded double ring resonator sensor, the ΔFSR of the sensing ring and reference ring has to be designed fairly small to get a large F. while a small ΔFSR means a flat envelope function of the total transmission spectrum, making it hard to determine the wavelength shift accurately. However, for the novel MZI-ring sensor, the ΔFSR of the MZI and the ring resonator can be designed relatively larger while still maintain almost a same level of sensitivity.

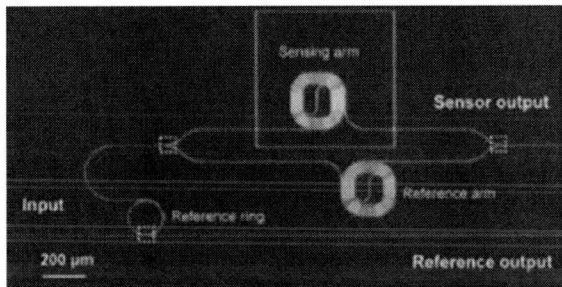

Fig.8 Optical microscope image of the cascaded MZI- ring sensor

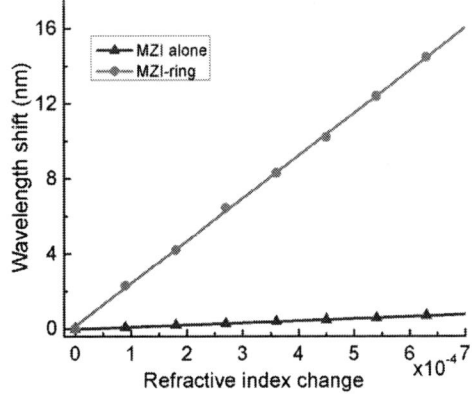

Fig. 9 Measured wavelength shift versus refractive index change of NaCl solutions with different concentrations

According to the wavelength peak shift in the measured transmission spectra of MZI-ring sensor with NaCl solution of different concentrations, the sensitivities for MZI sensor and cascaded MZI-ring sensor are 370nm/RIU and 19100nm/RIU respectively as shown in figure 9.

The wavelength shift of the MZI-ring sensor is about 14 times larger than that of the MZI sensor alone. To investigate the temperature dependence of the sensors, TEC was used to heat the sensor, the wavelength shifts versus temperature change are shown in figure 10. The measured results indicate that MZI-ring sensor is sensitive to the temperature, so using thermoelectric cooler (TEC) control the temperature is necessary in extremely low concentration detection.

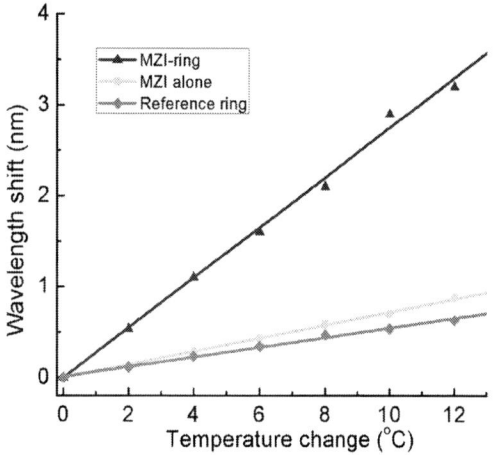

Fig. 10 Wavelength shift versus temperature change for reference ring, MZI and MZI-ring

After the standard surface functionalization on the sensing arm of the MZI, the sensor was used for monitoring the binding reactions between goat and antigoat IgG pairs. TEC maintained the sensor at 25℃ during the measurement. The measured sensor response is shown in the figure 11.

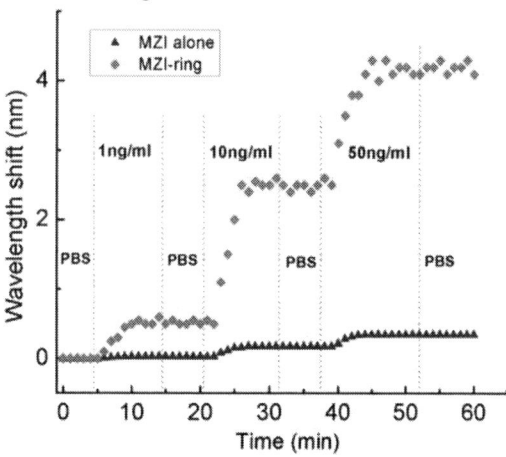

Fig. 11 Wavelength shift of the sensors with injection of antigoat IgG with different concentrations.

While the wavelength shift of the MZI-ring sensor is about 14 times larger than that of the MZI sensor, the wavelength measurement error also increased due to the peak broadening in the envelope function. The estimated lowest detectable concentration for the MZI sensor is about 0.29 ng/ml IgG with a minimal detectable wavelength shift of 10 pm. For the MZI-ring sensor, the reliable minimal detectable wavelength shift is around 50 pm using data fitting, which gives the lowest detectable concentration of 0.1 ng/ml IgG.

Although, the requirement of the spectral resolution is 5 times lower, the sensitivity of the MZI-ring is still about 3 times higher than that of MZI sensor. Thus, the sensor can detect extremely low concentration of IgG.

4. Conclusion

Cascaded double-ring sensor and MZI-ring sensor employing the Vernier effect have been proposed and experimentally demonstrated. The experimental results show that Vernier effect can increase the sensitivity dramatically. These SOI nanowire based high-sensitivity optical waveguide sensors are promising for real time measurements of molecule binding kinetics in biomedical and chemical applications.

Acknowledgment

This work was supported by the National High-Tech Research and Development Program of China (No. 2014AA06A504), the Science and Technology Department of Zhejiang Province (No. 2014C31030), Fundamental Research Funds for Central Universities (No. 2014QNA5018), and the National Natural Science Foundation of China (No. 61535010) , Zhejiang Provincial Natural Science Foundation of China (No.LY16F050001).

References

[1] B. Sepulveda, J.S. Del Rio, M. Moreno, F. Blanco, K. Mayora, C. Domínguez, L.M. Lechuga, "Optical biosensor microsystems based on the integration of highly sensitive Mach–Zehnder interferometer devices", .Opt.[A: Pure Appl. Opt., 8 S561-S566 (2006)..

[2] R.W. Boyd, J.E. Heebner, Sensitive disk resonator photonic biosensor, Appl. Opt., 40(31), 5742-5747 (2001)

[3] J.T. Kindt, R.C. Bailey, "Biomolecular analysis with microring resonators: Applications in multiplexed diagnostics and interaction screening", Curr. Opin.Chem.Biol., 17,818-826 (2013).

[4] L. Jin, M. Li, J.-J. He, "Highly-sensitive silicon-on-insulator sensor based on two cascaded micro-ring resonators with vernier effect", Opt. Commun., 284 (2011) 156-159.

[5] X. Jiang, Y. Chen, F. Yu, T. Tang, M. Li, J.-J. He, "High-sensitivity optical biosensor based on cascaded Mach-Zehnder interferometer and ring resonator using Vernier effect", Opt. Lett., 39(15), 6363-6365 (2014).

[6] L. Jin, M. Li, and J.-J. He, "Optical waveguide double-ring sensor using intensity interrogation with low-cost broadband source", Opt. Lett. 36(7), 1128–1130 (2011).

[7] Y. Chen, F. Yu, C. Yang, J. Song, L. Tang, M. Li, J.-J. He, "Label-free biosensing using cascaded double-microring resonators integrated with microfluidic channels", Opt. Commun. 344, 129-133 (2015).

Microlens Array Diffuser with Randomly Distributed Structure Parameters

Tianyi Guo[1,2], Chao Yu[1,2], Haifeng Li[*], Chen Su[1,2], Yinxu Bian[1,2] and Xu Liu[1,2]

[1]College of Optical Science and Engineering, Zhejiang University, Hangzhou, 310027, China

[2]State Key Laboratory of Modern Optical Instrumentation, Zhejiang University, Hangzhou, 310027, China

lihaifeng@zju.edu.cn

Abstract. With regard to optical diffusers, the configuration parameters is determined according to the need of application including beam shape, homogeneity, diffusing angle, transmittance etc. Differentiating with conventional optical diffuser design methods such as diffractive optical element (DOE) methods, we proposed a method to design microlens array (MLA) diffuser configured with random-distributed microlens parameter for generation of a large diffusing angle and uniform intensity distribution. And we discussed the relation between the beam homogeneity and random distribution type of structure parameters. Furthermore, in this article, the author validated that it was possible to splice duplicated small-scale unit into a large-scale diffuser screen with maintaining its scattering properties.

1. Introduction

Optical diffusers have attracted the attention from worldwide researchers for its application in the respects of display, beam-shaping and directional illumination, for example, the diffuser with large diffusing angle and uniform intensity distribution is widely used to achieve horizontal-parallax-only 3D display. Based on reflection or diffraction theory, optical diffusers redistribute incident light into target area, which is characterized by the parameters such as diffusing angle and projection distance. A lot of work have been done in diffractive optical element（DOE）diffuser. Parikka[1] and Mendez[2] have applied the DOE (diffractive optical elements) method to theoretically design optical diffusers. However, dealing with the introduced sampling interval properly and meeting strict processing precision makes fabrication of the diffuser more difficult. Frank C. Wippermann has proposed a novel setup that consists of a chirped microlens arrays (MLAs), which homogenizes the far-field intensity distribution, using reflective principle. Notwithstanding simple set-ups, the mapped structure of two-microlens arrays demands higher alignment precision[3].

Traditional scattering material, such as holographic diffuser and frosted glass, distributes beam into room because its homogenization capability is statistically related to the random distribution of micro-unit centres[4]. Meanwhile, the shape and emergent angle of diffusing beam depends on micro-unit's surface profile and aperture type. Thus, the single MLA diffuser whose micro-unit centres satisfies the random distribution has been proposed[5]. However, its diffusing angle is small with expensive fabrication cost.

Content from this work may be used under the terms of the Creative Commons Attribution 3.0 licence. Any further distribution of this work must maintain attribution to the author(s) and the title of the work, journal citation and DOI.
Published under licence by IOP Publishing Ltd

In this paper, we established one kind of microlens arrays model using optical simulation software LightTools and VBA macro language. Then, the structure parameters of microlens array model was distributed randomly. Due to random relation between parameters of microlens, beam through the diffuser will be deflected randomly in direction and quantity. Consequently, large-angle and uniform distribution of incident energy is constituted by randomly deflected beam in target area. Moreover, the relation was discussed that microlens deflection direction is mainly determined by the off-aperture extent of microlens centre. Also, it was discussed that which type random distribution results in uniform profile in target area. Finally, we focused on the possibility to splice duplicated small-scale diffuser units into a large-scale diffuser screen with maintaining its scattering properties.

2. Design principle

For the microlens array, with random distribution of its micro-unit scattering centres, it could make some contribution to a large angular uniform intensity distribution, as shown in Fig.1. For the reason that regarding emergent beam profile of every unit as Gaussian distribution in this scheme, the centre axis of emergent beam is exactly perpendicular to the target plane. Moreover, it's inevitable that beam energy is accumulated in the centre of the target plane to form a Gaussian-like energy distribution. Researchers made efforts on aspheric surface design of micro-units to improve non-uniform distribution. But the fabrication cost is expensive with little diffusing angle below ± 5 °.

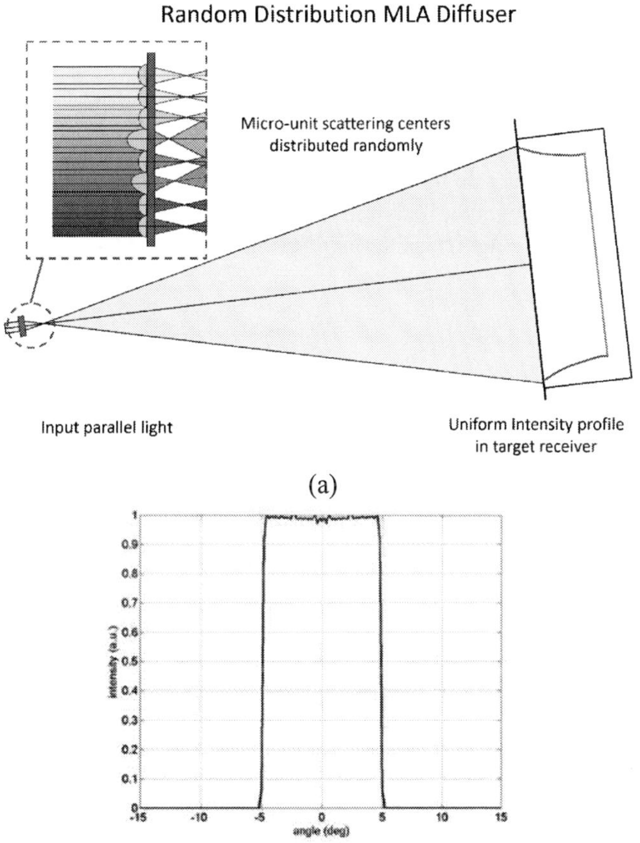

(a)

(b)

Fig. 1. (a) Ideal MLA diffuser configuration. (b) The angular distribution of one kind of MLA diffuser.

Fig.1(a) shows the far-field configuration of an ideal top-hat MLA diffuser, and Fig.1(b) is the angular distribution of one kind of MLA diffuser[6]. Assuming that the beam transmitting microlens corresponds to Gaussian distribution, thereby the emergent light intensity of MLA should be the intensity integration of all the Gaussian beams as indicated by Eq.(1),

$$I_x = \sum_{p=1}^{m} I_p = \sum_{p=1}^{m} A_p^2 \tag{1}$$

Where m is amount of micro-units in MLA diffuser. The letter p represents the index of micro-units from 1 to m. And I_x is the intensity distribution in target plane, while I_p is the intensity distribution of micro-units with index p. And A_p is the amplitude of micro-units emergent wave surface with index p.

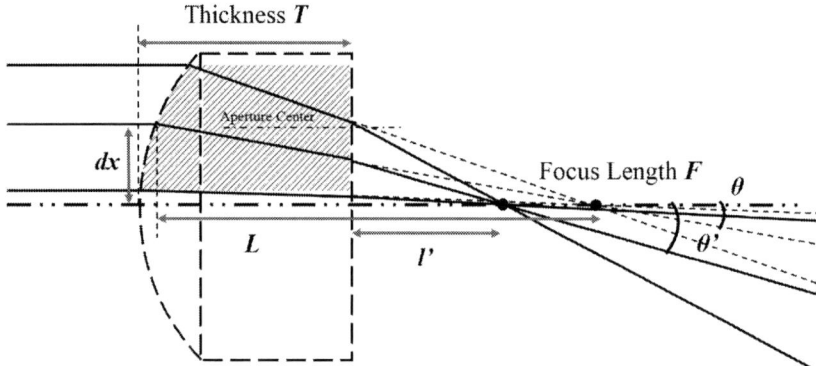

Fig. 2. Ideal microlens incident ray-path

With regard to an ideal lens model, as shown in Fig.2, optical ray path can be described using geometrical optics. For micro-units on the diffuser, their surface profile correspond to the specific part of an ideal lens surface which is symmetrical or asymmetrical (for example, as gray-shadow part shows). When the surface profile is asymmetrical, the micro-unit achieves outgoing beam with deflection. The first deflection angle $\tan\theta$ can be expressed as,

$$F = \frac{R}{n' - n} \tag{2}$$

$$\tan\theta = \frac{dx}{L} = \frac{dx}{F - (R - \sqrt{R^2 - dx^2})} \tag{3}$$

Where F and R indicates the focal length and the curvature radius of the microlens respectively. The distance dx is measured by length from the incident position of beam centre to the ideal lens optical axis, L is the axial distance from the position of incident beam centre to the lens focus point. Similarly, the twice deflection angle can be expressed as,

$$\tan\theta' = \frac{1}{l'} \times \left\{ dx - \tan\theta \cdot \left[T - (F - L) \right] \right\} \tag{4}$$

Where l' represents the back focal length of microlen, and T is the maximum thickness of the lens. It is found that the emergent beam deflection direction is determined by the off-aperture direction of microlens centre. Obviously, the deflection direction is opposite to the off-aperture direction of microlens centre. Following, deflection degree can be calculated by the extent of microlens off-aperture extent dx.

The microlens array diffuser with randomly distributed structure parameters is proposed to achieve homogeneity illumination. The random distribution of microlens structural parameters within a specific range causes beam deflection is randomly distributed in target area. The specific range of structural parameters is calculated by Eq. (2) (3) (4), including the radius of curvature, the micro-unit aperture, thickness, etc. The micro-units beam deflection of the proposed MLA diffuser is different in direction and quantity as the Fig. 3. Setting arbitrary 3*3 microlens as one group, the centre unit of each group (Blue part) will be surrounded by eight nearby units (Yellow part). Due to their different radii, resulting in partly surface overlapping of the centre unit, its beam deflection occurs varying in direction and degrees. Except micro-units in the edge of the diffuser, each unit conducts independently random deflection, which constitutes the uniform distribution of energy in target scattering zone.

In the Fig.3, it shows several situations that the emergent beam from centre unit of each group deflects in target area, (a) shows upwards beam-shaping comparing with (b) shows downwards beam-shaping. Moreover (c) and (d) shows different deflection characteristics of the micro-unit with randomly distributed structural parameters.

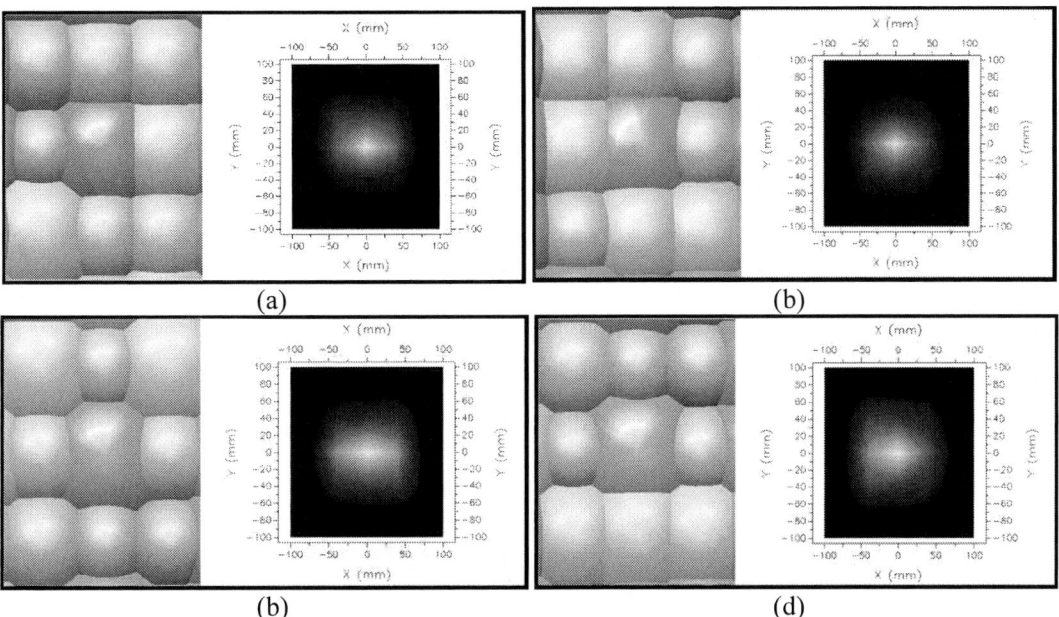

(a)

(b)

(b)

(d)

Fig. 3. Various beam deflection of center micro-unit (Blue part) surrounded by eight nearby micro-units (Yellow part) (a) Beam deflection upward. (b) Beam deflection downward. (c) Beam deflection towards right. (d) Beam deflection towards left.

3. Experiment

In the experiment, one model of standard microlens array was built by the optical simulation software LightTools. It concluded 30*30 microlens and the device size was 0.9mm*0.9mm. The unit size and microlens thickness was 30μm*30μm and 15μm respectively. Then, by using VBA macro language, we resetted the structure parameters like radius randomly distributed in the range from 25μm to 50μm. Actual micro-units structure characteristics was related with designed scattering beam shape. Simulation results about output beam of various designed MLA diffuser with randomly distributed structure parameters was obtained by LightTools, shown in Figure 4. The intensity profile in Y-axis of scattering beam shape is shown in below, and the diffusing angle in X-axis is $\pm 17°$.

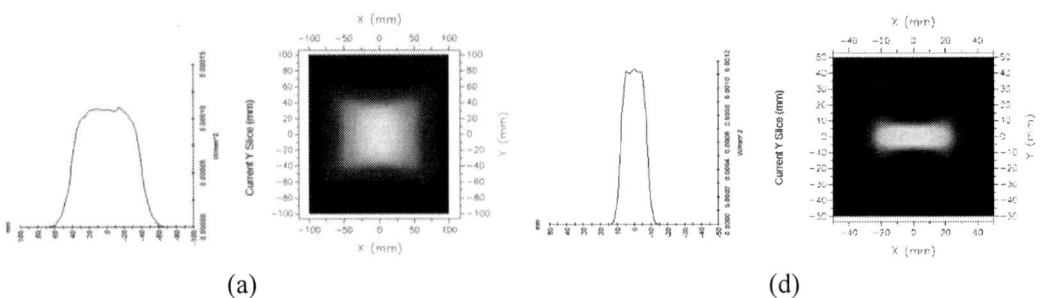

(a)   (d)

Fig. 4. The energy intensity profile of simulation result in Y-axis (Left), and the intensity distribution of diffusing shape (Right) from 200mm. (a) Rectangular-shape. (b) Slender rectangular-shape.

In order to verify the replicability of designed MLA diffuser, we confirmed the possibility that splicing duplicated small-scale diffuser units into a large-scale diffuser screen with maintaining its scattering properties. In this experiment, two MLA diffuser units A and B were connected together, with parallel light incident in the centre of Unit A and another one incident at the junction section of A and B. For the reason that the structure parameters accorded with random distribution, it made no difference to change the micro-units index of designed MLA diffuser. As the theory, the simulation results were shown in Figure 5. The intensity outline of the conjunction units is the same with the single one.

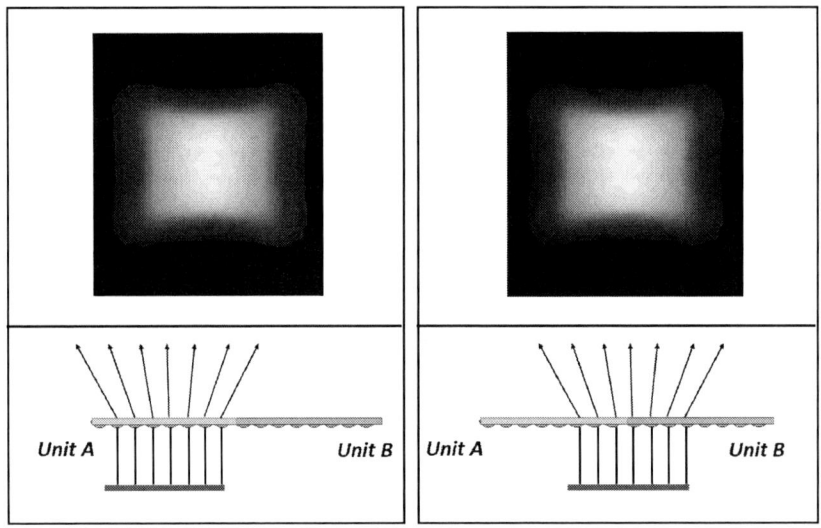

Fig. 5. Verification the scalability and replicability of designed MLA diffuser by the setup (Bottom) and the intensity (Top).

4. Conclusions

In this paper, based on the beam deflection characteristics of micro-units, the emergent beam deflection of designed MLA diffuser is randomly distributed in direction and quantity, which realizes a large-angle uniform distribution of the device scattering capability. Theoretically, the quantitative relation of beam deflection characteristics is discussed among the off-aperture extent of microlens centre, deflection degree and deflection direction. And we propose a design method of MLA diffuser with randomly distributed structure parameters. In addition, optical simulation software LightTools with Visual Basic Application macro language programming is used to model and simulate scattering effect of designed diffuser. Compared to general MLA diffusers, the designed diffuser expands diffusing angle to ± 17 °

in X-axis. In particular, it's possible to make larger diffuser device using the proposed design method by conjunction the duplicated diffusing units. In fabrication, the designed diffuser could be fabricated by lithography-melting process with high precision.

References
[1]   Parikka, M, Kaikuranta, T, Laakkonen, P, Lautanen, J, Tervo, J, Honkanen, M, et al. 2001 Deterministic diffractive diffusers for displays *Applied optics* **40**(14) 2239-2246.
[2]   Mendez, E R, Martinez-Niconoff, G C, Maradudin, A A, Leskova, T A 1998 Design and synthesis of random uniform diffusers *SPIE's International Symposium on Optical Science, Engineering, and Instrumentation (International Society for Optics and Photonics)* pp 2-13
[3]   Wippermann, F C, Zeitner, U D, Dannberg, P, Bräuer, A, Sinzinger, S 2007 Fly's eye condenser based on chirped microlens arrays *Photonic Devices+ Applications (International Society for Optics and Photonics)* pp 666309-666309
[4]   Wu, J H, Chen, J B, Tang, M X 2001 Design of the Directional Scatterer Composed of Random Gratings *JOURNAL OF OPTOELECTRONICS LASER* **12**(10) 1018-1021
[5]   Sales, T R, Schertler, D J, Chakmakjian, S 2006 Deterministic microlens diffuser for lambertian scatter *SPIE Optics+ Photonics (International Society for Optics and Photonics)* pp 629005-629005
[6]   Ducharme, A D 2007 Microlens diffusers for efficient laser speckle generation *Optics express* **15**(22) 14573-14579

Acknowledgments
This work is supported by National Basic Research Program of China (973 Program) (No. 2013CB328802), National High Technology Research and Development Program of China (863 Program) (No.2012AA011902) and National Natural Science Foundation of China (Grant No. 61177015).

Reflective Characteristics of Spatially-Bounded Laser Beam by Rugate coating

Zheng Guangwei

Information and Navigation College, Air Force Engineering University, 710077, Xi'an, Shannxi, China

Email: zgw198196@126.com

Abstract. Based on the transfer matrix analysis and discrete Fourier transform, the reflective characteristics of a spatially-bounded laser beam by a Rugate coating are investigated. The spatial intensity distribution and reflectance efficiency of the reflected beam are formulated. When the central vector of the laser beam satisfies the coating's Bragg law, the spatial intensity distributions of the diffracted beam with the change of the coating's parameters are simulated. The results show that with the decrease of the thickness, the increase of the period, or the decrease of the relative permittivity modulation, the spatial intensity distributions of the reflected beam in the incidence plane are broadened compared with those of incident beam, while those perpendicular to the incidence plane are not. And the reflectance efficiencies decrease. Among these three parameters, the relative permittivity modulation and thickness influence the reflectance characteristics a lot, while the period influences the reflectance characteristics a little. It means that when the thickness decreases, or modulation decreases, the Rugate coating has finer angular selective characteristics and less angular selectivity bandwidth. The results are instructive for the design and application of Rugate coating.

1. Introduction

Rugate coating is a type of thin films, the refractive index of which changes continuously along its thickness[1]. Compared with the traditional graded coatings, Rugate coating has many merits, such as no harmonious reflective bands, low stress between different layers, high laser damage threshold, and so on[2, 3]. The characteristics of Rugate coatings have been analyzed since 1960s[4]. However, with the development of modern coating's technique, Rugate coating's application has attracted lots of attention recently[5-7].

Spatial filter can improve the laser beam quality in the spatial domain, which is very important for its usage, such as the ignition laser beams in the Inertial Confinement Fusion system, and so on[8-9]. Nowadays, the most widely-used spatial filter is pinhole filter, the structure of which is two lens aligned coaxially with a pinhole plate placed in the con-focusing plane in order to select the required angular spectrum. Due to its focusing characteristics, pinhole filter cannot be adjusted easily to the application for the high power laser beam because of the higher beam intensity around the pinhole[10-12]. So the non-focusing methods have attracted a lot of attention. Due to the fine wave vector selectivity of the coatings, the non-spatial filtering for laser beam based on coatings has been put forward since 2000s[13,14]. The fine wave vector selectivity denotes the fine wavelength selectivity at some angular domain and the fine angular selectivity at some wavelength domain. So Rugate coatings may be a potential candidate for the pinhole filter, especially in the high power laser field, with its fine angular selectivity[15]. However, there

Content from this work may be used under the terms of the Creative Commons Attribution 3.0 licence. Any further distribution of this work must maintain attribution to the author(s) and the title of the work, journal citation and DOI.

Published under licence by IOP Publishing Ltd

is lots of attention paid to the reflectance or transmittance characteristics of the Rugate coating for the wavelength spectrum or angular spectrum, while there is little attention paid to the reflectance characteristics of the Rugate coating for the actual laser beam. Maybe the reflected laser beam is spatially deformed compared with the incident spatially-bounded laser beam by the Rugate coating. So it is necessary to calculate the reflectance characteristics of the Rugate coating for the real laser beam.

2. Configuration and Theoretical Analyses

Fig.1 shows the spatially-bounded laser beam incident into a Rugate coating.

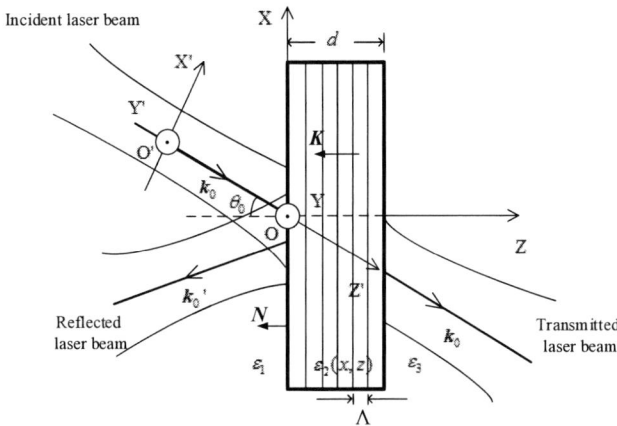

Fig.1. Schematic of spatially-bounded laser beam incident into a Rugate coating

In fig.1, the solid parallel lines stand for the coating's fringes. Λ denotes the period, d denotes the thickness. $\mathbf{K}$ denotes the coating's vector. Without any loss of generality, $\mathbf{K}$ is selected to be in the XOZ plane. $\varepsilon_2(\mathbf{r})$ is the coating's relative permittivity, which can be expressed as:

$$\varepsilon_2(\mathbf{r}) = \varepsilon_{20} + \varepsilon_{21}\cos(\mathbf{K}\cdot\mathbf{r}) \tag{1}$$

where ε_{20} and ε_{21} denote the relative mean dielectric permittivity and its modulation amplitude, respectively. $\mathbf{r}$ denotes the position vector, stands for (x, y, z). X' axis is in the plane of incidence, which is formed by the central wave vector $\mathbf{k}_0$ of the incident laser and the coating's front surface normal $\mathbf{N}$, while Y' axis is orthogonal to the plane of incidence. And Z' axis is parallel to $\mathbf{k}_0$. θ_0 is the incidence angle between $\mathbf{k}_0$ and Z axis. ε_1 and ε_3 are the relative permittivities in region 1 and region 3, respectively.

2.1. Decomposition of the spatially-bounded laser beam by discrete Fourier transforms

Based on the deduced reflectance results of a Rugate coating illuminated by monochromatic plane waves (MPWs) based on the transfer matrix analysis (TMA)[16,17], intuitively the spatially-bounded laser beam should be firstly transformed into linear combinations of MPWs based on the two-dimensional discrete Fourier transforms in spatial domains. And secondly the reflected and transmitted characteristics of each MPW can be achieved depending on the TMA. Finally based on the inverse discrete Fourier transform, the spatial intensity distributions of reflected laser beam can be formulated.

The field amplitude of incident laser beam (at $Z=0$) can be expanded into linear combinations of MPWs as:

$$E_v\left(x_{n_1}, y_{n_2}, z=0\right) = \sum_{m_1=-M_1/2}^{M_1/2-1}\sum_{m_2=-M_2/2}^{M_2/2-1} F_v\left(k_{x,m_1}, k_{y,m_2}, z=0\right)\exp\left[-j\left(k_{x,m_1}x_{n_1} + k_{y,m_2}y_{n_2} + k_{z,m_1,m_2,p}z\right)\right] \tag{2}$$

where M_1 and M_2 are the numbers of sampling points over the spatial intervals $-D_1/2 \le x_{n_1} \le D_1/2$, and $-D_2/2 \le y_{n_2} \le D_2/2$, in X and Y axis respectively. $k_{x,m_1} = m_1\left(2\pi/D_1\right)$ and $k_{y,m_2} = m_2\left(2\pi/D_2\right)$

are the wave vector components along X and Y axis, respectively. Furthermore, the z wave vector component for MPW $\left(m_1, m_2\right)$ can be achieved by $k_{z,m_1,m_2} = \left(k_0^2 \varepsilon_1 - k_{x,m_1}^2 - k_{y,m_2}^2\right)^{1/2}$. $k_0 = \omega_0/c$, c denotes the light speed in vacuum. Subscript v stands for x, y, or z, which means the polarization in the X, Y, or Z axis.

Furthermore, the corresponding monochromatic plane-wave spectrum coefficient for MPW $\left(m_1, m_2\right)$ can be performed by

$$F_v\left(k_{x,m_1}, k_{y,m_2}, z=0\right) = \frac{1}{M_1 M_2} \sum_{n_1 = -M_1/2}^{M_1/2-1} \sum_{n_2 = -M_2/2}^{M_2/2-1} E_v\left(x_{n_1}, y_{n_2}, z=0\right) \times \exp\left[j\left(k_{x,m_1} x_{n_1} + k_{y,m_2} y_{n_2} + k_{z,m_1,m_2} z\right)\right] \tag{3}$$

where $x_{n_1} = n_1 D_1 / M_1$, $y_{n_2} = n_2 D_2 / M_2$.

For MPW $\left(m_1, m_2\right)$, the incidence angle θ_{m_1,m_2} can be performed by

$$\theta_{m_1,m_2} = \cos^{-1}\left[k_{z,m_1,m_2} / \left(k_0 \varepsilon_1^{1/2}\right)\right] \tag{4}$$

2.2 Intensity distributions of the reflected laser beam in spatial domains

Assuming the incident laser beam is TE mode, and based on the deduced conclusions of TMA, the reflected amplitude of each MPW $\left(m_1, m_2\right)$ can be determined by

$$F_{yD}^R\left(k_{x,m_1}', k_{y,m_2}', 0\right) = E_R(\lambda_0, \theta) F_y\left(k_{x,m_1}, k_{y,m_2}, 0\right) \tag{5}$$

where $\lambda_0 = 2\pi/k_0$, $k_{x,m_1}' = k_{x,m_1}$, $k_{y,m_2}' = k_{y,m_2}$, $E_R(\lambda_0, \theta)$ denotes the reflected coefficient for each MPW based on TMA.

As a result, the amplitude of the reflected laser beam is expressed as:

$$E_{yD}^R\left(x_{n_1}, y_{n_2}, z\right) = \sum_{m_1 = -M_1/2}^{M_1/2-1} \sum_{m_2 = -M_2/2}^{M_2/2-1} F_{yD}^R\left(k_{x,m_1}', k_{y,m_2}', 0\right)$$
$$\times \exp\left[-j\left(k_{x,m_1}' x_{n_1} + k_{y,m_2}' y_{n_2} + k_{z,m_1,m_2}^R' z\right)\right] \tag{6}$$

where k_{z,m_1,m_2}^R' denotes the projection of $\boldsymbol{k}_{m_1,m_2}^R'$ into Z axis. And it can be expressed as:

$$k_{z,m_1,m_2}^R' = \begin{cases} -\left(k_0^2 \varepsilon_1 - k_{x,m_1}^2' - k_{y,m_2}^2'\right)^{1/2} & k_{x,m_1}^2' + k_{y,m_2}^2' \leq k_p^2 \varepsilon_3 \quad \textit{transmitting waves} \\ +j\left(k_{x,m_1}^2' + k_{y,m_2}^2' - k_0^2 \varepsilon_3\right)^{1/2} & k_{x,m_1}^2' + k_{y,m_2}^2' > k_p^2 \varepsilon_3 \quad \textit{evanescent waves} \end{cases} \tag{7}$$

So the intensity distribution of the reflected laser beam can be expressed as:

$$I_{sD}^R\left(x_{n_1}, y_{n_2}, z\right) = \left| \sum_{m_1 = -M_1/2}^{M_1/2-1} \sum_{m_2 = -M_2/2}^{M_2/2-1} F_{yD}^R\left(k_{x,m_1}', k_{y,m_2}', 0\right) \right.$$
$$\left. \times \exp\left[-j\left(k_{x,m_1}' x_{n_1} + k_{y,m_2}' y_{n_2} + k_{z,m_1,m_2}^R' z\right)\right] \right|^2 \tag{8}$$

And the reflectance efficiency is:

$$\eta_D^R = \frac{\displaystyle\sum_{m_1 = -M_1/2}^{M_1/2-1} \sum_{m_2 = -M_2/2}^{M_2/2-1} \left| F_{yD}^R\left(k_{x,m_1}', k_{y,m_2}', 0\right) \right|^2}{\displaystyle\sum_{m_1 = -M_1/2}^{M_1/2-1} \sum_{m_2 = -M_2/2}^{M_2/2-1} \left| F_y\left(k_{x,m_1}, k_{y,m_2}, 0\right) \right|^2} \tag{9}$$

3. Simulation results

The Gaussian laser beam has been widely used in lots of optical systems, so we select Gaussian laser beam as the incident spatially-bounded laser beam. For simplicity, the dispersion is not considered in the paper. The laser's normalized amplitude is expressed as

$$\mathbf{E}\left(x',y',z',t\right)=\frac{w_0}{w(z')}exp\left[-\left(x'^2+y'^2\right)\big/w^2\left(z'\right)\right]$$
$$\times exp\left\{j\left[\left(\omega_0 t-\mathbf{k}_0\cdot z'\right)-k_0\frac{\left(x'^2+y'^2\right)}{2R(z')}-\tan^{-1}\left(\frac{z'\lambda_0}{\pi w_0^2}\right)\right]\right\}\cdot\boldsymbol{y} \tag{10}$$

where w_0 is the beam waist radius. $\boldsymbol{y}$ is the unit vector along y axis.

$$w(z')=w_0\left\{1+\left[z'\lambda_0\big/\left(\pi\omega_0^2\right)\right]^2\right\}^{1/2} \tag{11}$$

$$R(z')=z'+\left(\pi\omega_0^2\big/\lambda_0\right)^2/z' \tag{12}$$

where $w(z')$ and $R(z')$ are the beam radius and the phase front curvature radius at the distance z' away from the beam waist center, respectively.

In the practical application, the highest diffraction efficiency is the key factor for the Rugate coating. So we just consider the condition that the central wave vector of the incident laser pulse satisfies the coating's Bragg's law, which is a requisite condition for its greatest reflectance efficiency of this kind of laser beam. And it can be expressed as:

$$2\left|\boldsymbol{k}_0\right|\varepsilon_{20}\cos\theta_0=\left|\boldsymbol{K}\right| \tag{13}$$

The other parameters used in the simulation are shown in tab.1.

Tab.1. Parameters and quantities in the simulation

Parameter	Quantity	Parameter	Quantity
λ_0	1053 nm	ε_1, ε_3	2.250
D_1, D_2	310 μm	M_1, M_2	310
w_0	50 μm	ε_{20},	2.250

3.1 Spatial intensity distributions of the reflected beam with the change of the relative dielectric permittivity modulation

Three groups of parameters are selected, with the relative dielectric permittivity modulation of 0.01 ppm, 0.02 ppm, or 0.03 ppm, respectively. The coating's thickness is 100 μm. The grating's period is 380 nm. The spatial intensity distributions of the reflected laser with the change of the relative permittivity modulation is shown in figure 2.

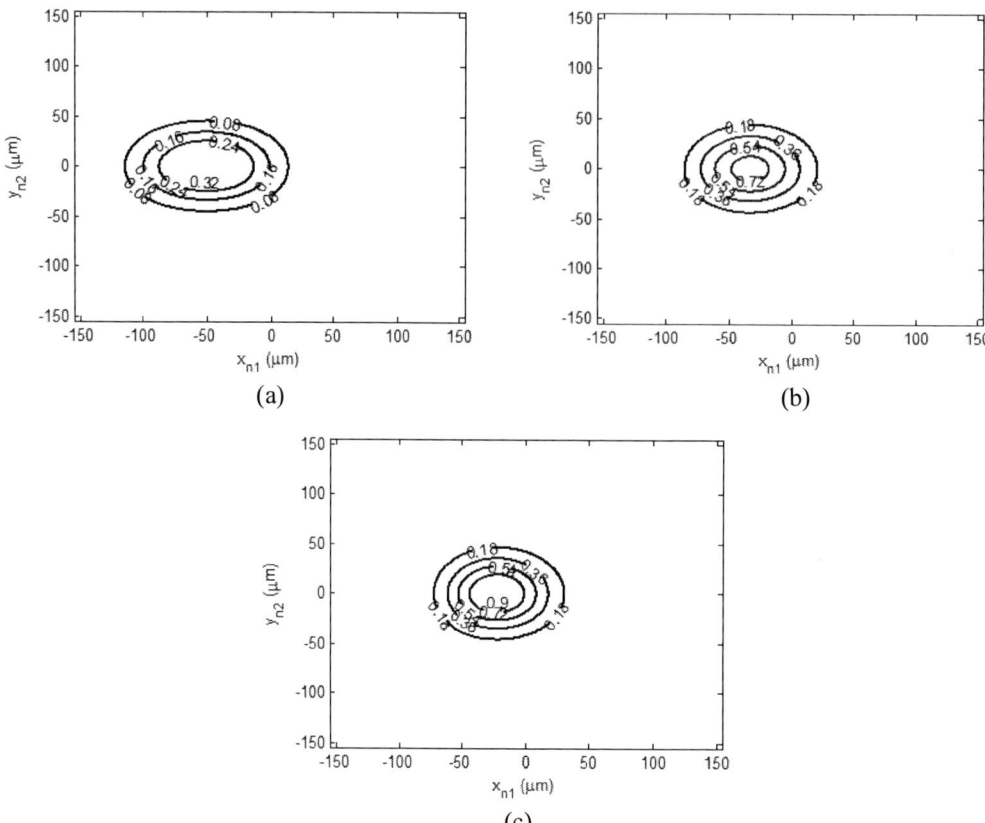

Fig.2. Spatial intensity distributions of reflected laser beam with the coating's relative permittivity modulation of (a) 0.01, (b) 0.02, and (c) 0.03.

From figure 2, we can deduce that with the relative permittivity modulation's increment, the intensity distributions of the reflected laser are getting more and more similar to that of the input laser in the spatial domain. It means that more decomposed MPWs are reflected at the higher relative dielectric permittivity modulation than those at the lower one. And when the relative dielectric permittivity modulation is low, the profile in X axis is widened greatly, which means some higher angular frequencies are not reflected. It is shown that the spatial width in the Y axis is approximately the same as that of the input one, which means that the Rugate coating has little effect on the wave vector components that are perpendicular to incidence plane. The reflectance efficiencies are 52.36%, 91.26%, and 98.96% in correspondence to the relative permittivity modulation of (a) 0.01, (b) 0.02, and (c) 0.03, respectively.

3.2 Spatial intensity distributions of the reflected beam with the change of the coating's thickness
In this part, three groups of coating's parameters are selected, with the coating's thickness of 50μm, 100μm, or 150μm, respectively. The relative dielectric permittivity modulations are fixed as 0.02. The grating's periods are 380nm. The spatial intensity distributions of the reflected laser with different gratings' thickness are shown in figure 3.

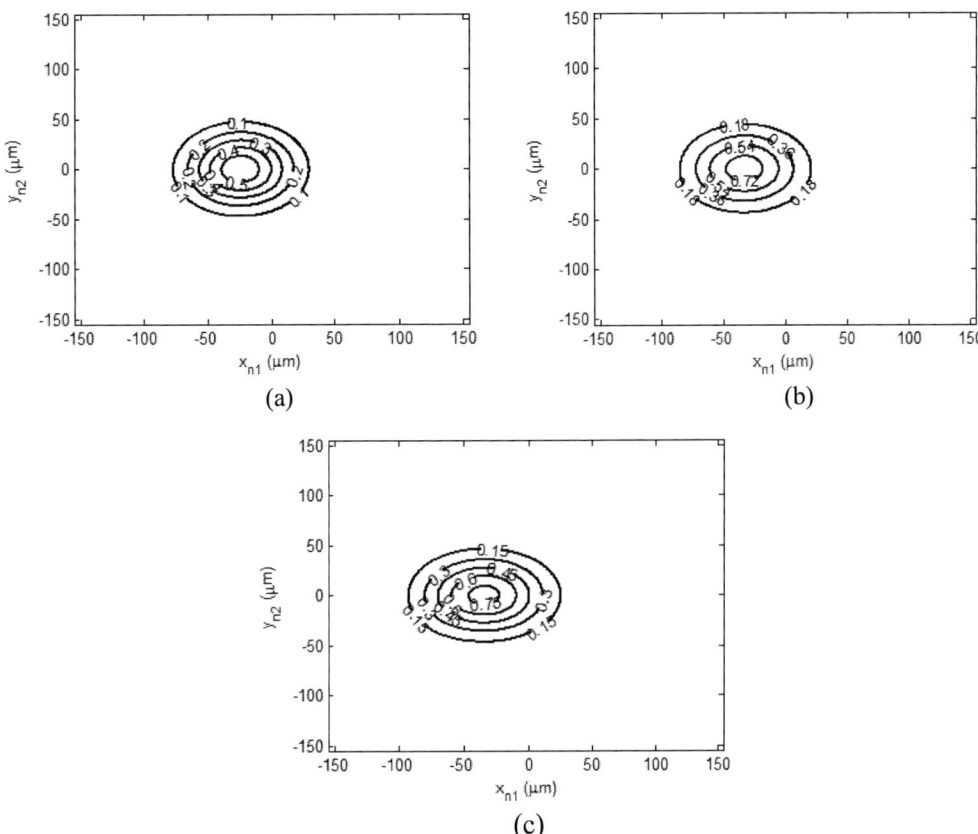

(a)

(b)

(c)

Fig.3. Spatial intensity distributions of reflected laser beam with the coating's thickness (a) $d = 50 \mu m$; (b) $d = 100 \mu m$; (c) $d = 150 \mu m$

From figure 3, we can deduce that with the thickness's increment, the intensity distributions of the reflected laser are similar to those of the input laser in the spatial domain firstly. And then it spreads along the X axis. It means that when the coating's thickness is small, the angular selectivity bandwidth is wide. Although at this situation, the diffraction efficiency is not high, most of the decomposed MPWs are reflected. With the thickness's increment, the diffraction efficiency gets higher. But due to the narrow angular selectivity bandwidth, the spatial intensity distribution of the reflected beam is broadened. The reflectance efficiencies are 60.18%, 91.26%, and 96.05% with respect to the thickness of 50μm, 100μm, and 150μm, respectively.

3.3 Spatial intensity distributions of the reflected beam with the change of the coating's periods

Three groups of coating's parameters are selected, with the coating's periods 360μm, 380μm, or 400μm, respectively. The relative dielectric permittivity modulations are fixed as 2500ppm. The grating's periods are 380nm. The spatial intensity distributions of the reflected laser with different gratings' thickness are shown in figure 4.

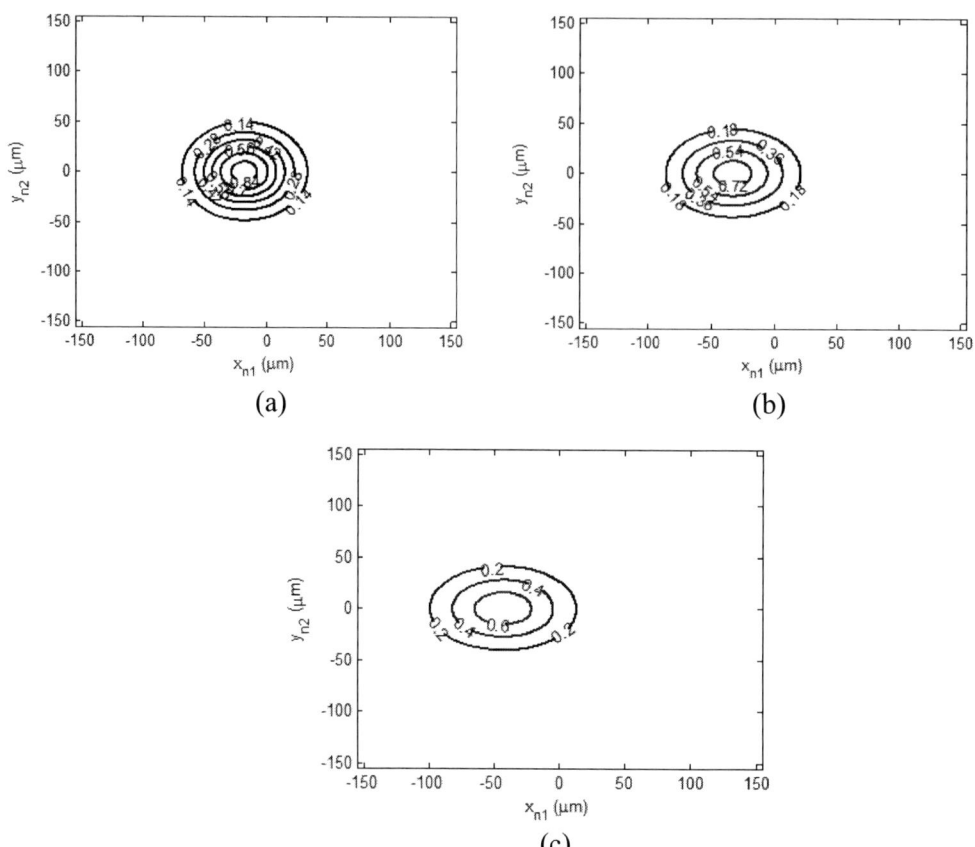

(a)

(b)

(c)

Fig.4. Spatial intensity distributions of reflected laser beam with the coating's period

(a) $\Lambda = 360\mu m$; (b) $\Lambda = 380\mu m$; (c) $\Lambda = 400\mu m$

From figure 4, we can deduce that with the period's increment, the intensity distributions of the reflected laser are broadened in the X axis compared with that of the input laser in the spatial domain. It means that less decomposed MPWs are reflected at the higher period than those at the smaller one. So the angular selectivity bandwidth is thinner with the higher period. The reflectance efficiencies are 89.46%, 91.26%, and 92.58% with respect to the period of 50μm, 100μm, and 150μm, respectively. It means that the period's change does not influence the diffraction efficiency very much.

4. Conclusions

We analyze the reflective characteristics of a Rugate coating illuminated by a spatially-bounded laser beam. Besides the reflectance efficiency, the spatial intensity distribution is put forward with the change of the different coating's parameter. The results show that with the decreasing of the grating's period, increasing of its thickness, and increasing of its relative dielectric permittivity modulation, the spatial intensity distribution of the reflected laser are similar to those of the input pulse. dAnd the reflectance efficiencies increase. Among these three parameters, the relative permittivity modulation and thickness influence the reflectance characteristics a lot, while the period influences the reflectance characteristics a little. It means that when the thickness decreases, or modulation decreases, the Rugate coating has finer angular selective characteristics and less angular selectivity bandwidth. The results are instructive for the design and application of Rugate coating.

Acknowledgement

The author acknowledges the support of National Natural Science Foundation of China (No.61205002).

References

[1] Bertrand G. Bovard 1993 Rugate filter theory: an overview *Appl. Opt* **32** 5427

[2] W. H. Southwell 1988 Spectral response calculations of rugate filters using coupled-wave theory *Appl. Opt* **28** 2949

[3] T. Herffurth et al 2014 Roughness and optical losses of rugate coatings *Appl. Opt* **53** A351

[4] Erwin Delano 1967 Fourier synthesis of multilayer filters *JOSA* **57** 1529

[5] Andy C. van Popta et al 2004 Gradient-index narrow-bandpass filter fabricated with glancing-angle deposition *Optics Letters* **29** 2545

[6] Xinbin Cheng et al 2008 Gradient-index optical filter synthesis with controllable and predictable refractive index profiles *Optics Express* **16** 2315

[7] P. V. Usik et al 2009 Spatial and spatial-frequency filtering using one-dimensional graded-index lattices with defects *Optics Communications* **282** 4490

[8] J. T. Hunt et al 1978 Suppression of self-focusing through low-pass spatial filtering and relay imaging *Appl. Opt* **17** 2053

[9] Liu Daizhong et al 2006 Design and application of a laser beam alignment system based on the imaging properties of a multi-pass amplifier *Chinese optics letters* **4** 601

[10] Liu Daizhong et al 2006 Design and application of a laser beam alignment system based on the imaging properties of a multi-pass amplifier *Chinese optics letters* **4** 601

[11] A. K. Potemkin et al 2007 Spatial filters for high-peak-power multistage laser amplifiers *Appl. Opt* **46** 4423

[12] Bonghoon Kang et al 2010 Optimization of the input fundamental beam by using a spatial filter consisting of two apertures *Journal of the Korean Physical Society* **56** 325

[13] Ivan Moreno and J. Jesus Araiza 2004 Thin-film optical filters for spatial frequencys *Proc. of SPIE* **5524** 409

[14] Ivan Moreno et al 2005 Thin-film spatial filters *Optics Letters* **30** 914

[15] Luo Zhaoming et al 2010 Low-pass rugate spatial filters for beam smoothing *Optics communications* **283** 2665

[16] M. G. Moharam and T. K. Gaylord 1982 Chain-matrix analysis of arbitrary-thickness dielectric reflection gratings *JOSA* **72** 187

[17] David J. Mccartney 1989 The analysis of volume reflection gratings using optical thin-film techniques *Optical and quantum electronics* **21** 93

A Novel Atomic Force Microscope with Multi-Mode Scanner

Chun Qin, Haijun Zhang, Rui Xu, Xu Han and Shuying Wang

State Key Laboratory of Modern Optical Instrumentation, Zhejiang University, Hangzhou 310027, People's Republic of China, Email: zhanghj@zju.edu.cn

Abstract. A new type of atomic force microscope (AFM) with multi-mode scanner is proposed. The AFM system provides more than four scanning modes using a specially designed scanner with three tube piezoelectric ceramics and three stack piezoelectric ceramics. Sample scanning of small range with high resolution can be realized by using tube piezos, meanwhile, large range scanning can be achieved by stack piezos. Furthermore, the combination with tube piezos and stack piezos not only realizes high-resolution scanning of small samples with large-scale fluctuation structure, but also achieves small range area-selecting scanning. Corresponding experiments are carried out in terms of four different scanning modes showing that the AFM is of reliable stability, high resolution and can be widely applied in the fields of micro/nano-technology.

1. Introduction

In recent years, micro and nano technology have become one of the frontier of technological development. Atomic force microscope (AFM) has been proved to be a powerful and versatile tool [1] to study the surface of nanomaterials in various research fields [2-3]. In an AFM, a micro cantilever with a micro-fabricated tip is used to scan the sample surfaces, and the deflection of the micro cantilever is measured to detect the distance between the tip and the sample surface [4-8].

Currently, the majority of conventional AFMs only have a single scanner and a single scanning mode to realize small scanning range of small sample with high resolution [9-10], which cannot meet the growing demand. In order to achieve more functions at the same time, in this paper, a new type of AFM with multi-mode scanner is developed. With some special designs, the AFM can provide some different scanning forms. So it can satisfy the specific need of various samples under different situations.

2. Method and Working Principle

2.1. System components

The overall schematic diagram of the AFM with multi-mode scanner is shown in Figure 1. The system consists of an AFM probe, a scanning and feedback controlling circuit, and a personal computer (PC) with AFM software system. The AFM probe is of an elaborate design including cantilever and tip, measured sample, the XYZ scanner and optical paths, etc. The scanner is made up of tube piezos and stack piezos combined in X, Y and Z directions. The AFM employs sample-scan mode with micro-cantilever fixed to realize the given functions.

Content from this work may be used under the terms of the Creative Commons Attribution 3.0 licence. Any further distribution of this work must maintain attribution to the author(s) and the title of the work, journal citation and DOI.

Published under licence by IOP Publishing Ltd

AOM2015

Journal of Physics: Conference Series **680** (2016) 012009

IOP Publishing

doi:10.1088/1742-6596/680/1/012009

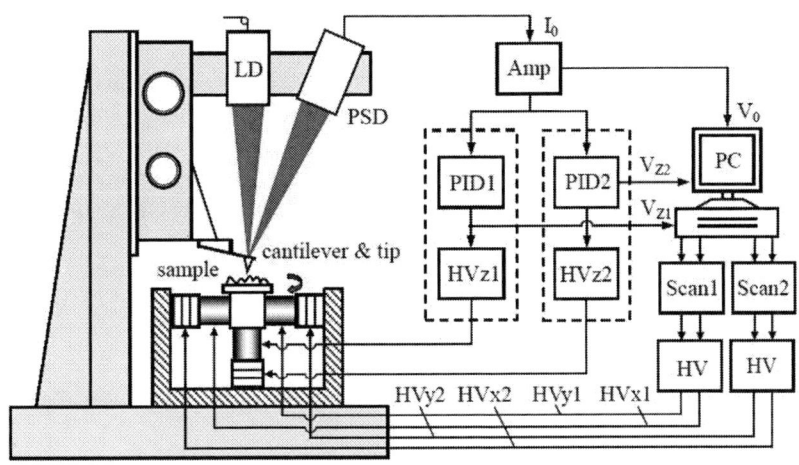

Figure 1. Scheme of the multi-mode scanner AFM

2.2. Working principle
The specially designed AFM can provide at least four scanning forms as shown in Figure 2.

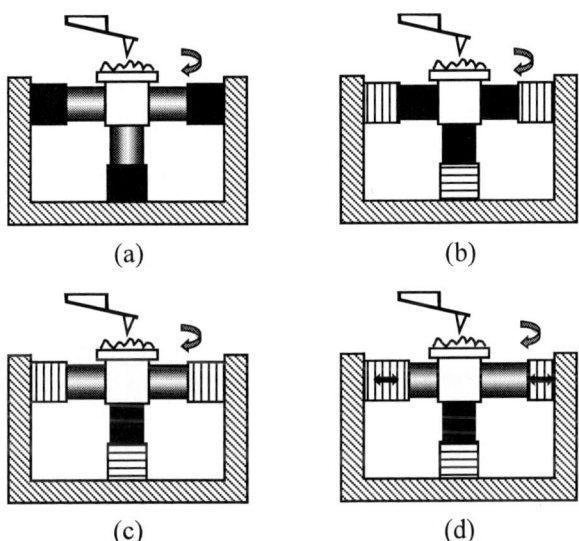

Figure 2. Scanning forms of the multi-mode scanner AFM. (a) small range scanning mode with tube-piezos, (b) large range scanning mode with stack-piezos, (c) combined mode with stack-piezo feedback and tube-piezos scanning, (d) combined mode with stack-piezos area selecting and tube-piezos scanning.

The first scanning form is shown in Figure 2 (a), the stack piezos are regarded as rigid body when uncharged. Like most traditional AFMs, the small scanning range of small sample with high resolution and high speed can be realized by using three orthogonal tube piezos for XY scanning and Z feedback control. The resolution can reach 0.2 nm laterally and 0.1 nm vertically.

In Figure 2 (b), the second scanning form can realize a wider range scanning of the sample by using three orthogonal stack piezos for XY scanning and Z feedback control. In this situation, the tube

39

piezos are rigid body because they are not energized. At the same time, this scanning form can keep a comparatively high resolution both laterally and vertically.

In Figure 2 (c), the third scanning form is a combined one to use stack-piezo for Z feedback and tube-piezos for XY scanning. This form is suitable for scanning small samples with large-scale fluctuation in height. Meanwhile, as shown in Figure 2 (d), small range area-selecting scanning can be realized using two stack piezos in XY directions. First of all, in order to select the target area, a certain size of voltage is applied in the stack piezos in XY directions according to the requirements and remains unchanged. After that, the small range scanning with high accuracy for target area can be realized by using the tube piezos. Moreover, like the first form, the resolution of this form can also reach nanometer level.

3. Experiment

In order to verify the performance of the multi-mode scanning AFM, corresponding experiments are carried out in terms of four different scanning modes.

For small scanning range of small sample, we choose the first scanning form in Figure 2 to realize high accuracy and high speed. The AFM image of nanoimprint structure on polycarbonate film (shown in Figure 3) is scanned under the first scanning form. The distribution of the sample can be clearly seen.

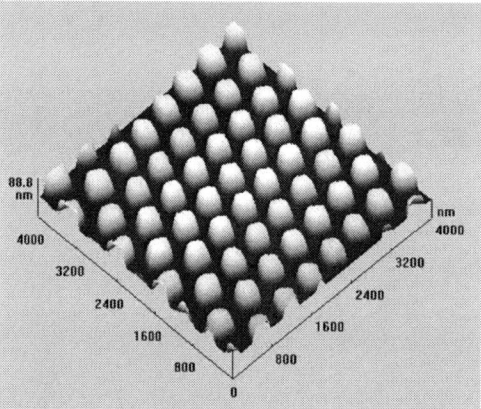

Figure 3. The AFM image of nanoimprint structure on polycarbonate film

When the sample is a larger one and needs greater scanning range, we choose the second scanning form. The AFM image of two-dimensional lattice structure is shown in Figure 4.

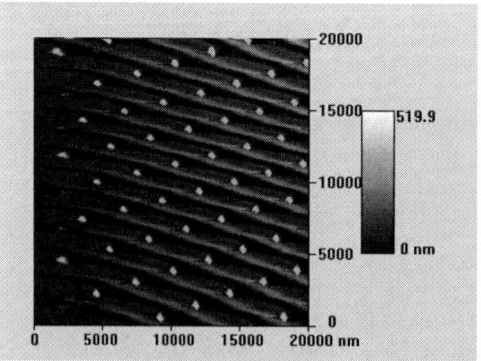

Figure 4. The AFM image of two-dimensional lattice structure

For small sample with large-scale fluctuation structure in height, the third scanning form in Figure 2 can satisfy the requirement. The AFM image of large-scale fluctuation structure on a sapphire glass substrate is shown in Figure 5 (a). In order to see the full image of our interesting area inside the marked squares, the fourth scanning form is selected. As shown in Figure 5 (b), the target area can be seen completely and clearly.

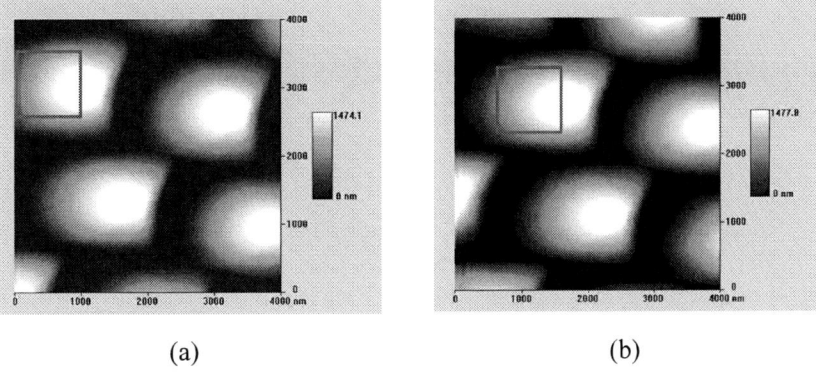

(a)   (b)

Figure 5. The AFM images of large-scale fluctuation structures on a sapphire glass substrate

4. Conclusion
A new type of AFM with multi-mode scanner is developed, and it can be applied to the scan images of different samples. Experiments show that the AFM is of reliable stability and can provide different scanning forms according to the actual requirements. Compared with the conventional AFMs, the novel AFM owns more optimized performance, better repeatability and has a potential of satisfying much wider applications in various research fields.

References
[1] Yamanaka K, Tsuji T, Noguchi A and Koike T. Nanoscale elasticity measurement with in situ tip shape estimation in atomic force microscopy. 2000, *Rev. Sci. Instrum,* **71(6):** 2403-08
[2] Wu Jia, Deng Xiao, Zhang Yun, Wang Lan, Tian Bingqiang and Xie Bijun. Application of atomic force microscopy in the study of polysaccharide. 2009, *Agricultural Sciences in China,* **8(12):** 1458-65
[3] Kwon E Y, Kim Y T and Kim D E. Investigation of penetration force of living cell using an atomic force microscope. 2009, *J. Mech. Sci. Tech,* **23(7):** 1932-38
[4] Zhang Haijun, Zhang Dongxian and Shi Yang. A horizontal atomic force microscope and its applications. 2005, *Nanophotonics, Nanostructure, and Nanometrology,* **5635:** 511-520
[5] Kwon J, Hong J and Kim Y S. Atomic force microscope with improved scan accuracy, scan speed, and optical vision. 2003, *Rev. Sci. Instrum,* **74(10):** 4378-83
[6] Xie Zhigang, Fu Xia and Zhang Dongxian. Study on a novel atomic force microscopy for the detection of large-size samples. 2009, *Acta Optica Sinica,* **29(s2):** 327-330
[7] Zhang Dongxian, Zhang Haijun and Lin Xiaofeng. Atomic force microscope in liquid with a specially designed probe for practical application. 2005, *Rev. Sci. Instrum,* **76(5):** 3705-1
[8] Liu Mingyue, Zhang Haijun and Zhang Dongxian. A compact CCD-monitored atomic force microscope with optical vision and improved performances. 2013, *Microscopy research and technique,* **76(9):** 931-935
[9] Schitter G, Astrom K J, Demartini B E and Hansma P K. Design and modeling of a high-speed AFM-scanner. 2007, *IEEE Transactions on Control Systems Technology,* **15(5):** 906-915
[10] Dedkov V G, Dedkova E G. Contact atomic force microscopy of biological tissues. 2010, *Technical Physics Letters,* **36(2):** 130-132

Photoacoustic sensor of Temperature with Linear-shaped light source

Yuanyuan Peng, Shulian Wu, Dongqing Peng, Zhifang Li and Hui Li[*]

Key Lab of OptoElectronic Science and Technology for Medicine, Ministry of Education; Fujian Provincial Key Lab of Photonic Technology, College of Photonic and Electronic Engineering, Fujian Normal University, Fuzhou, Fujian 350007, People's Republic of China.

E-mail: hli@fjnu.edu.cn

Abstract A method of temperature measurement based on photoacoustic technique with its noninvasive, real-time, high precision is a new type technology. A cylindrical lens was applied to improve the signal to noise ratio of the detector system in this paper. Then, the relation of photoacoustic signal and temperature were discussed, the corresponding image was obtained. The result indicated that the photoacoustic pressure amplitude presented a good linear relationship at temperature range from 20 ℃ to 50 ℃. The study demonstrated that the photoacoustic temperature measurement approach is feasible.

1. Introduction

A temperature measurement method with real-time, noninvasive and accurate is necessary for biomedicine research, especially on the process of photothermal therapy and cooling therapy. At present, some promising techniques of temperature measurement have been developed, including infrared thermography, ultrasonography, and MRI techniques. Infrared thermography method is the real-time temperature measurements with the accuracy of 0.1℃. However, this method is only applicable from surface to superficial of tissue [1]. Ultrasonography is capability of monitoring in real time with good resolution and has relatively low cost, but lacks of high accuracy on temperature measurement [2-5]. MRI is able to provide the images with high resolution and contrast. However, it has limitations associated with long acquisition time and high cost [6].

Laser optoacoustic technique for monitoring temperature is a new type technology with its noninvasive, real time, high precision. Photoacoustic technique has applied on monitoring the process of heating, coagulation, freezing, hypothermia and cooling in biological tissues with its high contrast and resolution [7-11]. But the many applications have not considered the effects of photoacoustic signal caused by irradiated by the pulsed laser. In our study, a cylindrical lens was applied in photoacoustic system. Photoacoustic signal with and without a cylindrical lens were compared to study, and then cylindrical lens was used to change the way of pulsed laser irradiation to improve the signal to noise ratio (SNR) of the photoacoustic imaging system. Finally, vitro pork liver was used for experiment of photoacoustic monitoring temperature on the best SNR of photoacoustic system.

2. Theoretical background

Laser optoacoustic imaging technique is based on thermoelastic mechanism of pressure wave generation. Pressure of thermoelasticity rises with absorbing non-scattering medium upon stress-confined irradiation conditions is defined as [12].

$$P(z) = (\beta c^2 / C_p) \mu_a F(z) = \Gamma \mu_a F(z) = \Gamma \mu_a F_0 e^{-\mu_a z} \tag{1}$$

where, β [1/°C] is the thermal expansion coefficient; c [cm/s] is the speed of sound; C_p [J/g·°C] is the heat capacity at constant pressure; μ_a [1/cm] is the absorption coefficient; $F(z)$ [J/cm²] is the laser fluence; F_0 [J/cm²] is the incident laser fluence ; $\Gamma = \beta c^2 / C_p$ is the Grüneisen parameter. The factor $e^{-\mu_a z}$ represents exponential attenuation of the optical radiation in the medium.

According to equation (1), optoacoustic pressure is proportional to the Gruneisen parameter, fluence, and absorption coefficient of the medium. The exponential slope of optoacoustic signal determines by the absorption coefficient of a nonscattering medium .

For high scattering tissues, the exponent is determined by the effective attenuation coefficient μ_{eff} [1]:

$$\mu_{eff} = \sqrt{3\mu_a(\mu_a + \mu_s')} \tag{2}$$

where $\mu_s' = \mu_s(1-g)$ [1/cm] is the reduced scattering coefficient, μ_s [1/cm] is the scattering coefficient, and g [dimensionless] is the anisotropy factor of the tissue.

Thus, Optoacoustic pressure distribution in a biological tissue with attenuation coefficient μ_{eff} can be expressed:

$$P(z) = \Gamma \mu_a k F_0 e^{-\mu_{eff} z} \tag{3}$$

where k is the parameter resulting from multiple scattering in tissue and is dependent on the absorption and scattering coefficients.

Since the Grüneisen parameter of tissue is linearly dependent on temperature in the temperature range from 20 to 52 °C[13].

$$\Gamma = A + BT \tag{4}$$

where A and B are constants and T is the temperature. one can rewrite equation (3) as follow:

$$P(z) = [A + BT(z)] k \mu_a F_0 e^{-\mu_{eff} z} \tag{5}$$

Using this equation one can obtain:

$$T(z) = C + DP(z)/P(z)_{T=T_0} \qquad (6)$$

where $T(z)$ is the temperature distribution in tissue, $P(z)_{T=T_0}$ is optoacoustic pressure profile recorded at initial temperature T_0 (before heating). C and D are parameters that are dependent on tissue properties. Therefore, the temperature distribution during hyperthermia process will be reconstructed through recording and analysing the temporal optoacoustic pressure profile .

3. Materials and Method

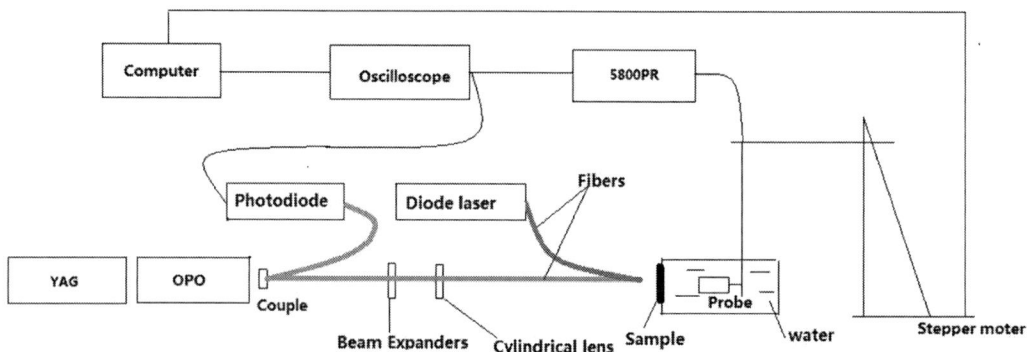

Fig 1. experimental setup

3.1 sample
Fresh excised pork liver with 60mm x 50 mm x 25mm was used for the experiments.

3.2 system
The experimental setup was shown on Fig 1. Nd:YAG laser with wavelength 532nm, pulse width 5ns, repetition frequency 10Hz was used for optoacoustic wave generation. Diode laser with wavelength 810nm was used for treatment heat source.

The light emitted from Nd:YAG laser was divided into two beams. One beam was received by the photodiode, and the signal was displayed on the digital oscilloscope which connected with the computer. The other beam was irradiated on the sample with linear spot after penetrating the beam expanders and cylindrical lens. The thermocouples was inserted the place where the continuous laser emitted from Diode laser irradiated in the center of a line-shaped light.The sample and water were separated by transparent plastic wrap. The ultrasonic signals were detected by ultrasound transducer (5MHz) on the other side of irradiated surface. The signals received by ultrasound transducer were transmitted to ultrasonic pulse generator/receiver (5800PR) and then displayed on the digital oscilloscope. In addition, the stepper motor was applied to realize scanning imaging of sample.

3.3 method
In case of keeping the same conditions during the experimental, the intensities of photoacoustic

signals of the pork liver with and without a cylindrical lens were detected at the same environment with 16 ℃. And then, the signals of two-dimensional scanning of the pork liver without and with a cylindrical lens were scanned, respectively. Then, the 2D images were reconstructed by the signal, which obtained from without and with cylindrical lens.

Finally, the pork liver were heated intermittently when the thermocouple temperature reached at 20 ℃, 25 ℃,30 ℃, 35 ℃,40 ℃, 45 ℃,50 ℃, respectively. During this process, the thermocouple temperature changed slightly with 2D scanning by the stepper motor. All above scans are along the line-shaped light with 0.1mm scanning steps and 27mm scanning length.

4. Results

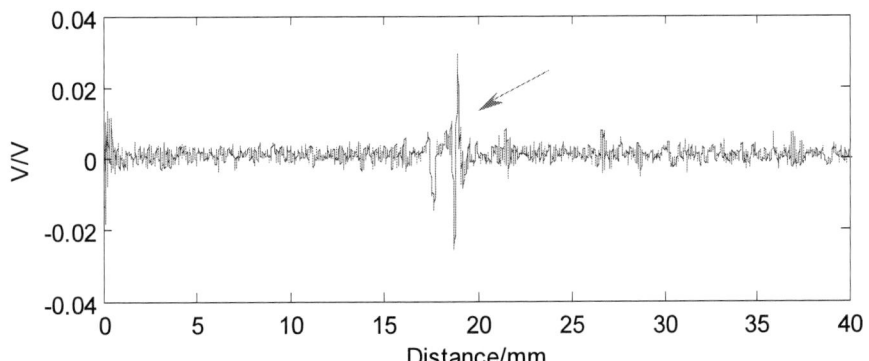

Fig 2. photoacoustic signal of without cylindrical lens

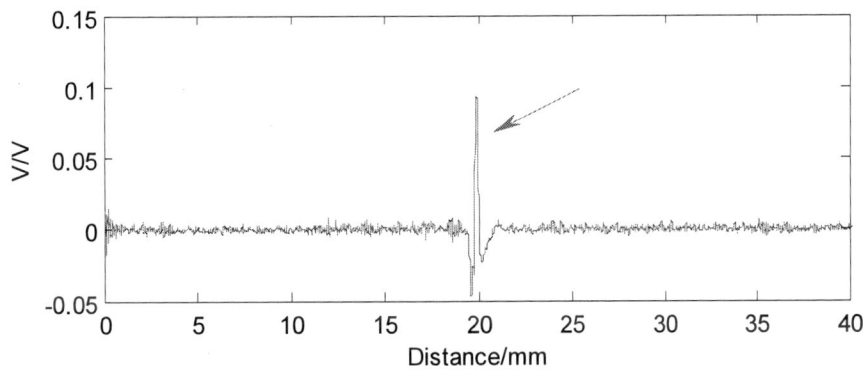

Fig 3. photoacoustic signal with cylindrical lens

The case without and with cylindrical lens were displayed in Fig 2 and Fig 3, respectively. The photoacoustic signal were shown by the arrow. The SNR of photoacoustic signal without cylindrical lens is 12.835, while it is 53.289 with cylindrical lens . The results indicated that the signal to noise ratio of the photoacoustic system will be greatly improved and the quality of photoacoustic images also will be obtained with the cylindrical lens.

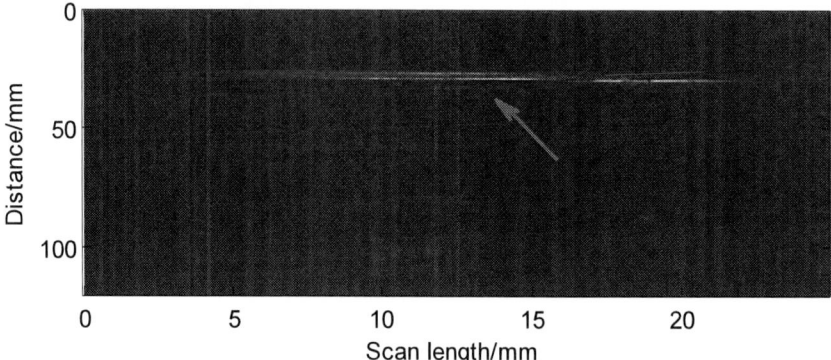

Fig 4. 2D scanning image of the pork liver without cylindrical lens(surface light source) at 16 ℃.

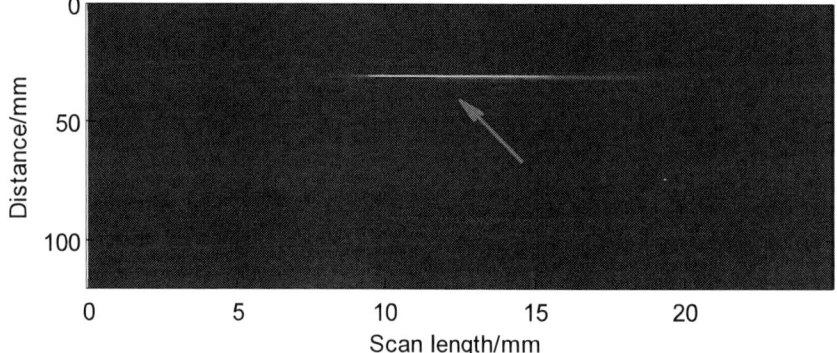

Fig 5. 2D scanning image of the pork liver with cylindrical lens(surface light source) at 16 ℃.

The case 2D scanning image of the pork liver without and with cylindrical lens (surface light source) at 16 ℃ were displayed in Fig 4 and Fig 5, respectively. The photoacoustic signals were shown by the arrows. X-axis represents scanning length, and Y-axis represents the distance from the ultrasonic transducer to the detected signal. The light distribution is more concentrated and photoacoustic signal is stronger with cylindrical lens. The results showed that the way of line-shaped light irradiation with cylindrical lens with its focusing function has higher utilization efficiency of source in the same laser energy, compared with the way of being irradiated by large area uniform light,the defect of insufficient light intensity in per unit area in the case of uniform illumination was overcome in the way with cylindrical lens.

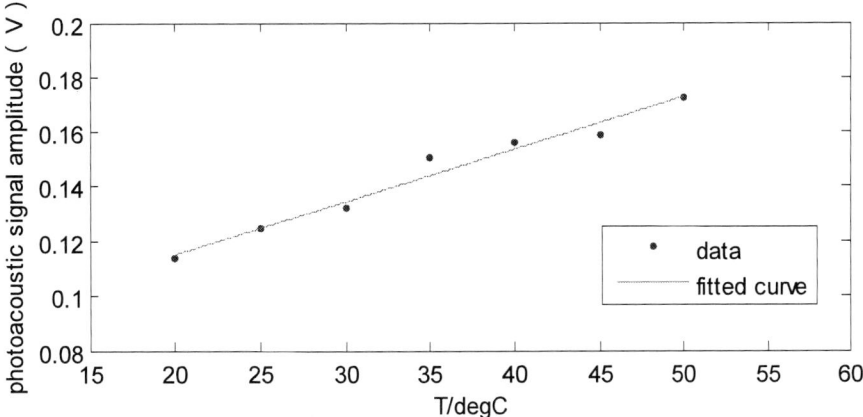

Fig 6. photoacoustic signal amplitude versus temperature

The linear light was uniform in a small enough area using the cylindrical lens .The photoacoustic signal amplitude versus temperature was displayed in Fig 6. Concretely the maximum photoacoustic signal amplitude was extracted from 2D scanning data sets of the pork liver with cylindrical lens (surface light source) at 20 ℃, 25 ℃,30 ℃, 35 ℃,40 ℃, 45 ℃,50 ℃, respectively. The values were 0.1137 V, 0.1245V, 0.1320V, 0.1503V, 0.1562V, 0.1588V, 0.1726 V. From the fitted curve, we can see that with the increasing temperature, the photoacoustic signal that excited by the same laser energy becomes stronger with the temperature from 20 ℃to 50 ℃.

5. Conclusion

In this paper, photoacoustic signal with and without a cylindrical lens were compared to study,and then fresh vitro pork liver was used for experiment of photoacoustic monitoring temperature. The results demonstrated that the SNR of photoacoustic signal can be improved by cylindrical lens. The light distribution is more concentrated and photoacoustic signal is stronger with cylindrical lens. The temperature and the photoacoustic pressure amplitude presented a good linear relationship at temperature range from 20 ℃ to 50 ℃. The photoacoustic technique monitoring temperature can potentially be a valuable method for temperature measurement.

Acknowledgment

This work was supported by the National Natural Science Foundation of China (61178089, 81201124), the Natural Science Foundation of Fujian Province (No.2014J01226).

Reference

[1] Welch A J and Gemert M J C V 1995 Optical-thermal response of laser-irradiated tissue *Plenum Press NY.*

[2] Gilbert J C, Onik G M,Hoddick W K, Rubinsky B 1985 Real time ultrasonic monitoring of hepatic cryosurgery *Cryobiology* **22** 319-30.

[3] Laugier P and Berger G 1993 Assessment of echography as a monitoring Technique for cryosurgery *Ultrasonic Imaging* **15** 14–24.

[4]   Seip R,VanBaren P,Cain C A,Ebbini E S 1996 Noninva-sive real-time multipoint temperature control for ultrasound phased array treatments *IEEE Transactions on Ultrasonics, Ferroelectrics, and Frequency Control* **43** 1063- 1073.

[5]   Maass-Moreno R,Damianou C A 1996 Noninvasive temperature estimation in tissue via ultrasound echo-shifts *Part II. In vitro study. J. Acoust. Soc. Am* **100** 2522-2530.

[6]   Matsumoto R, Selig A M, Colucci V M, Jolesz F A 1993 MR monitoring during cryotherapy in the liver: predictability of histologic outcome *Journal of Magnetic Resonance Imaging* **3** 770-6.

[7]   Esenaliev R O, Larina I V, Larin K V, Motamedi M, Karabutov A A, Oraevsky A A 1999 Laser optoacoustic technique for real-time measurement of thermal damage in tissues *Proceedings of SPIE - The International Society for Optical Engineering* **3594** 98-109.

[8]   Larina I V, Larin K V, Motamedi M, Esenaliev R O 2000 Optoacoustic monitoring of freezing and hypothermia of tissue and tissue phantoms *Proceedings of SPIE - The International Society for Optical Engineering* **4001** 361-367.

[9]   Larin K V, Larina I V, Motamedi M, Esenaliev R O 2000 Monitoring of temperature distribution in tissues with optoacoustic technique in real time *Proceedings of SPIE - The International Society for Optical Engineering* **2000** 311-321.

[10]   Larina I V, Motamedi M 2002 Optoacoustic laser monitoring of cooling and freezing of tissues *Quantum Electronics* **32** 953–958.

[11]   Larin K V, Larina I V, Esenaliev R O 2005 Monitoring of tissue coagulation during thermotherapy using optoacoustic technique *Journal of Physics D Applied Physics* **38** 2645–2653.

[12]   Gusev VE, Karabutov AA 1993 Laser Optoacoustics *ALP Press New York.*

[13]   Esenaliev R O, Oraevsky A A, Larin K V, Larina IV, Motamedi M 1999 Real-time optoacoustic monitoring of temperature in tissues *Proc. SPIE* **3601** 268-275.

Design of high resolution panoramic endoscope imaging system based on freeform surface

Qun Liu, Jian Bai, and Yujie Luo

State Key Laboratory of Modern Optical Instrumentation, Zhejiang University, Hang Zhou, 310027, P.R.China

E-mail: bai@zju.edu.cn

Abstract. This paper introduces a novel endoscope design based on the panoramic annular staring imaging technology. This design utilizes a single optical system to realize both panoramic observation and local high resolution on a single sensor. The freeform surface is employed to improve the image quality and reduce system volume. The design results based on the commercial optical design software package ZEMAX, indicate that this optical system is able to acquire an excellent image quality with a modulation transfer function above 0.6. Compared with the traditional ones, this novel endoscope design with wide FOV is likely to decrease the diagnostic time dramatically and improve the lesion detect rate considerably.

1. Introduction

Wide field endoscopic imaging technology is pivotal in the area of both medical and industrial endoscope. However, current methods toward surrounding observation of inner surface of a tube, are mostly rely on devlating prism or complicated mechanical structure [1-4]. Obviously, these approaches doomed to bring large aperture or long length, which are not fit for detections of small size tube or body cavity. Panoramic Annular Lens (PAL) is a kind of compact optical structure with wide field-of-view (FOV) and small distortion [5]. The PAL design provides panoramic views and sharp imaging quality which means it is very suitable for endoscope system.

Endoscopic objectives adopting PAL structure have been reported in previous literature [6-9]. However, there is an inevitable blind area in the centre of the existing PAL's image plane, which brings visual discomfort and pixels waste. In this study, a PAL-based endoscope design is put forward, which is able to achieve both panoramic view and local high resolution. This system takes full use of the pixels in blind area on the image surface of conventional PAL systems, and solves the contradiction between wide FOV and high resolution.

In order to solve the simultaneous imaging problem of the axisymmetric part and the non-axisymmetric part, freeform surface is employed in this optical structure, so that it has several outstanding features such as simple structure, wide FOV, small size, light weight and high resolution. Compared with the traditional large FOV endoscope, this novel endoscope imaging system is likely to decrease the diagnostic time observably and improve the rate of lesion detecting signally. Moreover, this system has a great value of application in many fields, such as the robot vision, pipeline detection, medical equipment and aerospace. Consequently, the basic research of this system is of profound significance.

Content from this work may be used under the terms of the Creative Commons Attribution 3.0 licence. Any further distribution of this work must maintain attribution to the author(s) and the title of the work, journal citation and DOI.

Published under licence by IOP Publishing Ltd

2. Principle of Design

2.1. PAL Optics

Figure.1 gives the characteristics of a PAL optical system .The cylindrical object space of the PAL has a vertical field angle β which is rotational symmetric around the z axis, that is the optical axis of the system. The area from α to the z axis is a blind space which is not engaged in imaging [10]. The PAL block consists of two reflective surfaces marked by slashes in figure.1 and two refractive surfaces. The optical path can be represented in the following way. Rays leaving the object are refracted into the PAL block from surface 1 and reflected off the rear mirrored surface 2. Then they travel forward in the lens and contact the front mirrored surface 3. Reflected back, the rays exit the PAL block from the rear surface 4. After leaving the PAL, the divergent rays enter a relay lens that corrects and balances aberrations of the PAL block. In the end, an annular image comes into being on a Charge Coupled Device (CCD) sensor.

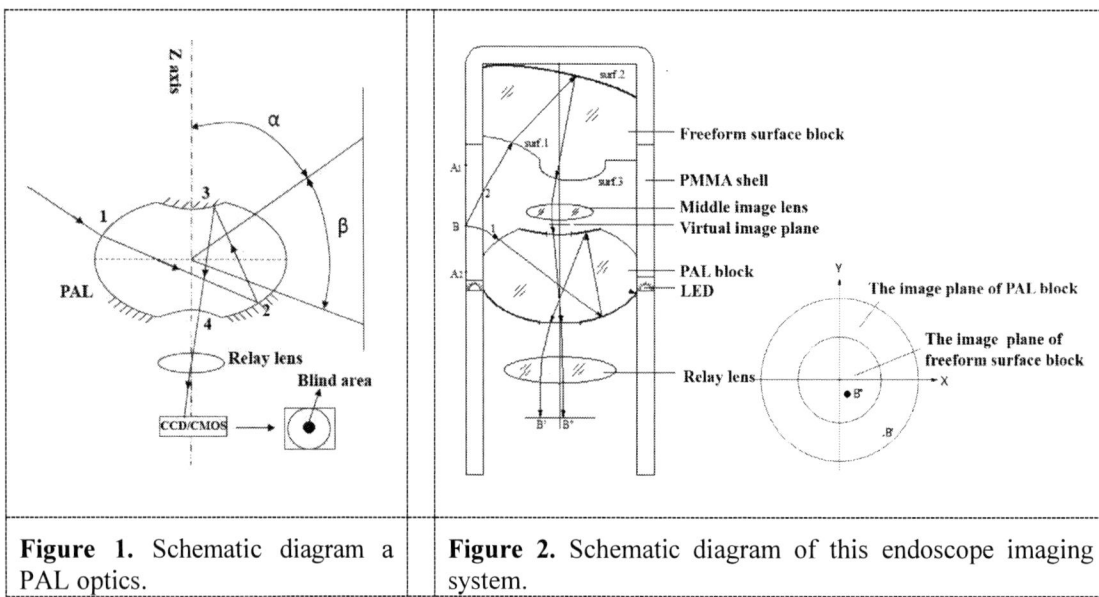

Figure 1. Schematic diagram a PAL optics.

Figure 2. Schematic diagram of this endoscope imaging system.

This work based on a miniaturized panoramic annular imaging optical system takes use of freeform surface lenses to achieve the dual function of panoramic survey and key areas magnification. The PAL block of this system images 360 degrees surrounding the optical axis of the lens to an annular image plane, and the freeform surface block in front of the PAL block amplifies the local area of the PAL's FOV, which uses the blind area of the annulus image plane for imaging. The schematic diagram of the design is shown in figure 1. As performed in the left part of figure 2, the front and back surfaces of PAL are all coated with annular reflective film (for the inside light reflection), while its interior is a transmissive region. The light from the freeform surface block passes through the transmissive area of the PAL. The relay lens forms rays from the PAL block and the freeform surface block to converge and image on the CCD sensor. The right part of figure 2 gives the distribution of image, in which the image of freeform surface block is in the center of the annular image plane of the PAL block. The material of the system shell is selected as optical plastic mainly due to surrounding view. White LEDs located at the joint of endoscopic shell and digestive wall are used to illuminate the body tissue attaching to the optical plastic surface.

In the process of design, the entire system can be divided into two subsystems, one is a PAL subsystem for side view, and the other is a high resolution subsystem for local magnification. The first step is to design the PAL subsystem to find the middle virtual image surface; secondly, designing the high resolution subsystem; finally optimizing the two subunits together.

The resolution of the CCD sensor used in this design is 648*488. The size of pixel is 7.4um, and its spectrum is in the visible range (0.486μm −0.656μm). The design is optimized by a standard optical design software package (Zemax). It achieves good image quality and has a modulation transfer function (MTF) above 0.6, which is within the cutoff frequency of CCD/complementary metal-oxide semiconductor (CMOS) sensors. All the optical design details are described here.

3. PAL subsystem

For the purpose of miniaturization, in our demonstration design the surface 1 and 2 are aspheric surfaces and the material of the PAL block is Polymethyl Methacrylate (PMMA). Figure.3 shows the structure of the PAL subsystem. The largest caliber is 10mm, the PMMA shell is 1mm, and total length is 25mm.The effective focal length is -0.819mm, and its back work distance is 3mm. FOVs of this system are shown by object height specifically, considering the gastrointestinal tract attaches to the endoscope shell during diagnosing. In a PAL system, lateral aberrations, such as lateral chromatic, coma, and others, are major factors impacting the imaging quality, when considering annular imaging [10]. The relay lens system is just for imaging and correcting aberrations. The doublet is applied to reduce the spherical aberration.

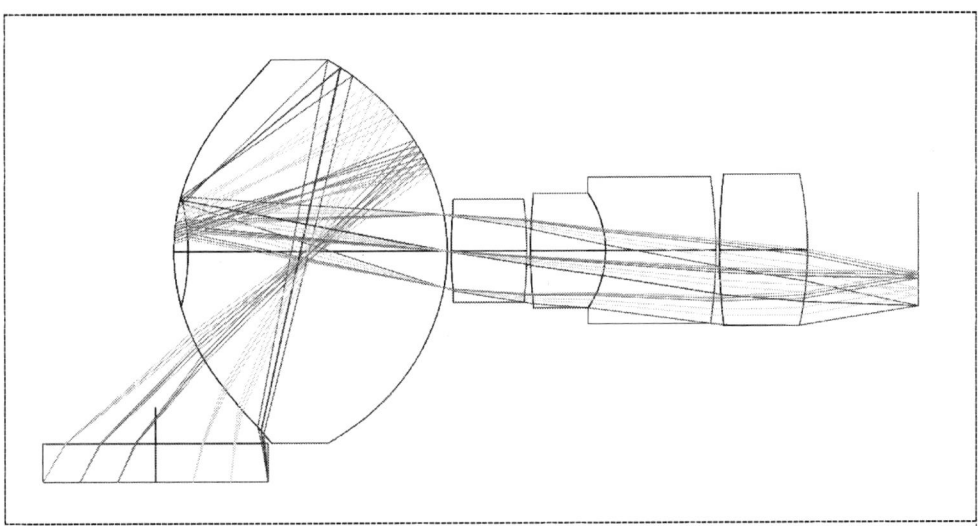

Figure 3. Structure of the PAL subsystem. The different colors represent the different FOVs, which are -3mm,-2mm,-1mm, 1mm, 2mm, and 3mm.

Figure.4 is the graphic report of the PAL subsystem. The MTF effectively evaluates the imaging quality. In this system, the spatial cutoff frequency of the CCD sensor is 70lp/mm. As figure.4(c) shows, the MTF is above 0.8 @70lp/mm, which is very close to the diffraction limit. What's more, the aberration affected the image quality in FOV is balanced well, as shown in figure.4.

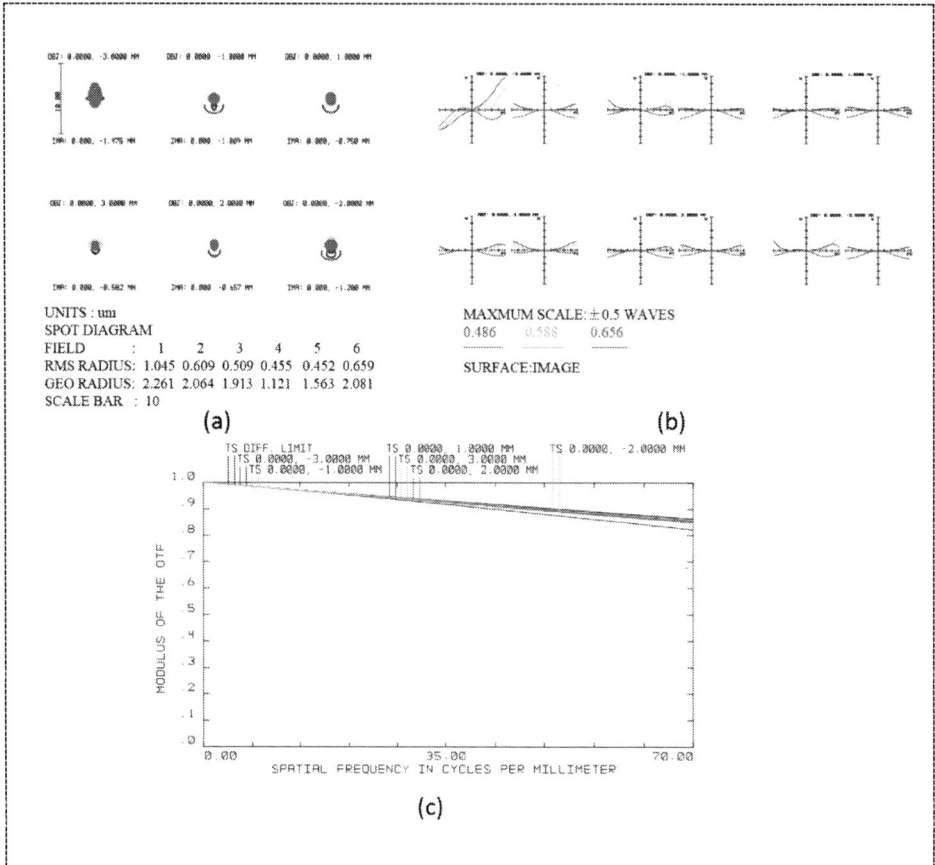

Figure 4. Graphic report of the PAL subsystem: (a) spot diagram, (b) OPD diagram, and (c) MTF diagram.

4. High resolution subsystem

The high resolution subsystem can be given based on the blind area on the image plane and relay lenses of the PAL subsystem, which is used to magnify the target area in the panoramic FOV. Image of the high resolution subsystem is just in the blind area of the PAL's image plane, which means the idle pixels coming in handle.

During the process of optical design, first of all, the known blind area in the image plane can be seen as the object of the high resolution subsystem to design reversely to get a preliminary basic structure easily. An axisymmetric model is used to build an all spherical initial structure, and then the freeform surfaces and mirror are added to turn optical axis and implement asymmetric design. The magnification of cylindrical lens is different in radial direction and axis direction. Therefore a cylindrical lens settled before the PMMA shell is capable of imaging the columnar gastrointestinal tract to a plane, for its character of different magnifications in two perpendicular direction. Figure.5 shows the structure of the high resolution subsystem. The distance between the mirror and the shell is 5mm.The FOV is represented by object height. As the object plane is a cylindrical surface, both the direction of x axis and y axis should be taken into consideration when setting the field data. It is telecentric in image space (which means the exit pupil is located at infinity) for uniform distribution of illumination on the image surface. The transmission part of PAL block and the relay lenses between PAL and image plane are also included in this unit, with the same optical parameters as in PAL subsystem. Lenses near the mirror are all aspheric surfaces, and are made of PMMA in terms of correcting aberration and easy fabrication. Aspheric surface can help to increase the freeform of design. The FOV of this subsystem is just the edge of that of the PAL subsystem. And its image fills the blind area of the PAL subsystem. The magnifying factor is -0.88 which is much bigger than -0.15 of PAL subsystem. The resolution of object space is up to 0.008mm.

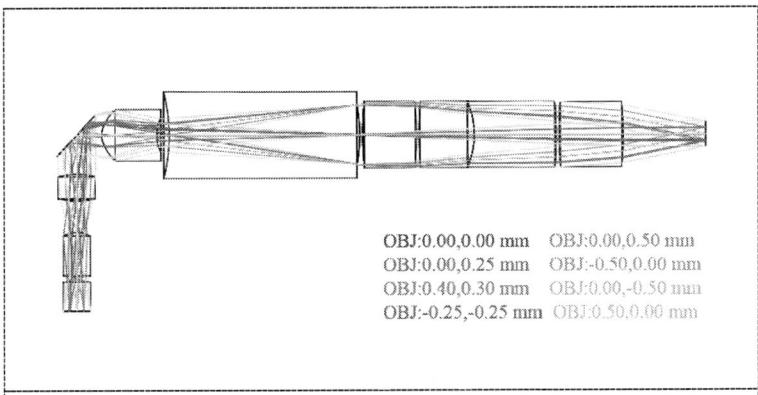

Figure 5. Structure of the high resolution subsystem. The different colours represent the different FOVs.

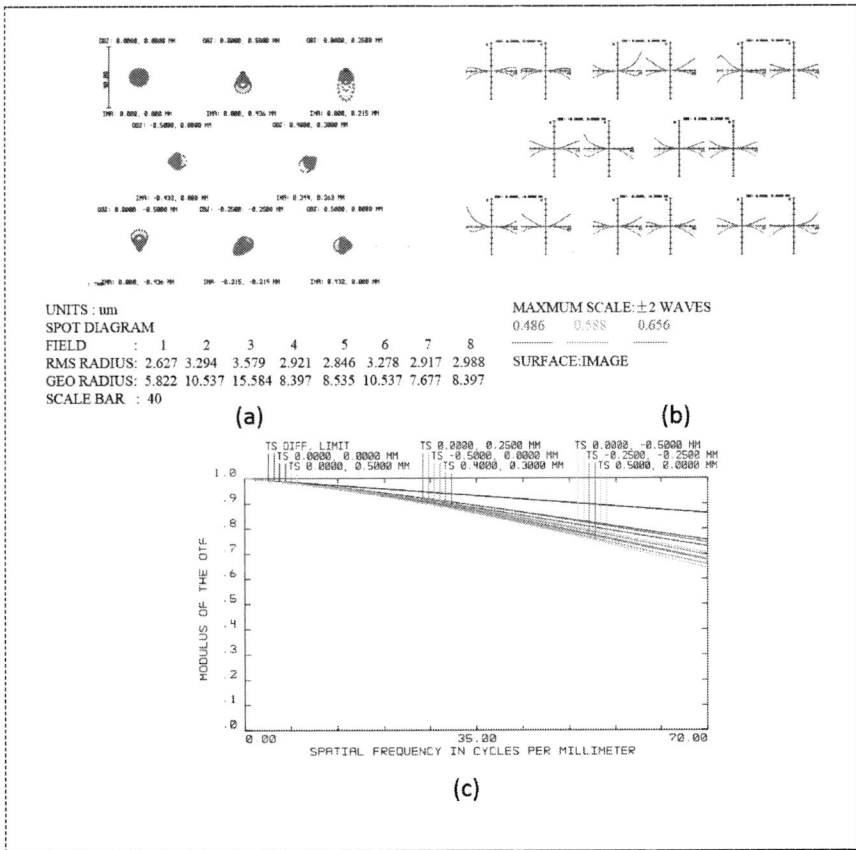

Figure 6. Graphic report of the high resolution subsystem: (a) spot diagram, (b) OPD diagram, and (c) MTF diagram.

Figure.6 gives the graphic report of high resolution subsystem. As figure.6(c) shows, the MTF is above 0.6 @70lp/mm, which is higher than the resolution requirement of the CCD sensor. Further, the aberration affecting the image quality in FOV is balanced well, as shown in figure.6.

5. Results

The final important step is combining the two subsystems together into an integrated endoscope. The Multi-Configuration function in ZEMAX software is able to optimize two different units simultaneously. The two subsystems have different FOV, image height and entrance pupil, but they share the same locations of stop and image surface. A little adjustments to the high resolution subsystem can help to solve the inevitable light blocking problem caused by the small displacement between the cylindrical lens and the shell, but the image quality may be impacted slightly.

After repeated optimization, we get the optical structure of the whole endoscope, as illustrated in figure.6, and the optical parameters is shown in Table.1. The results show that the imaging quality of this system is very well balanced and the volume size meets the requirement of endoscopy. Freeform surface lenses help to solve the simultaneous imaging problem of the axisymmetric part and the non-axisymmetric part. In addition, the results of FOV and object space resolution shows that this endoscope can realize panoramic survey and local high resolution. The tolerance result analysed by ZEMAX indicates that the stability of this system meets the requirements of engineering processing.

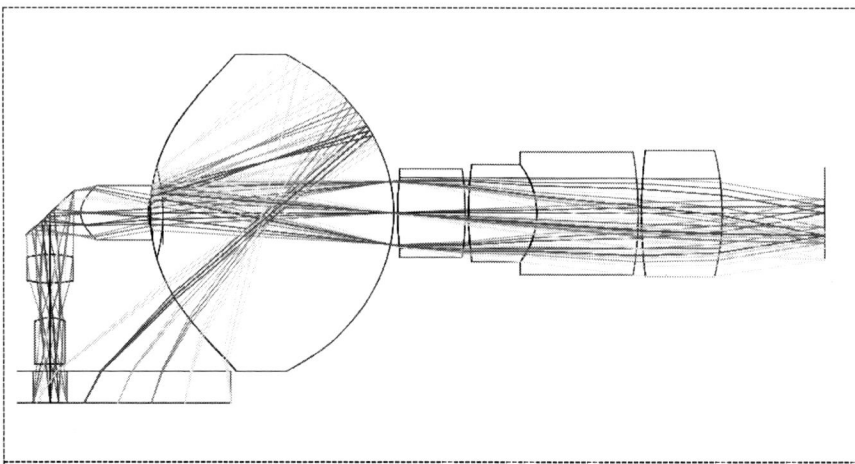

Figure 6. Structure of the whole endoscope optical design. The different colours represent the different FOVs.

Table 1. Optical parameters of PAL subsystem and high resolution subsystem

Optical parameters	PAL subsystem	High resolution subsystem
Spectral range	486.1nm~656.3nm	486.1nm~656.3nm
FOV	$(60° \sim 97.5°) \times 360°$	$(-10.2° \sim 10.2°) \times 20.4°$
Finite object distance	6mm	6mm
EFL	-0.82mm	2.12mm
Working F/#	2.8	2.65
Object space resolution	0.047mm	0.008mm
MTF (full field)	MTF>0.8 @70lp/mm	MTF>0.6 @70lp/mm
Spot size	RMS radius<1um	RMS radius<4um
OPD	±0.5 waves	±2 waves
Image plane	Radius =1.5mm	
Volume	Total length =23.5mm, Diameter =10mm	

Conclusion

The high resolution panoramic endoscope imaging system based on freeform surface is proposed in this work, which achieves both panoramic viewing and partial magnification. In comparison to the traditional endoscopes with wide FOV, this design gives possibility to reduce the diagnostic time and improve the lesion detect rate without regular rotation. More importantly, this system has potential use in robot vision, pipeline detection, medical equipment and aerospace. This basic research is of great importance for the further fabrication.

Reference

[1] E. Kobayashi, I. Sakuma, K. Konishi, M. Hashizume and T. Dohi 2004 A robotic wide-angle view endoscope using wedge prisms *Surg Endosc* **18** 1396-1398

[2] Eric L. Hale, Nathan Jon Schara and Hans David Hoeg, WIDE ANGLE FLEXIBLE ENDOSCOPE, US Patent: 20120116158A1

[3] Patrice Roulet, Pierre Konen and Mathieu Villegas 2010 360° endoscopy using panomorph lens technology *Photonics West SPIE* **7558** 75580T-1

[4] Rysuke Sagawa, Takurou Sakai, Tomio Echigo, etc, 2008 Omnidirectional Vision Attachment for Medical Endoscopes, the 8th Workshop on Omnidirectional Vision, Camera Network and Non-classical Cameras-OMNIVIS, Marseille: France

[5] Z. Huang, J. Bai and X. Y. Hou 2012 Design of panoramic stereo imaging with single optical system *Opt. Expres* **20**: 6085~6096

[6] John A. Gilbert, Donald R. Matthys and C. M. Lindner 1992 Endoscopic inspection and measurement *SPIE* **1771** Application of Digital Image Processing XV,

[7] Donald R. Matthys, John A. Gilbert and Pal Greguss October 1991 Endoscopic measurement using radial metrology with digital correlation *Optical Engineering* **30** 1455-1460

[8] Roy Chih Chung Wang, M. Jamal Deen, David Armstrong, and Qiyin Fang June 2011 Development of a catadioptric endoscope objective with forward and side views *Journal of Biomedical Optics* **16** 066015

[9] Zong-Ru Yu, Cheng-Fang Ho, Annie Liu, etc. 2012 Design and development of bi-directional Viewer *SPIE* **8486** 848613

[10] Dong Hui, Mei Zhang, Zheng Geng, etc. July 2012 Designs for high performance PAL-based imaging systems *APPLIED OPTICS* **51** No. 21/20

AOM2015   IOP Publishing
Journal of Physics: Conference Series **680** (2016) 012012   doi:10.1088/1742-6596/680/1/012012

Full-aperture long focal-length measurement based on divergent beam

Jia Luo*, Jian Bai, and Kaiwei Wang
State Key Laboratory of Modern Optical Instrumentation, Zhejiang University,
Hangzhou 310027, China

E-mail: 11330045@zju.edu.cn

Abstract. A new method for long focal-length measurements is proposed. In this method, we employ divergent beam and two Ronchi gratings of different periods, as the alternative to collimated beams and two identical gratings in traditional method, to realize achieve higher accuracy. Moreover, with divergent beam, the full-aperture measurement is easily realized when detect large diameter lens. The experiments demonstrated the proposed method features high accuracy and repeatability.

1. Introduction

Ultra-long focal-length lens (UFL) is now being widely used in large optical systems. It is of great significance to develop a fast, real-time measurement system to satisfy the stringent requirement for ultra-precision long focal-length measurement. However, accurately measurement is still a great challenge because it is difficult to find the exact position of the focus and the long measurement light-path can be easily affected by the disturbance of environment [1, 2]. Some methods have been proposed in previous literatures to deal with this problem. Meshcheryakov et al. [3] acquired a measured focal-length of 25 m by inserting an optical wedge into the light-path, which changed the position of a luminous slit, achieving an accuracy of 0.1%. Moreover, DeBoo and Sasian [4, 5] applied a Fresnel-zone hologram with a precision better than 0.01% when measured a 9 m focal-length lens. This technique worked well when applied to large, slow lenses; but in terms of lenses with large curvature, it is difficult to fabricate suitable hologram and the measurement precision is limited by optical lithography. Besides, W. Zhao et al [1, 6] proposed a laser differential confocal technique for long focal-length measurement in 2009. They used a differential confocal focusing system (DCFS) to measure the variation in position of DCFS focus with and without a UFL and then obtain the UFL focal-length from the distance between the two focuses. The precision is about 0.01%, which is depending on the reference lens.

The key for improving the measurement precision is shortening the light path and improving the focusing precision. Those above methods have reached high precision. But the stringent requirement for environmental temperature, air disturbance and vibration are hardly achieved. To overcome these difficulties, the moiré interferometry method, based on Talbot effect and the moiré technique, have been reported in recent years [7-12]. The research discussed this method in details which employed collimated rays, two equal-period gratings, and the scanning system. The sensitivity of moiré interferometry can be easily tuned by varying the crossed angle between the two gratings, thus high measurement accuracy can be achieved. This method greatly shorten the testing light path, however, in

---

Content from this work may be used under the terms of the Creative Commons Attribution 3.0 licence. Any further distribution of this work must maintain attribution to the author(s) and the title of the work, journal citation and DOI.

Published under licence by IOP Publishing Ltd

most case the scanning system induced inevitable cumulative errors, making an enormous impact on the accuracy.

In this paper, we propose and demonstrate a novel, feasible and accurate method to overcome these problems. In order to realize full-diameter measurement, we employ divergent beam and two Ronchi gratings of different periods. In this way, the collimation process and scanning system are avoided. The stability and accuracy of the measurement are improved and the time for testing is tremendously reduced accordingly. Furthermore, the whole measurement can be easily achieved through the stripe matching of the moiré patterns, depending on neither precise adjusting mechanism nor troublesome manual adjustment. The experiments confirm the validity and high accuracy of this method.

As follow, the principle of stripe matching and simple description of measurement system will be presented in Section 2; Section 3 provides the experiment process and results; The conclusion will be demonstrated in Section 4.

2. Method

The schematic representation of optical path in our method is shown in Figure (1). We can see from Fig. (1-a) there are two gratings set parallel with each other, we denote the two gratings as G1 and G2 with periods p_1 and p_2. The length of optical path along the optical axis between grating G1 and grating G2 is set as the distance Z, which be referred to as Talbot distance . After pass through grating G1, the monochromatic divergent wave produces a Talbot image G1'. As the wave is divergent, the image G1' is a magnified Talbot image and the period of G1' is expressed as p_1'. Then the Talbot image G1' is superimposed on the second grating G2, creating the image of moiré patterns.

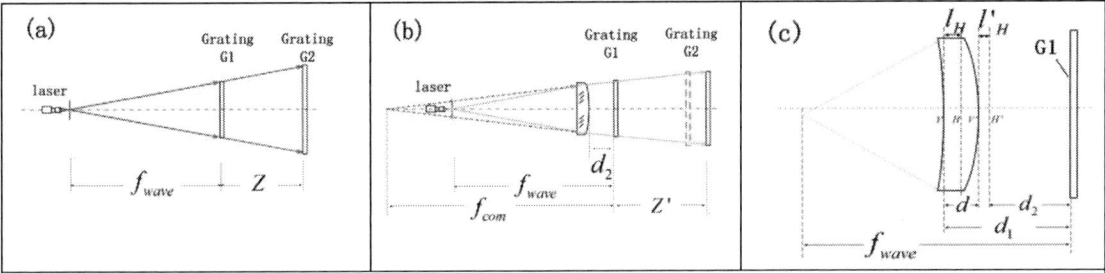

Figure 1. : (color online) Schematic representation of our method. (a) Focal-length of divergent beam, (b) Test lens is inserted, the combined focal-length of new wave is represented as the red dash line. (c) Parameters of test lens

According to the geometricaloptics, shown in Fig.1 (a), the focal-length of the divergent wave satisfies the following relation:

$$\frac{f_{wave}}{f_{wave} - Z} = \frac{p_1}{p_1'} \tag{1}$$

Here f_{wave} is negative, regarded as the focal-length of the beams illuminated on the grating G1. The optimal Talbot distance Z is determined by moving grating G2 along the optic axis.

The tilt angle α between the patterns and the x-axis can be calculated out by computer with the help of Fourier transform algorithm, and the relationship in the image of moiré patterns can be given by

$$\tan \alpha = \frac{\dfrac{p_2}{p_1'} - \cos\theta}{\sin\theta} \tag{2}$$

In which θ is the crossed angle between two gratings in x-y plane.

Then, as shown in Fig.1 (b), we insert the test lens into the light path in front of the first grating G1and get a new combined divergent beam. The focal-length of this new beam is called .Now the period of the image G' (Talbot image of grating G1) on the second grating G2 is changed. The relationship about the new moiré patterns can be given by

$$\tan \alpha'' = \frac{\dfrac{p_2}{p_1''} - \cos\theta}{\sin\theta} \tag{3}$$

Here α'' is the tilt angle of new patterns, p_1'' is period of new image G1''. This allows us to rewrite Eq. (2) as:

$$\frac{f_{com}}{f_{com} - Z'} = \frac{p_1}{p_1''} \tag{4}$$

In Eq. (4), Z' is the new Talbot distance. Moving the Grating G2 along the z-axis until the Z' is equal to Z. According to Eq. (2) and Eq. (4) the period is p_1'' equal to p_1' at this point. Hence we get:

$$f_{com} = \frac{Z'}{Z} f_{wave} \tag{5}$$

The diagram of the lens is shown in Fig.1 (c). The lens parameters are got by the preliminary measurement, the focal-length of the measured lens can be given by:

$$\begin{cases} l_H = \dfrac{-dr_1}{n(r_2 - r_1) + (n-1)d} \\[2mm] l'_H = \dfrac{-dr_2}{n(r_2 - r_1) + (n-1)d} \\[2mm] d_2 = d_1 - d - l'_H \\[2mm] f_{lens} = 1/(\dfrac{1}{f_{com} + d_2} - \dfrac{1}{f_{wave} + d_1 - l_H}) \end{cases} \tag{6}$$

Where, l_H and l'_H indicate the distance between principal plants and surfaces of the lens. r_1, r_2 and n are radius of two surfaces and refractive index of the lens, respectively. The distance d_1 from front surface of the lens to grating G1 can be measured precisely by grating ruler.

3. Experiment and Result

Figure 2 shows schematic diagram of experiment setup. An infrared laser with wavelength 1053nm was used in our experiment. The parallelity of test lens and two gratings was ensured by an auto collimation. The focal-length of divergent beam was measured by grating ruler with 10um precision, and an electric displacement platform was used to move the grating G2, the precision is also 10um. The accuracy of the inclination angle and the crossed angle, which has been discussed in [15, 16, 17], are 0.005° and 0.003° respectively.

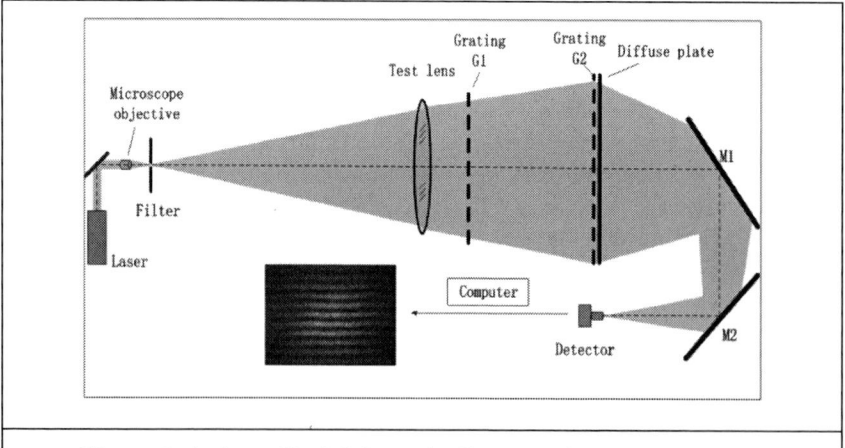

Figure 2. (color online) Schematic diagram of experiment setup.

The experiment steps are as follow. First we measured the distance between laser and the first grating G1 i.e. the focal-length of divergent beam by a grating ruler. The distance between two gratings was also obtained with the help of the ruler. Tilt angle of the moiré pattern was calculated by computer as α. After that we inserted the test lens into the light path and obtained the distance between the lens and grating G1. The corresponding new Talbot distance was determined when the tilt angle of dynamic fringe pattern achieved α (by moving grating G2 with the help of displacement platform). Finally, the focal-length of the lens could be computed out according to Eq. （6）.

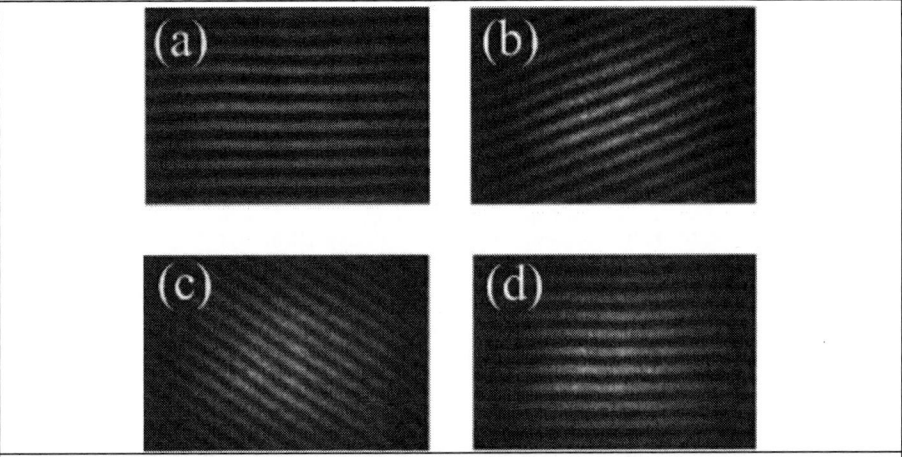

Figure 3. Fringe patterns recorded with CCD camera. Images (a) without lens (b) at in-corresponding Talbot distance with respect to test lens. (c) at out-of-applicable Talbot distance with respect to test lens. (d) at at-corresponding Talbot distance with respect to test lens.

Figure 3 shows the moiré pattern obtained with the system. Fig. 3(a) is the fringes taken without lens. Fig. 3(b) and (c) are the fringes when the grating G2 was at in- and out-of-corresponding Talbot distance, respectively. Fig. 3(d) is the fringes at at-corresponding Talbot distance with respect to test lens can be captured for calculation.

Table 1. Measurement results of proposed method.

No.	Nominal $f(mm)$	Measured $f(mm)$	$\Delta f(mm)$	$\Delta f / f$ (%)
1	13500	13498.7	-1.3	-0.009
2	13500	13485	-15	-0.111
3	13500	13479.3	-20.7	-0.153
4	13500	13489.5	-10.5	-0.077
5	13500	13492.5	-7.5	-0.055
6	13500	13468.4	-31.6	-0.234
7	13500	13450.6	-49.4	-0.365
8	13500	13488.3	-11.7	-0.086
9	13500	13487.9	-12.1	-0.089

Table 1 shows the results of the measurement undertaken with 10 lenses of same focal length. To reduce the effect of air turbulence and electronic noise, 10 continuous measurement values were averaged for each measurement point. Comparing experiment results with the nominal values, the relative error of the measurement of focal-length of lenses is better than 0.36%. Hence, the measured focal-length of the lenses agrees well with the nominal values.

Table 2. Measurement results of knife-edge test and Hartmann approach.

No.	Knife-edge (mm)	Error (%)	Hartmann (mm)	Error (%)
1	13192	0	-	-
2	13179	-0.098	-	-
3	13195	0.022	-	-
4	13197	0.037	-	-
5	13191	-0.007	13453	0.01
6	13169	-0.174	13426	-0.19
7	13156	-0.272	13409	-0.32
8	13167	-0.189	-	-
9	13177	-0.113	-	-

Further experiments were performed to verify the new method works with high accuracy. The lenses we tested above were also measured by the knife-edge test and Hartmann approach. Because of experimental conditions, only 3 pieces of lenses were tested by Hartmann approach. And different wavelength beam were employed in different approaches, hence the nominal values of these lenses are different (13192mm in knife-edge test and 13452mm in Hartmann approach). Table 2 shows the measurement results of knife-edge test and Hartmann approach and Fig. 4 shows the relative errors of each approach. The blue, red and green curves represent the relative error of the knife-edge test, Hartmann approach and our system, respectively. Comparison between these three systems is demonstrated in Fig. 5. Good agreement can be seen from the two figures. The relative error between three systems in Fig. 5 is better than 0.18%, indicating the high relative precision of our system. Comparison of these systems reveals that the long focal-length can be measured accurately by using the proposed method.

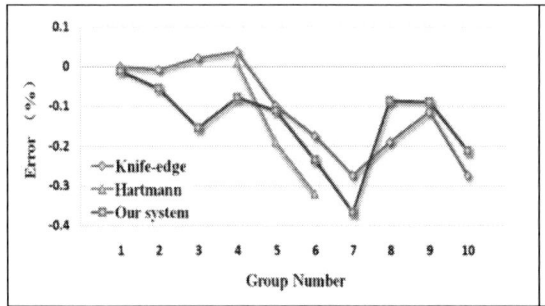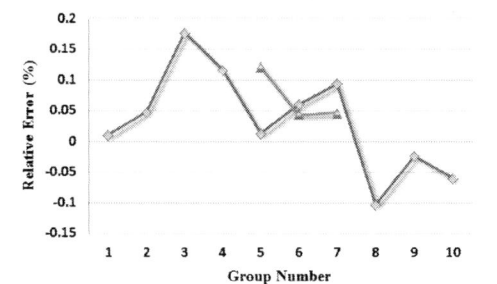

Figure 4. (color online) Results of three approaches. The blue, red and green curves represent the relative error of the knife-edge test, Hartmann approach and our system, respectively.

Figure 5. (color online) Results for comparison. The blue curve represents the relative error between knife-edge test and our system, and the green curve represents the relative error between Hartmann approach and our system

The stability is also important for a precision measurement system. To check it out, some other independent measurements were carried out. A 13500mm focal-length lens was fixed in our system for 24 hours. As shown in Fig. 6, 20-groups measurements with time interval of one hour are demonstrated. Each spot in Fig. 6 denotes the average values of 10 continuous measurements within each group. The standard deviation of those measurements values is 0.0557mm, better than 0.0004%.

It is clearly that our system is of high accuracy and insensitive to the testing environment. Both the sound validity and stability of the proposed new method can be told obviously.

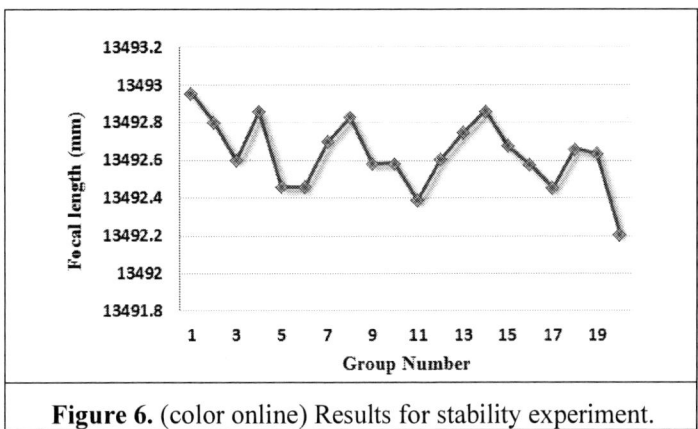

Figure 6. (color online) Results for stability experiment.

4. Conclusion

A novel method of long focal-length measurement based on stripe matching is proposed. In this method, a divergent beam illuminating two parallel gratings produces moiré fringes on the second grating. Then, added a test lens in front of the first grating, a new moiré fringes is produced. The two moiré fringes are matched by moving the second grating and the corresponding focal-length can be easily obtained by the measurement of the distance between gratings. In the experiments, the novel method shows good performance in testing accuracy, and both the sound validity and stability of the proposed new method can be told obviously.

Acknowledgements

This work was supported by Research Center of Laser Fusion and State Key Laboratory of Modern Optical Instrumentation, Zhejiang University. The authors are gratefully to L. Chai and Q. Li for their encouragement and Y. Jiang, F. He for discussion

Reference

[1] P. Singh, M. S. Faridi, C. Shakher, and R. S. Sirohi, "Measurement of focal-length with phase-shifting Talbot interferometry,"Appl. Opt. **44**, 1572–1576 (2005).

[2] V. I. Meshcheryakov, M. I. Sinel'nikov, and O. K. Filippov, "Measuring the focal-lengths of long-focus optical systems,"J. Opt. Technol.**66**, 458–459 (1999).

[3] B. DeBoo and J. Sasian, "Precision focal-length measurement technique with a relative Fresnel-zone hologram,"Appl. Opt.**42**, 3903–3909 (2003).

[4] B. DeBoo and J. Sasian, "Novel method for precise focal-length measurement," in International Optical Design Conference, 2002 OSA Technical Digest Series (Optical Society of America, 2002), paper IMCS5.

[5] W. Zhao, R. Sun, L. Qiu, and D. Sha, "Laser differential confocal radius measurement," Opt. Express. **18**, 2345–2360 (2010).

[6] Y. Nakano and K. Murata, "Talbot interferometry for measuring the focal-length of a lens,"Appl. Opt.**24**, 3162–3166 (1985).

[7] Y. Nakano and K. Murata, "Measurements of phase objects using the Talbot effect and Moiré techniques," Appl. Opt. **23**, 2296–2299 (1984).

[8] Y. Nakano and K. Murate, "Talbot interferometry for measuring the small tilt angle variation of an object surface," Appl. Opt. **25**, 2475–2477 (1986).

[9] Y. Nakano, "Measurements of the small tilt-angle variation of an object surface using Moiré interferometry and digital image processing," Appl. Opt. **26**, 3911–3914 (1987).

[10] C.-W. Chang and D.-C. Su, "An improved technique of measuring the focal-length of a lens," Opt. Commun. **73**, 257–262

[11] K. V. Sriram, M. P. Kothiyal, and R. S. Sirohi,"Direct determination of focal-length by using Talbot interferometry,"Appl.Opt.**31**, 5984–5987 (1992).

[12] J. Stoer and R. Bulirsch, Introduction to Numerical Analysis (Springer, 2002).

[13] C. Sun, Y. Shen, J. Bai, X. Hou, and G. Yang, "The precision limit analysis of long focal-length testing based on Talbot effect of Ronchi grating," Acta Photon. Sin.**33,** 1214–1217(2004).

[14] Changlun H, Jian B, and Xiyun H. "The measurement of long focal length based on Talbot effect of Ronchi grating," Acta Optica Sinica, 2002, **22**(11): 1328-1330.

[15] C. Hou, J. Bai, and X. Hou, "Novel method for testing the long focal length lens of large aperture," Opt. Lasers Eng. **43**(10), 1107–1117 (2005).

[16] X. Jin, J. Zhang, J. Bai, et al. "Calibration method for high-accuracy measurement of long focal length with Talbot interferometry," Appl. Opt. **51**(13): 2407-2413 (2012).

[17] J. Luo, J. Bai, and J. Zhang. "Long focal-length measurement using divergent beam and two gratings of different periods," Opt. Express. **22**(23): 27921-27931 (2014).

Three-Photon Luminescence of Gold Nanorods Excited by 1040 nm Femtosecond Laser for High Contrast Tissue and *In Vivo* Imaging

Shaowei Wang[1], Xinyuan Zhao[2], Hequn Zhang[1], Fuhong Cai[1,3,*], and Jun Qian[1]

[1]State Key Laboratory of Modern Optical Instrumentions, Centre for Optical and Electromagnetic Research, Zhejiang University, Hangzhou, China 310058;
[2]Bioelectromagnetics Laboratory, School of Medicine, Zhejiang University, Hangzhou, China 310058; [3]Suzhou WiHealth Information Technology Co. Ltd, Changshu, China 215500.

*Corresponding author: caifuhong@zju.edu.cn

Abstract. Gold Nanorods (GNRs) with tunable aspect ratios can strongly absorb and scatter light in the NIR region due to their localized surface plasmon resonance (LSPR) property, and have been demonstrated to exhibit strong plasmon enhanced multi-photon luminescence (MPL) with brightness many times stronger than the conventional organic chromophores. In this study, we synthesized GNRs with longitudinal LSPR peak at 1036 nm to match our home-built light source 1040 nm femtosecond laser, which locates in the "optical window" where the tissue absorbs relatively little light. PEGylated GNRs with great biocompatibility were intravenously injected through the tail vein into mice. Excited by 1040 nm laser, the GNRs exhibit bright three-photon luminescence (3PL) signals while circulating in the blood vessels. The use of GNRs as bright contrast agents for 3PL imaging of mouse ear blood vessels *in vivo* was demonstrated. And GNRs targeted in tissues can be excited by 1040 nm laser and could be clearly visualized with no autofluorescence background. These results indicated that 3PL of GNRs is very promising for deep *in vivo* bio-imaging and assessing the distribution of GNRs in tissues with high contrast.

1. Introduction

Gold nanoparticles display many unique optical properties including localized surface plasmon resonance (LSPR), which makes them ideal enhancement agents for imaging, sensing and theranostics in biological systems [1]. Gold Nanorods (GNRs) with well-defined shapes and sizes are very attractive for their plasmon resonant absorption and scattering in the NIR region, making them promising probes for *in vitro* and *in vivo* imaging [2]. GNRs have been demonstrated to exhibit strong plasmon enhanced multi-photon luminescence (MPL) with brightness many times stronger than the conventional organic chromophores [3,4].

In this paper, three-photon luminescence of high aspect ratio gold nanorods with longitudinal LSPR peak at 1036 nm excited by 1040 nm femtosecond laser was investigated and its applications for high

---

Content from this work may be used under the terms of the Creative Commons Attribution 3.0 licence. Any further distribution of this work must maintain attribution to the author(s) and the title of the work, journal citation and DOI.

Published under licence by IOP Publishing Ltd

contrast tissue and *in vivo* imaging were also demonstrated. Due to the bright three-photon luminescence (3PL) of PEGylated GNRs excited by 1040 nm laser, GNRs targeted in tissues could be clearly visualized with negligible autofluorescence background. The results demonstrated that 3PL imaging could be used to assess the distribution of GNRs in tissues with high contrast. After intravenously injected through the tail vein into mice, 3PL of GNRs can be excited and visualized in mice ear blood vessels, suggesting that GNRs are promising in *in vivo* imaging.

2. Materials and Methods

2.1 Synthesis of GNRs

GNRs with the longitudinal LSPR peak at 1036 nm were synthesized using an improved seed-mediated method proposed by Xingchen Ye [5]. Briefly, 0.2 mL of 25 mM $HAuCl_4$ and 5 mL of 0.2 M CTAB were mixed in 5 mL water solution and 0.6 mL of 0.01 M ice-cold sodium borohydride was quickly injected. The solution was stirred for 2 min and kept for 2 h before use. 250 mL of GNRs growth solution containing 3.5 g CTAB and 0.617 g NaOL and 0.5 mM $HAuCl_4$ was prepared in warm water (30 °C). 12 mL of 4 mM $AgNO_3$ solution was added and the mixture solution was stirred for 60 min. 2.6 mL of HCl (37 wt. %) and 0.625 mL of 64 mM ascorbic acid were added. Afterwards, 0.4 mL of seed solution was injected into the growth solution and left undisturbed at 37 °C for 12 h to let GNRs grow.

2.2 Surface modification of GNRs

50 mL of the as-prepared GNRs were centrifuged twice at 6000 rpm for 10 min, and then the precipitate was dispersed in 25 mL of 2 mg/mL PEG (SH-PEG-CH$_3$, MW = 5000) aqueous solution. The solution was stirred magnetically for 12 h. The final PEGylated GNRs solution was washed twice and dispersed in 1 × PBS solution for use.

2.3 Tissue slices preparation

PEGylated GNRs (in 200 µL 1 × PBS, 30 nM) were intravenously injected into mice and the control group were treated with the same amount of saline. The liver was excised at 0.5 h post-injection and fixed in 4% paraformaldehyde solution. The samples were embedded in paraffin and sectioned at 50 µm thickness and mounted on slides.

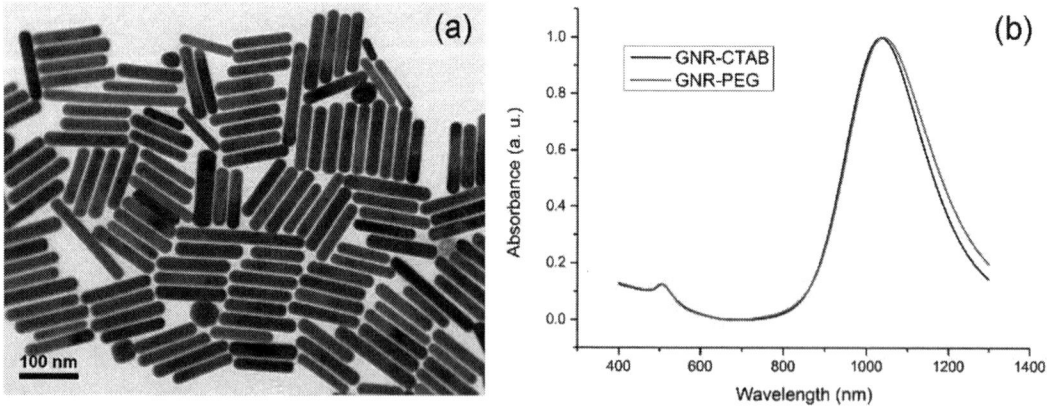

Figure 1. (a) Representative TEM image of GNRs; (b) Normalized UV-vis-NIR absorption spectra of GNRs (GNR-CTAB) and PEGylated GNRs (GNR-PEG).

3. Results and Discussion

3.1 Characterizations of GNRs

Transmission electron microscope (TEM) image of GNRs were captured by a JEOL JEM-1200EX microscope at 160 kV (Figure 1a). The GNRs are ~ 15 × 100 nm, with aspect ratio = ~ 6.5. As shown in Figure 1b, the absorption spectra of GNRs and PEGylated GNRs were measured using a Shimadzu UV3600 UV-vis-NIR scanning spectrophotometer. The longitudinal LSPR peak of PEGylated GNRs showed a little red shift compared with the as-prepared GNRs, indicating a successful coating of PEG polymer on GNRs. In addition, the absorption spectrum did not broaden after the PEG modification and the optical properties of GNRs were maintained for the imaging.

3.2 3PL imaging of GNRs in liver tissue

To access the biodistribution of GNRs in liver tissue, 1040 nm fs laser beam was focused onto the sample by a 20 × objective lens (NA = 1.00) with a power of 10 mW. A short pass filter (492 nm) and a long pass filter (590 nm) were used to collect 3PL and 2PL signals from GNRs and liver tissue, respectively. As shown in Figure 2a and 2b, PEG-GNRs accumulated in the mouse liver due to the reticuloendothelial system (RES) and appeared as bright dots in the sliced tissue images (white circle in Figure 2a). When excited by 1040 nm laser, GNRs exhibit plasmon-enhanced multi-photon luminescence, including three-photon luminescence (3PL) and two-photon luminescence (2PL) [6,7]. The autofluorescence from liver was clearly observed in the 2PL channel (> 590 nm) and disappeared in the 3PL channel (< 492 nm), whereas the 3PL from GNRs can be visualized with super high contrast and little autofluorescence background. As a control, no bright dots of GNRs were observed in images of the tissue explanted from mice liver injected with saline (Figure 2c and 2d). These results demonstrated that bright 3PL of GNRs could be excited by 1040 nm femtosecond laser and this feature enables us to access the biodistribution of GNRs in tissues with high contrast.

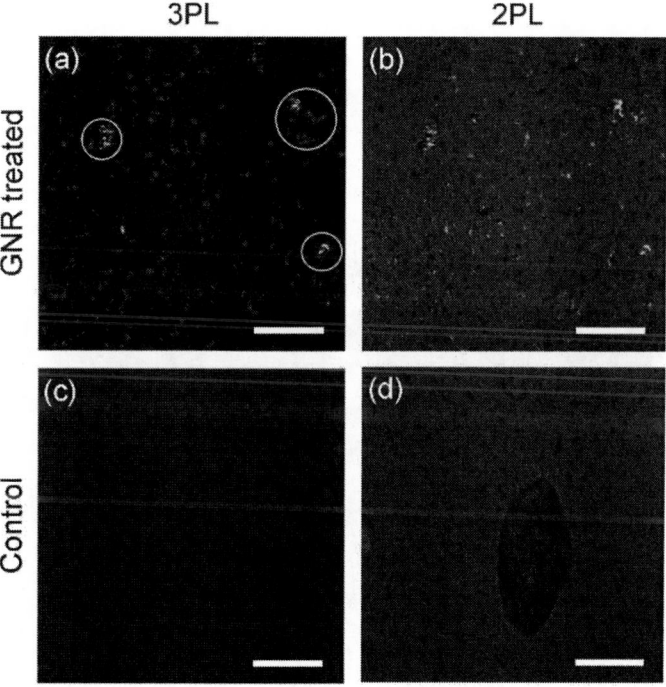

Figure 2. Multi-photon luminescence (3PL and 2PL) images of tissue slices harvested from mice liver injected with PEGylated GNRs (2a and 2b) and saline (2c and 2d). λ_{ex} = 1040 nm, emission signals collected within < 492 nm for 3PL channel and > 590 nm for 2PL channel. The white circles indicate the aggregates of PEG-GNRs in liver tissue. Scale bars: 50 μm.

3.3 In vivo 3PL imaging of GNRs in mouse ear blood vessels

PEGylated GNRs (in 200 µL 1 × PBS, 30 nM) were intravenously injected into mice and the mice were anesthetized and placed on a petri dish with one ear attached to the coverslip [3]. The 1040 nm fs laser beam was focused by a water-immersion objective lens (20 ×, NA = 1.00) onto the earlobe immersed by water and 3PL signals were collected using a filter within < 492 nm. An 80 µm-deep stacks for 3PL imaging was taken with 5 µm step depth. As shown in Figure 3, 3PL of PEG-GNRs flowing in the blood vessels excited by 1040 nm laser at various depths could be clearly observed. Aside from major veins and arteries, the small capillaries could also be visualized with the GNRs. This result demonstrated that GNRs hold great promise to serve as a three-photon agent for intravital vasculature imaging.

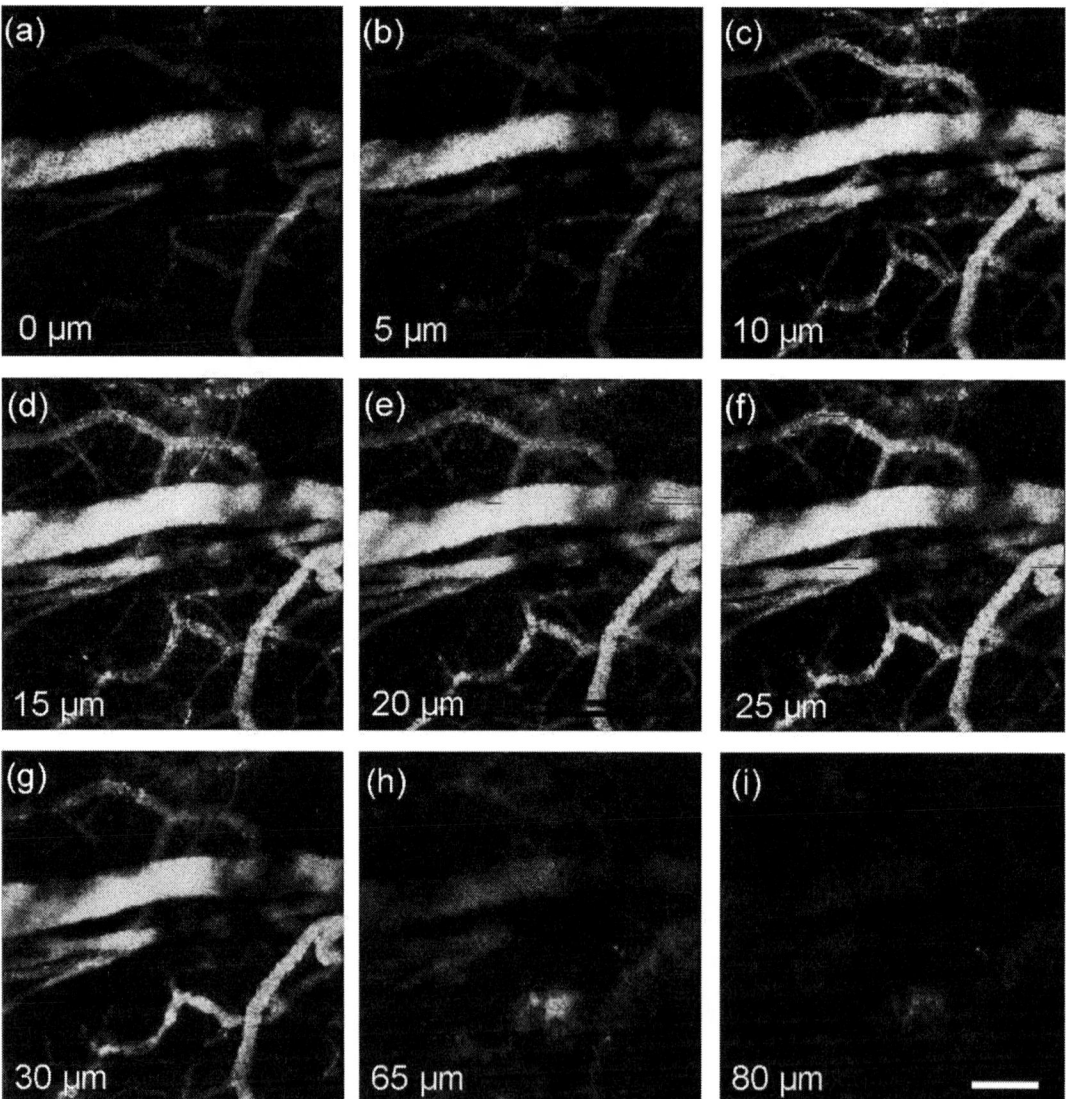

Figure 3. Intravital 3PL images of PEG-1036GNR-stained mouse ear blood vessels. (a)-(i) Images at various vertical depths of mouse ear skin. λ_{ex} = 1040 nm. Signal collected within < 490 nm. Scale bars: 100 µm.

4. Conclusions

In summary, high aspect ratio gold nanorods with the longitudinal LSPR peak at 1036 nm were synthesized and their three-photon luminescence excited by 1040 nm laser were observed in *in vitro* and *in vivo* imaging. PEGylated GNRs were intravenously injected through the tail vein into mice and flowed *via* blood circulation. Bright 3PL imaging of GNRs accumulated in liver tissue clearly illustrated the distribution of GNRs in major organs with high contrast. In addition, 3PL of GNRs in *in vivo* imaging of mouse ear blood vessels was demonstrated. Excited by longer wavelength laser, which penetrates deeply into tissues, 3PL imaging of GNRs holds great potential in high contrast and deep *in vivo* imaging.

Acknowledgment

This work was supported by National Basic Research Program of China (973Program; 2013CB834704 and 2011CB503700), the National Natural Science Foundation of China (61275190 and 91233208), the Program of Zhejiang Leading Team of Science and Technology Innovation (2010R50007), the Fundamental Research Funds for the Central Universities. Shaowei Wang is grateful to Mr. Kai Wu for help with animal experiments.

References

[1]   Li N, Zhao P, Astruc D 2014 Anisotropic Gold Nanoparticles: Synthesis, Properties, Applications, and Toxicity *Angew. Chem. Int. Ed.* **53** 1756-1789

[2]   Huang X, Neretina S and El-Sayed M A 2009 Gold nanorods: from synthesis and properties to biological and biomedical applications *Adv. Mater.* **21** 4880-4910

[3]   Wang H, Huff T B, Zweifel D A, He W, Low P S, Wei A and Cheng J 2005 In vitro and in vivo two-photon luminescence imaging of single gold nanorods *Proc. Natl. Acad. Sci. U.S.A.* **102** 15752-756

[4]   Durr N J, Larson T, Smith D K, Korgel B A, Sokolov K and Yakar A B 2007 Two-photon luminescence imaging of cancer cells using molecularly targeted gold nanorods *Nano Lett.* **7** 941-945

[5]   Ye X, Zheng C, Chen J, Gao Y and Murray C B 2013 Using binary surfactant mixtures to simultaneously improve the dimensional tunability and monodispersity in the seeded growth of gold nanorods *Nano Lett* **13** 765-771

[6]   Hu K W, Liu T M, Chung K Y, Huang K S, Hsieh C T, Sun C K, Yeh C S 2009 Efficient near-IR hyperthermia and intense nonlinear optical imaging contrast on the gold nanorod-in-shell nanostructures. *J Am Chem Soc.* **131** 14186-187.

[7]   Wang S, Xi W, Cai F, Zhao X, Xu Z, Qian J and He S 2015 Three-Photon Luminescence of Gold Nanorods and Its Applications for High Contrast Tissue and Deep In Vivo Brain Imaging *Theranostics* **5** 251-266

Classification and recognition of texture collagen obtaining by multiphoton microscope with neural network analysis

Shulian Wu, Yuanyuan Peng, Liangjun Hu, Xiaoman Zhang and Hui Li*

Key Lab of OptoElectronic Science and Technology for Medicine, Ministry of Education; Fujian Provincial Key Lab of Photonic Technology, College of Photonic and Electronic Engineering, Fujian Normal University, Fuzhou, Fujian 350007, People's Republic of China.

E-mail: hli@fjnu.edu.cn

Abstract Second harmonic generation microscopy (SHGM) was used to monitor the process of chronological aging skin in vivo. The collagen structures of mice model with different ages were obtained using SHGM. Then, texture feature with contrast, correlation and entropy were extracted and analysed using the grey level co-occurrence matrix. At last, the neural network tool of Matlab was applied to train the texture of collagen in different statues during the aging process. And the simulation of mice collagen texture was carried out. The results indicated that the classification accuracy reach 85%. Results demonstrated that the proposed approach effectively detected the target object in the collagen texture image during the chronological aging process and the analysis tool based on neural network applied the skin of classification and feature extraction method is feasible.

1. Introduction

Skin aging is an important issue in dermatology and cosmetic science. It involves chronological aging and photoaging process [1]. Chronological aging is the natural aging process induced by the passage of time. In terms of skin health, chronological aging causes visible changes that appear in skin with time, usually starting around the age of 25 years. The degeneration of collagen is a major factor in age-related dermal alteration. Chronological aging occurs as the natural regenerative processes begin to slow; a slower turnover of the surface skin cells follows [2]. Currently, the process of cutaneous aging and its underlying molecular mechanism and biochemical have attracted worldwide attention [1, 3-4]. A second harmonic generation microscope (SHGM) [5] is a powerful tool has been applied on the study of biological tissues [6-8]. SHGM is a nondestructive imaging tool that enables the clear visualization of non-centrosymmetric structural proteins, such as myosin, thyroid tissue, and collagen, without fixation, sectioning, dehydration, and using exogenous dyes or stains [8]. The age-related structural and morphological changes in collagen result in optical parameter alterations, causing an optical contrast that can be detected in vivo using SHGM. Many researchers have paid increased attention to structural and morphological changes. However, they mainly focused on qualitative analyses of the altered skin structure [9].

Texture refers to the properties that represent the surface or structure of an object and is defined as something consisting of mutually related elements. Texture analysis is an important and useful method for

---

Content from this work may be used under the terms of the Creative Commons Attribution 3.0 licence. Any further distribution of this work must maintain attribution to the author(s) and the title of the work, journal citation and DOI.

Published under licence by IOP Publishing Ltd

processing visual information. The spatial distribution of texture gray levels is a function of spatial variation in pixel intensities. It has a number of perceived qualities that play an important role in describing texture. This spatial distribution is useful in a variety of applications and has been the subject of intense studies by many researchers [10]. Therefore, quantitatively characterizing skin collagen, which has complex geometrical and local optical properties, is an important but difficult task.

Till now, few studies had extracted quantitative texture information from SHG images to evaluate collagen alteration during the aging process in vivo. This study aims to evaluate collagen texture using SHG images combined with mathematical processing to assess the chronological aging stages. SHGM was used to obtain collagen morphology during the chronological process *in vivo*. A quantitative analysis was established throughout the gray level co-occurrence matrix (GLCM) [10-12] between collagen alteration and aging progression to determine aging characteristics. The results show that combined SHG and GLCM is feasible to determine the main impact on age-related skin. It is important to understand the process of aging process and establish preventive regimens that would slow aging progression.

2. Materials and methods

2.1 Animal model

Forty matured Kunming mice with 8, 16, 50, and 60 weeks old, respectively, were selected as animal models in this study and were provided by the Animal Center of Fujian Medical University. All studies involving mouse tissues were approved by the Institutional Animal Care and Use Committee at Fujian Normal University. The mice were kept in a humid environment at a constant temperature and a 12 h light/dark cycle. They were fed a standard diet and given water. Before the experiment, each mouse was anaesthetized using intraperitoneal phenobarbital injection, and their dorsal hair was removed [13]. The mouse dorsal skin was stabilized in a certain fold chamber, covered with a cover slip, and observed by SHGM.

2.2 Second harmonic generation microscopy system

SHGM system used in this study has been described previously [8, 13]. Briefly, it is a commercialized technique based on the combination of an inverted microscopy with Axiovert 200 microscope of Zeiss LSM510 META laser scanning microscopy and a Coherent Mira 900-f mode-locked femtosecond Ti:sapphire laser (110 fs, 76 MHz), with tunable wavelength ranging 700 nm to 980 nm. The excitation wavelength at 850 nm was employed and the average output power was limited at 10 mW irradiated on the sample spot. An objective with Plan-Neofluar×10 (NA=0.3) was used to detect the SHG signal. All images are of 512×512 pixels with a size of 921 μm × 921 μm. SHG signal was detected at 425 nm in center with bandwidth of 20 nm, which was generated and collected at the sample focal plane.

2.3 Gray level co-occurrence estimate analysis

The images of collagen structure were processed by GLCM, which was reprogrammed and computed using Matlab 7.0. Spatial gray level co-occurrence estimates image properties related to second-order statistics, which is defined as the likelihood of observing a pair of gray values occurring at the endpoints of a dipole with a random length placed in the image at a random location and orientation. Haralick [14] suggested the use of GLCM, which has become one of the most well-known and widely used texture features. The greatest advantages of GLCM are its computational ease and powerful features. The G×G

GLCM Pd for a displacement vector d=(dx, dy) is defined as follows. The entry (i, j) of P is the occurrence number of the gray level pairs i and j, which are a distance d apart.

$$P(i,j,d,\theta) = \left\{ [(x,y),(x+\Delta x, y+\Delta y)] \middle| \begin{array}{l} f(x,y)=i, f(x+\Delta x, y+\Delta y)=j; \\ x=0,1,...,N_x-1; y=0,1,...N_y-1 \end{array} \right\} \quad \text{(1)}$$

First, an original texture image D is requantized into an image G with reduced number of gray levels Ng, with a typical value of either 16 or 32. Then, GLCM is computed from G by scanning the intensity of each pixel and its neighbor, defined by displacement d and angle θ. d can take a value of 1, 2, 3...n, whereas θ is limited to 0℃, 45℃, 90℃, and 135℃. The P(i, j | d, θ) of GLCM is the second-order joint probability density function P of the gray level pairs in the image for each element in the co-occurrence matrix, which is obtained by dividing each element with Ng. Finally, scalar secondary features are extracted from this co-occurrence matrix [10]. The common statistical features used in this study were contrast, correlation, and entropy, which were described following.

The contrast feature, which measures intensity contrast as a function of pixel distance, is related to the collagen structure with detached fibrils [14].

$$f_1 = \sum_{i,j} P(i,j)^2 \quad \text{(2)}$$

Contrast shows the definition and the degree of texture depth of the groove pattern of the image. The value of contrast signifies the distribution of the matrix pixel. Matrix pixel distribution is uniform in the large contrast value, and vice versa.

Correlation is the process of passing the mark w by the image matrix fcorrelation in the manner proposed by [14].

$$f_{correlation} = \frac{\sum_i \sum_j ij P_d(i,j) - \mu_x \mu_y}{\sigma_x^2 \sigma_y^2} \quad \text{(3)}$$

where
$$\mu_x = \sum_i i \sum_j P_d(i,j) \quad \mu_x = \sum_j j \sum_i P_d(i,j)$$

$$\sigma_x^2 = \sum_i (i-\mu_x^2) \sum_j P_d(i,j); \quad \sigma_y^2 = \sum_j (j-\mu_y^2) \sum_i P_d(i,j)$$

μ_x, μ_y are mean values and σ_x^2, σ_y^2 are variances.

The correlation of the images reflects the similarity on a direction of the image texture area and is the linear correlation measure of the local gray level in the image. This correlation is a measure of the dependence between two different pixel values. In particular, the correlation feature indicates a relation of intensity measure as a function of pixel distance, which is used to assess the similarity of matrix pixel horizontally and vertically.

Entropy is the complexity feature of an image matrix, and it mainly reflects the texture grayscale randomness distribution of the image. Entropy is a measure of the amount of information in an image and is defined as follows [14]:

$$f_{entropy} = \sum_i \sum_j P_d(i,j) \log_2 P_d(i,j) \tag{4}$$

When the image has many fine textures, the number of P (i, j) is approximately equal, and thus the value of entropy is larger, and vice versa. When the image has less number of fine textures, the number P (i, j) differs greatly, resulting in a smaller entropy value of the images.

These features were obtained using a combination of different displacements and angles. The calculated displacements of numerical parameters ranged from 0 to 60 pixels (0-152.4 µm) in the horizontal direction of each image.

Ten images from region of interest, with an area of 921 µm × 921 µm in each stage of each mouse, were selected for quantitative analysis. The experimental results were analyzed by statistic test with SPSS 15.0 software (SPSS Inc., USA). Statistical significances of the data in different types ages were evaluated by T-test which was used to determine whether there were any significant differences between the means of two independent groups. The differences were considered statistically significant when the P values obtained from T-test analysis were less than 0.05.

3. Results and discussions

3.1 Collagen structure

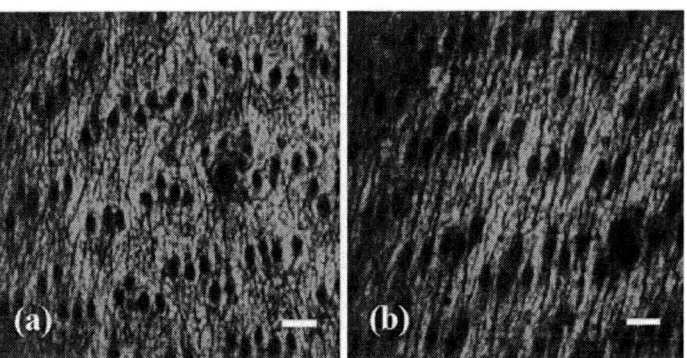

Figure.1 the morphological structure of collagen obtained by SHGM. The images are 512×512 pixels with an area of 921 µm × 921 µm, scale bar=100µm. (a) The 8 weeks skin; (b) the 60 weeks skin.

As skin at various ages has different epidermis thicknesses and optical parameters [15], collagen under the epidermis of different skin types may stay in different skin layers, and it may result in false analysis. To better determine the characteristics of collagen, different sections of collagen from different depths of the dermis layer with the strongest collagen intensity were investigated. This process guarantees obtain the best layer of collagen in different skin types. Figure. 1 partially shows the SHG images that visualize the collagen structure status at different depths where the SHG signals are the strongest in different aging processes.

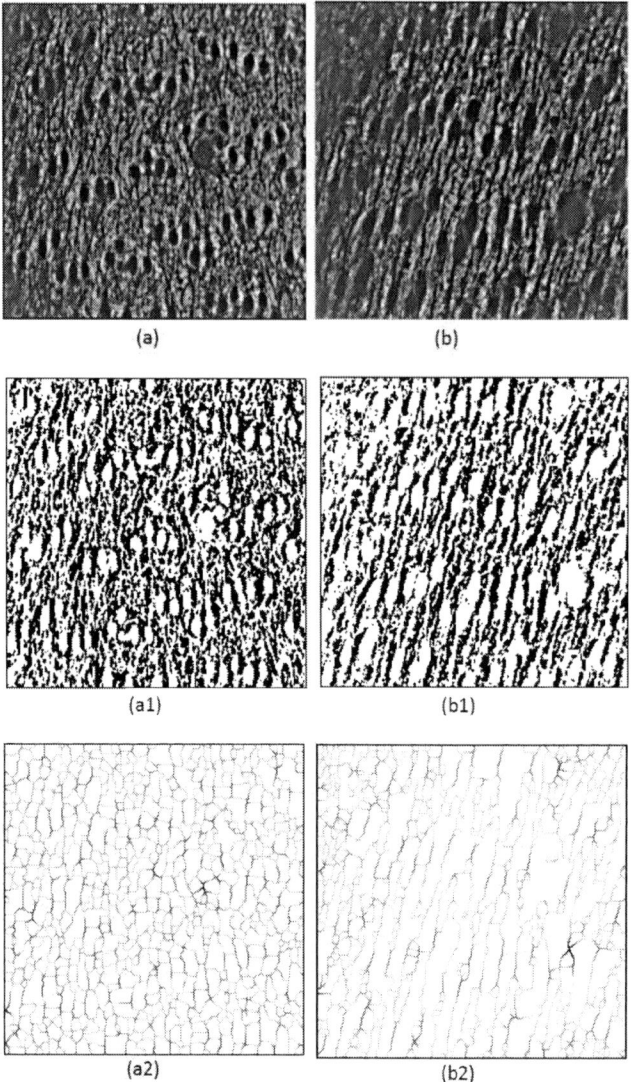

Figure.2 the (a) and (b) are the gray images of mice skin of 8 and 60 weeks; The (a1) and (b1) are the make binary images; the (a2) and (b2) are the voronoi processing images.

Then, the images were processed to perceptual intuition. Band-pass-filter was used to improve the visualizing degree. The results were displayed in Figure.2. The (a) and (b) are the gray images of mice skin of 8 and 60 weeks, respectively; The (a1) and (b1) are the make binary images; the (a2) and (b2) are the voronoi processing images. From the voronoi processing images, we could see that the quantity of polygonal in normal skin were much more that other states of chorological skin. The areas of polygonal were uniformity size in 60 weeks skin. With age, the quantity of polygonal decreased, and the size difference of areas were very big.

3.2 texture feature

The parameter of contrast extract from SHG imaging using GLCM method. The results of skin contrast at various ages are shown in Figure. 3. From the principle, if the contrast value sharply increases as the pixel distance, the collagen matrix is distinct. While, if this value remains constant along with the pixel distance, then the collagen matrix has a less defined fibrillar contrast structure.

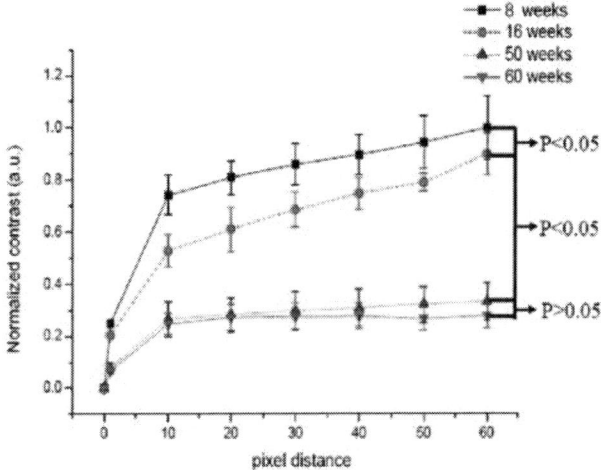

Figure.3 the normalized value of contrast with pixel distance in different age skin of images extract by GLCM.

From figure. 3, the contrast value in 8-week increased sharply with the pixel distance, similar to that in 16 weeks. In the 50- and 60-week skin textures, the value gradually increased as the pixel distance increased. At a distance of 20 pixels, the contrast value in a young skin slowly increased, and the chronological aging skin became stable and the values fluctuated less. Therefore, the contrast at 20 pixels is defined as the collagen contrast in this study. The results show that the contrast value in young skin was larger than that in chronological aging skin. The data were statistically significant in two types of skin groups. This strong significance showed that the collagen intensity in young skin was higher than that in chronological aging skin. Moreover, the result indicated that a loss of fine structure led to less contrast in collagen fibrils. The collagen texture of young skin was distinct with a great difference in the pixel matrix, whereas collagen texture was obscure with a homogeneous pixel matrix in aging skin. Therefore, contrast values may be used to determine the intensity of collagen during the chronological aging process.

The relationship between correlation features and pixel distance in skin at different ages is shown in Figure. 4. As far as our study is concerned, if the correlation value gradually decreases as the pixel distance increases, the collagen matrix is distinct. Conversely, when this correlation value remains at a constant value with the pixel distance, then the collagen matrix has less defined fibrillar correlation structure.

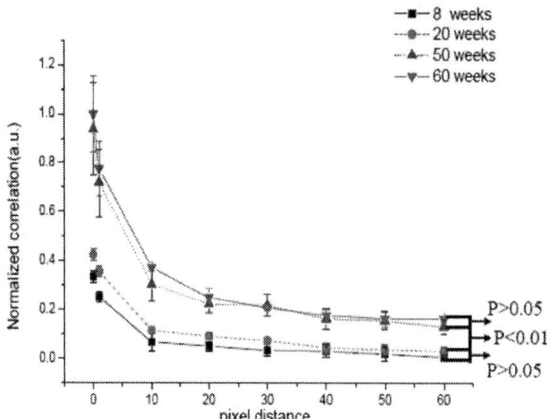

Figure.4 the normalized value of correlation with the pixel distance at various age skin gotten by GLCM.

The values of correlation that decreased along with the pixel in all ages are shown in Figure. 3. At a distance of 20 pixels, the correlation values became stable and remained steady. Thus, the correlation number at 20 pixels is defined as the collagen correlativity in this study. This definition of correlation indicates that a loss of fine structure can result from great collagen correlativity. Thus, the collagen fibril correlation sharply decreased with distance in aging skin, implying that the texture was obscured and had a big matrix at this skin state. Moreover, the collagen correlation value in chronological aging skin was larger than that in young skin, indicating that aging skin collagen texture had faint fibrils and a uniform pixel matrix. The value of collagen correlation was steady with distance in young skin, indicating that the collagen texture was distinct with a great difference in pixel matrix. Thus, the value of collagen correlation could be used to quantitatively estimate the similarity among various skin types. The extent of the similarity depended on the correlation value.

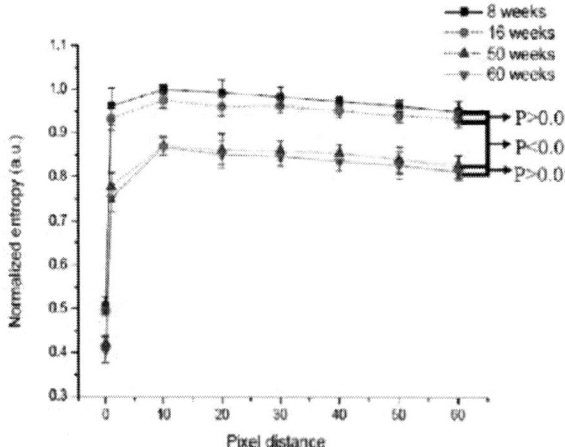

Figure.5 the normalized value of entropy with pixel distance in different age skin from SHG image obtained by GLCM.

The entropy values in skin at different ages are shown in Figure. 5. The entropy values were also stable at a distance of 20 pixels. Thus, these values of entropy are also defined as the collagen entropy in this study. Collagen entropy increased as distance in both young and chronological aging skin. However, the entropy value in young skin was greater than that in aging skin. This result indicated that the young skin collagen texture had many fine textures and linear fibrils, whereas the aging skin had a less fine texture. Thus, texture was complex in young skin and simple in chronological aging skin. Therefore, entropy value may be used to characterize the complexity of skin.

3.3 Classification skin type by neural network

Table.1 The network training results of mice skin images

Serial number	Theory output			Actual output			Classific-ation
a_1	0	0	0	0.005	0.011	0.039	younger
a_2	1	0	0	1.007	0.012	-0.045	older

The texture features extracted from the gray level co-occurrence matrix were classified by the neural network tool [16, 17]. First, to establish the feature that is confirmed by the known statue. Let the feature vector of skin texture as X= (CON, COR, ENT), that is X=(X1, X2, X3). Then input the normalization data in the NNtool. The output of NNtool is Y=(Y1, Y2, Y3), the output result means the skin statue. Using the neural network toolbox of matlab6.5, setting the relation parameters, and then training the network and classification the skin texture features. The training sample is about fifty, and the prediction samples about fifty, with the results of classification as shown in table.1.

The theory output were set 0, 0, 0 corresponding CON, COR and ENT in younger skin texture feature. And the actual outputs were 0.005, 0.011, and 0.039 of these parameters. And the theory output were set 1, 0, 0 corresponding CON, COR and ENT in older skin texture feature. And the actual outputs were 1.007, 0.012, and -0.045 of these parameters. It means that the third parameter entropy is a litter big in two status ages. The results indicated that the classification accuracy reach 85%. The result shows that the neural network tool of Matlab was applied to train the texture of collagen in different statues during the aging process is feasible.

4. Conclusions

In conclusions, GLCM was used to characterize the images of the texture features of skin at various ages obtained from SHGM. Three main parameters, namely, contrast, correlation, and entropy, were investigated to determine the intensity, similarity, and complexity of the skin during the chronological aging process, respectively. The results obtained from GLCM showed the distinct features of young skin, which were reflected through the great difference in the pixel matrix of the collagen. The collagen fibril contained abundant microgrooves in young skin, indicating complexity in terms of collagen texture distribution. By contrast, in chronological aging skin, the image of the collagen was blurred and had a uniform pixel matrix. The fine collagen decreased and the thick texture increased, indicating that the collagen distribution had a more definite orientation. The neural network tool of Matlab was applied to train the texture of collagen in the two statues of skin and the classification accuracy reach 85%. the results indicated that the approach effectively detected the target object in the collagen texture image during the chronological aging process and the analysis tool based on neural network applied the skin of classification and feature extraction method is feasible.

Acknowledgment

This work was supported by the National Natural Science Foundation of China (61178089), the Natural Science Foundation of Fujian Province (2014J01226) and Foundation of Fujian Educational Committee (JA14093).

References

[1] Farage M A, Miller K W, Elsner P and Maibach H I 2007 Structural characteristics of the aging skin: a review *Cutaneous and ocular toxicology* **26** 343-57

[2] Fisher G J, Kang S, Varani J, Bata-Csorgo Z, Wan Y, Datta S and Voorhees J J 2002 Mechanisms of photoaging and chronological skin aging *Archives of dermatology* **138** 1462-70

[3] Chung J H, Seo J Y, Choi H R, Lee M K, Youn C S, Rhie G-e, Cho K H, Kim K H, Park K C and Eun H C 2001 Modulation of skin collagen metabolism in aged and photoaged human skin in vivo *Journal of Investigative Dermatology* **117** 1218-24

[4] Cua A, Wilhelm K-P and Maibach H 1990 Elastic properties of human skin: relation to age, sex, and anatomical region *Archives of Dermatological Research* **282** 283-8

[5] Denk W, Strickler J H and Webb W W 1990 Two-photon laser scanning fluorescence microscopy *Science* **248** 73-6

[6] Jiang X, Chen S, Chen J, Zhu X, Zheng L, Zhuo S and Wang D 2011 Monitoring process of human keloid formation based on second harmonic generation imaging *Laser Physics* **21** 1661-4

[7] Perry S W, Burke R M and Brown E B 2012 Two-photon and second harmonic microscopy in clinical and translational cancer research *Annals of biomedical engineering* **40** 277-91

[8] Zhuo S, Wu G, Chen J, Zhu X and Xie S 2012 Label-free imaging of goblet cells as a marker for differentiating colonic polyps by multiphoton microscopy *Laser Physics Letters* **9** 465

[9] Koehler M J, Hahn S, Preller A, Elsner P, Ziemer M, Bauer A, König K, Bückle R, Fluhr J W and Kaatz M 2008 Morphological skin ageing criteria by multiphoton laser scanning tomography: non‑invasive in vivo scoring of the dermal fibre network *Experimental dermatology* **17** 519-23

[10] Honeycutt C E and Plotnick R 2008 Image analysis techniques and gray-level co-occurrence matrices (GLCM) for calculating bioturbation indices and characterizing biogenic sedimentary structures *Computers & Geosciences* **34** 1461-72

[11] Clausi D A 2002 An analysis of co-occurrence texture statistics as a function of grey level quantization *Canadian Journal of remote sensing* **28** 45-62

[12] Soh L-K and Tsatsoulis C 1999 Texture analysis of SAR sea ice imagery using gray level co-occurrence matrices *Geoscience and Remote Sensing, IEEE Transactions on* **37** 780-95

[13] Wu S, Li H, Yang H, Zhang X, Li Z and Xu S 2011 Quantitative analysis on collagen morphology in aging skin based on multiphoton microscopy *Journal of biomedical optics* **16** 040502--3

[14] Haralick R M, Shanmugam K and Dinstein I H 1973 Textural features for image classification *Systems, Man and Cybernetics, IEEE Transactions on* 610-21

[15] Wu S, Li H, Zhang X and Li Z 2013 Optical features for chronological aging and photoaging skin by optical coherence tomography *Lasers in medical science* **28** 445-50

[16] Ripley B D 1996 *Pattern recognition and neural networks*: Cambridge university press)

[17] Yoshida T and Omatu S 2000 Pattern recognition with neural networks. In: *Geoscience and Remote Sensing Symposium, 2000. Proceedings. IGARSS 2000. IEEE 2000 International*: IEEE) pp 699-701

The optical design of highly efficient cold shield in IR detector based on ASAP

Xianjing Zhang[1], Jian Bai[1,*], and Shuang Yin[2]

[1] State Key Laboratory of Modern Optical Instrumentation, Zhejiang University, Hangzhou, 310027, P.R.China
E-mail: bai@zju.edu.cn

[2] Kunming Institute of Physics, Yunnan, 650223, P.R.China

Abstract. In order to obtain higher stray radiation suppression of a cold shield and enhance the cold shield efficiency of cooled infrared system, different configurations of the vanes inside cold shield are proposed. The effects of these configurations of vanes on stray radiation suppression are simulated and analysed by Advanced Systems Analysis Program (ASAP software), which not only trace ray by Monte Carlo method but also simulate the scattering of each surface. According to the analysis of results, vane plays an important role in the improvement of cold shield efficiency and the suppression of stray radiation. The cold shield with 6 vanes characterized by rounded-square hole, whose configuration is calculated based on the principle of laying out a set of vanes, suppresses 99.5% more stray radiation comparing to the original one without any vanes. However, considering that the cold shield with 6 vanes is complicated and difficult to fabricate, the cold shield with 3 vanes owing rounded-square hole is selected, since its stray radiation suppression level is acceptable.

1. Introduction

In infrared system, stray radiation is the radiation that is received by the detector [1,2], but does not originate from the conjugate object plane [3]. It will reduce signal to noise ratio(SNR) and may damage the optical components inside, especially in thermal imaging systems. Nowadays, since the resolution of detector reaching diffraction limit, higher level stray radiation suppression is imperative to enhance cold shield efficiency [4]. In cooled infrared system, cold shield is of great importance in decreasing the amount of stray radiation incident into the detector and increasing the SNR of system. 100% cold shield efficiency will be achieved when the cold shield is on the exit pupil, indicating no out-of-field thermal radiation is propagating to the detector. However, considering surface scattering and spontaneous radiation of internal objects, it is impossible to achieve 100% cold shield efficiency. Furthermore, with the goal of obtaining a compact structure, most of the infrared systems get far below 100% cold shield efficiency, and some are even less than 50% [5]. In this paper, different configurations of vanes are proposed to improve cold shield efficiency of the cooled infrared system.

2. Theory

According to radiation transfer formula [7], scattering radiation from an infinitesimal area propagates to an infinitesimal detector area. And the radiation flux $d\Phi_c$ received by this infinitesimal detector area can be calculated by the following formula:

$$d\Phi c = d\Phi s(\theta_i, \phi_i) BRDF(\theta_i, \phi_i; \theta_c, \phi_c) GCF_{sc} \pi \tag{1}$$

Where $d\Phi s(\theta_i, \phi_i)$ is the radiation flux scattered from the infinitesimal source area as a function of scattering angle θ_i and azimuth ϕ_i; $BRDF(\theta_i, \phi_i; \theta_c, \phi_c)$ is the bidirectional reflectance distribution function of the scattering source surface, θ_i, ϕ_i are reflecting angle and reflecting azimuth respectively; $GCF_{sc}\pi$ is called geometrical configuration factor, which represents the projected solid angle from scattering source area to the detector area.

Apparently, in equation (1), the radiation flux $d\Phi_c$ received by infinitesimal detector area can be reduced by decreasing the three terms on the right hand of the equation.

- Decrease $d\Phi s(\theta_i, \phi_i)$: reduce the spontaneous radiation from internal objects by lowering the temperature of workspace [8].
- Decrease $BRDF(\theta_i, \phi_i; \theta_c, \phi_c)$: improve the absorption of surfaces by the process of nigrescence or choosing a special material with high absorption and low emittance.
- Decrease $GCF_{sc}\pi$: block the transmission paths of stray radiation by adding an aperture or some vanes inside the system.

Once one of the above three factors reaching zero, the detector plane will not be illuminated by stray radiation. Actually, no material can achieve 100% absorption, and spontaneous radiation would always exist no matter how low the temperature of workspace is. Obviously, $GCF_{sc}\pi$ is the only factor that could equal to zero so it is the key point in this study.

The energy of the radiation can be reduced when the radiation is scattered or absorbed. When the absorption of one surface reaches 90%, the attenuation coefficient would be 0.1^n (n is the times that radiation hits surfaces before it gets to detector). Obviously, the more times stray radiation is scattered or absorbed, the greater its energy would be attenuated.

3. Modeling
There are three different parts as shown in this section, presenting the modeling of the cold shield system.

3.1. Scattering model
The inside walls of cold shield and the surfaces of vanes are both considered as Lambertian diffusers after the process of nigrescence. When each stray radiation hits surface of internal walls or vanes, additional radiation with different directions is generated, based on the Lambertian scattering properties. In ASAP software, these can be simulated precisely. In addition, children radiation that is split from parent radiation after hitting the surface, can also be traced in this software.

3.2. Modeling of radiation source
A disk radiator placed at the entrance of cold shield is regarded as radiation source. Each point on the disk is considered as a point source emitting hemispherical radiation that can be catalogued as stray radiation and imaging radiation. Spherical coordinates are selected to describe the discrete radiation illuminated from each point source in the convenience of calculation. The whole system is axis-symmetrical, and so is the imaging information on the detector. Hence, only the point sources in first quadrant are necessary to simulate in consideration of efficiency. The imaging of other three quadrants could be calculated, thus the total flux distribution on the detector is obtained by making a superposition.

3.3. Modeling of vanes
Vanes can be used to reduce the amount of radiation that is reflected or scattered from the internal walls of an infrared system. As shown in the upper part in figure 1, out-of-field radiation would be reflected from the internal walls and split into several radiation, while in the lower part of figure 1, stray radiation is obscured by vanes and could not reach the detector directly.

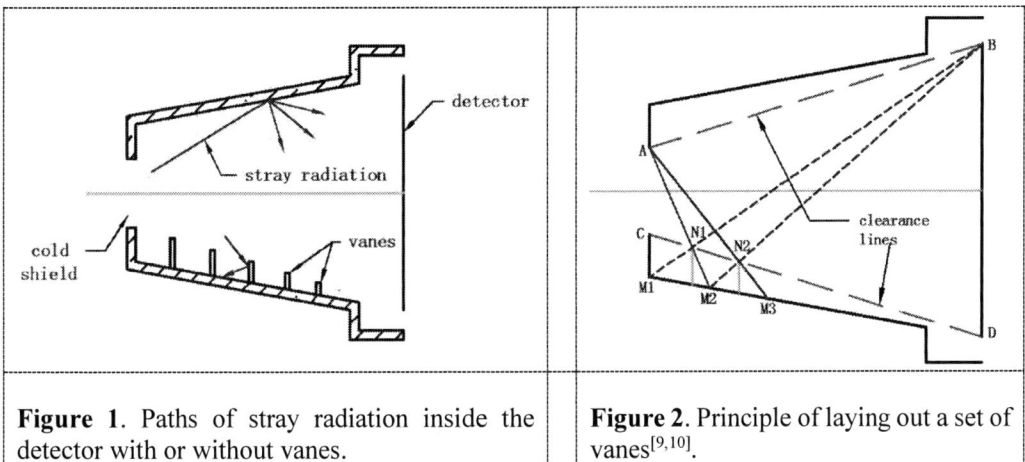

Figure 1. Paths of stray radiation inside the detector with or without vanes.

Figure 2. Principle of laying out a set of vanes[9,10].

Therefore, some configurations of vanes inside the system are proposed to minimize the stray radiation reaching the detector [6]. The crucial point to making efficient utilization of vanes is to arrange them in positions so that no out-of-field radiation can illuminate to any part of the detector directly [11]. The principle of laying out a set of vanes is illustrated in figure 2.

The necessary clearance space is indicated by the hidden lines CD and AB from the rim of the cold shield to the edge of the detector. Vanes cannot be intruded in the clearance space without blocking part of the radiation from the desired field of view. The dashed line M_1B is the critical line where the extraneous radiation begins from the point on the internal wall to the rim of the detector. The first vane is placed at the intersection of the dashed line M_1B and the hidden clearance line CD. The solid line AM_2 indicates the path of radiation from the rim of the cold shield to the internal wall. The area between first vane and the solid line AM_2 is shadowed and cannot be seen by any part of the detector. The dashed line M_2B is another critical line, and the second vane is set up at the intersection of the dashed line M_2B and the hidden clearance line CD. These procedures are repeated until no radiation from internal walls could illuminate to the detector directly.

However, if the effect of stray radiation suppression reaches the basic requirements, the number of vanes added inside the detector should be reduced and the complexity of cold shield should be reduced in consideration of cost and the influence of diffraction.

4. Simulation and Results

In this study, kinds of configurations are designed and simulated by ASAP software, including cold shields with several annular vanes (shorted as AN), with vanes which have rounded-square hole (shorted as RS) and the original one without vanes. The former two configurations are showed in figure.3, in which detector is marked as red, vanes are marked as green, and cold shied is marked as gray.

In this simulation, the internal walls of cold shield and the surfaces of vanes are regarded as Lambertian scattering bodies with the absorption of 93%. The thickness of each vane, which is ought to be about 0.1 millimetres, is being ignored during this simulation since it would have little influence on the results. Moreover, the absorption of detector is supposed to be 100% while it could not be achieved actually. The detector is partitioned into 251×251 multi-pixel squares for the convenience of analysis, by each the flux received can be obtained if it is necessary. And the flux of stray radiation is supposed to be 1000, while the total flux of whole radiation is 100000.

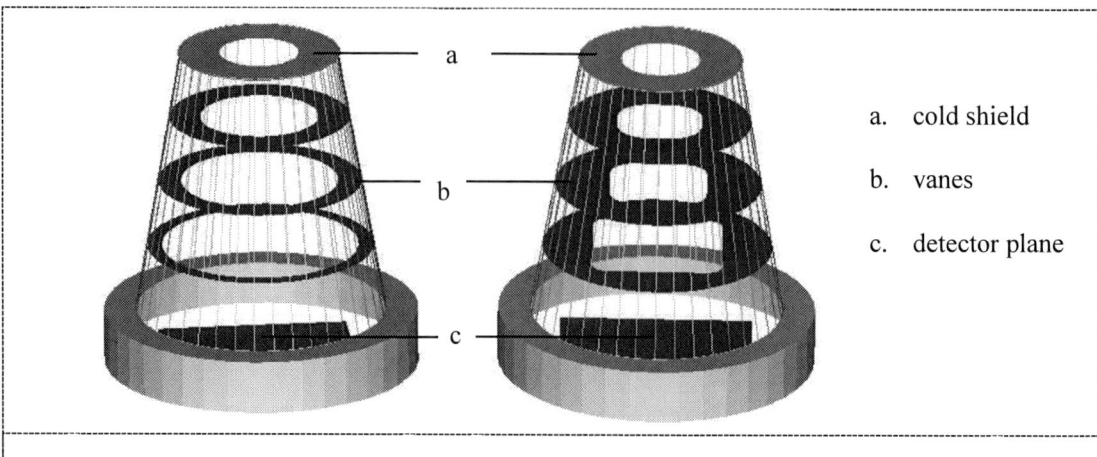

a. cold shield

b. vanes

c. detector plane

Figure 3. configurations of cold shields added with annular vanes and rounded-square vanes.

The result of stray radiation received by the detector is showed as Table 1(A, B, C means the different arrangements of the vanes' positions; example: 3RS vanes(A) is shorted for 3 rounded-square vanes in Configuration A). According to the design principle mentioned in section 3.3, configuration C is obtained, including the number of vanes and the position each vane laid. In consideration of manufacturing cost and the influence of diffraction, configuration A and configuration B, both containing only 3 vanes, are proposed and simulated. And the cold shield without vanes is also considered for comparison.

Table 1. the stray radiation illuminate to the detector in each cold shield added with different configurations of vanes based on the simulation results in ASAP software

Configuration	Max Flux/square	Min Flux/square	RMS Flux/square	Total Flux
6 RS vanes(C)	0.0092	3.84E-07	0.0011	1.1716
6 AN vanes(C)	0.3392	1.67E-06	0.0070	1.6848
3 RS vanes(A)	0.7336	1.60E-06	0.0274	6.6723
3 RS vanes(B)	0.4927	1.57E-06	0.0161	4.2756
3 AN vanes(A)	0.7999	5.22E-06	0.0406	14.5104
3 AN vanes(B)	0.4930	1.02E-05	0.0228	9.8298
No vanes	3.3422	1.58E-04	0.1716	234.1491

According to the results in table 1, detector in the cold shield added with vanes receive less illumination compared to the original cold shield and the more vanes are added into the cold shield, the less stray radiation illuminate into the detector. Moreover, vane with a rounded-square hole gets better effect on stray radiation suppression than the annular one. In the situation of using 3 vanes, configuration B does a better job than configuration A as a whole. Among these simulated flux distributions of different configurations, configuration of 6 RS vanes reduces 99.9% stray radiation to reach the detector, which suppresses 99.5% more compare to the original one without any vanes.

After a superposition and interpolating operation, the flux distributions of 6 AN vanes(C) and 6 RS vanes(C) are showed in figure 4 and figure 5 respectively. As is shown in figure 5, the flux of every multi-pixel square is lower than 0.008, which decreases two order of magnitude in comparison to the maximum one in figure 4. In figure 4, the flux received by the four corners of detector is much greater than which received by the other places of the detector. It illustrated that annular vane does not block out-of-field radiation completely and there is some stray radiation illuminate to the rim of the detector plane directly. Obviously, the configuration of 6 RS vanes is better at stray radiation suppression.

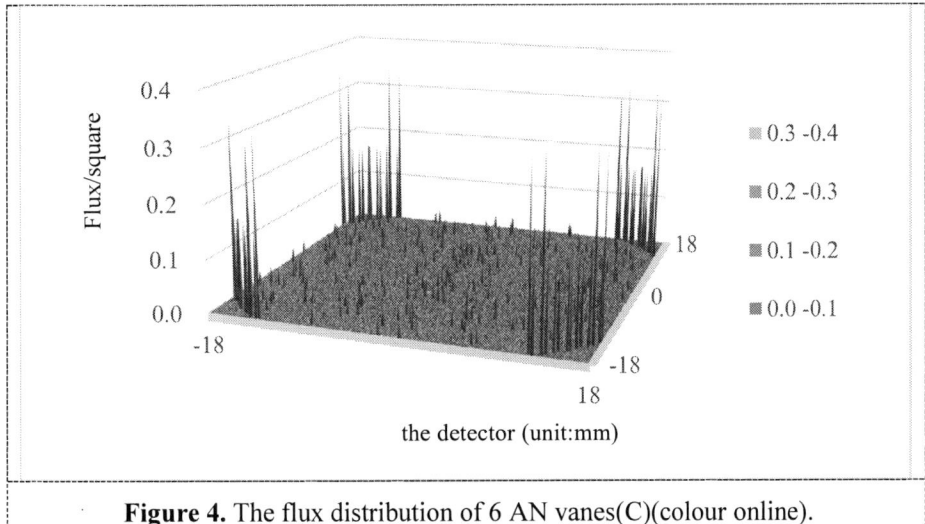

Figure 4. The flux distribution of 6 AN vanes(C)(colour online).

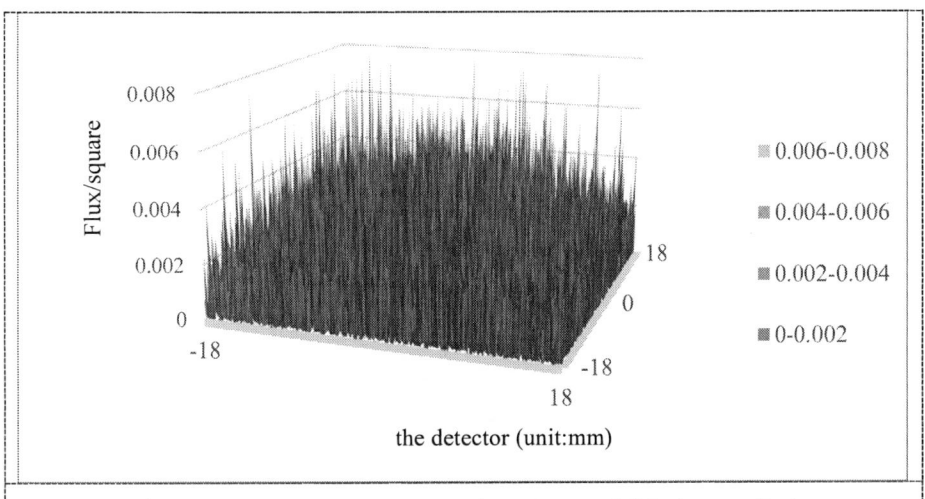

Figure 5. The flux distribution of 6 RS vanes(C)(colour online).

Considering that the cold shield with 6 vanes is complicated to fabraicate and would have more influence on diffraction ,though the cold shield with 6 RS vanes has a perfect effect on stray radiation suppression, it is wise to decrease the number of vanes as a widespread product. Based on the results in table 1, the configuration of 3 RS vanes(B), which suppress 98% more stray radiation comparing to the original one without any vanes, has an advantage on blocking stray radiation comparing to the other comfigurations of 3 vanes. The flux distribution of 3 RS vanes is showed in figure 6. Hence, a compare among 6 RS vanes, 6 AN vanes and 3 RS(B) vanes is made. The configuration of 6 RS vanes get far less stray radiation on detector while the the gap between 6 AN vanes and 3 RS(B) vanes on stray radiation suppression is not wide. Therefore, the cold shield added with 3 RS vanes(B) is choosen as a compromise proposal.

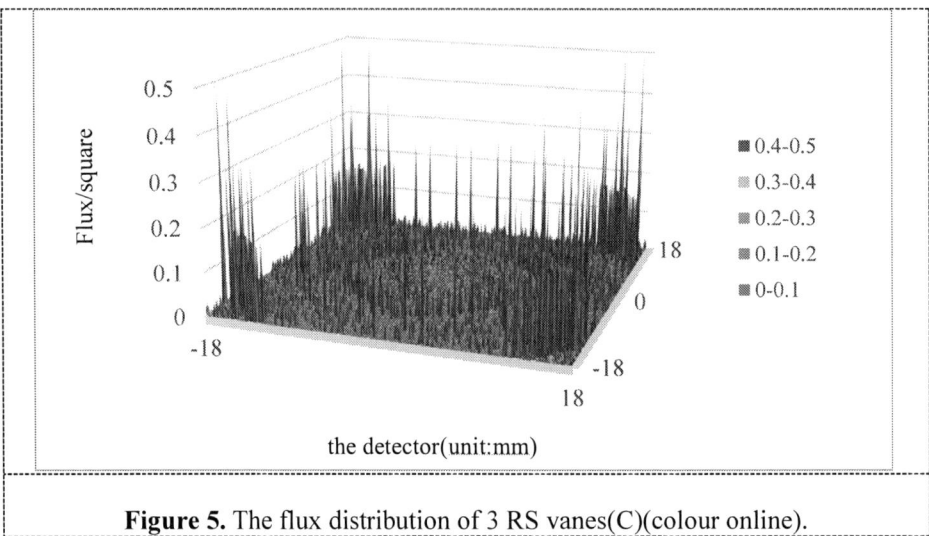

Figure 5. The flux distribution of 3 RS vanes(C)(colour online).

5. Conclusion

In this paper, several configurations of vanes inside cold shield are designed and simulated, and the results presented in this paper are summarized as follows:

- Vane plays an important part in minimizing background radiation and enhancing cold shield efficiency of cooled infrared systems.
- Vane with a rounded-square hole is more effective on stray radiation suppression than the annular vane.
- Cold shield added with 6 RS vanes(C) has the best effect on stray radiation suppression, but the one added with 3 RS vanes(B) is selected in consideration of cost and the influence of diffraction.

Acknowledgment

This study is supported by Kunming Institute of Physics.

Reference

[1] Pancrazzi M, Vives S, Landini F, Guillon C, Escolle C and Garcia J. 2013 Optimization of baffle configuration for stray light reduction *Proc. SPIE Optical Engineering Applications, International Society for Optics and Photonics (California, US, 25 August 2013)* pp 886205-886205

[2] Jinxing N, Shuheng S and Renkui Z 2011 Analysis to stray radiation of infrared detecting system *Proc. International Symposium on Photoelectronic Detection and Imaging, International Society for Optics and Photonics (Beijing, China, 24 May 2011)* pp 81931H-81931H

[3] Scholl M S, Páez-Padilla G 1997 Using the y, y-bar diagram to control stray light noise in IR systems *Infrared phys. & techn.* **38** 25-30

[4] Breault R P. 1977 Problems and techniques in stray radiation suppression *Proc. SPIE/SPSE Technical Symposium East. International Society for Optics and Photonics (Reston, 18 April 1977)* pp 2-23

[5] Kaiyu Y, Ning J, Ling C and Man X 2012 Calculation of energy distribution in image plane for all shapes of cold shields *Infrared Laser Eng.* **7** 009

[6] Breault R P. 1995 Control of stray light *Handbook of Optics* **1** 38-1

[7] Xia W, Sun W, Xiaokun W and Yanjin L 2012 Study on stray light suppression in IRFPA Dewar *Conf. 6th International Symposium on Advanced Optical Manufacturing and Testing Technologies, International Society for Optics and Photonics (Xiamen, China, 26 April 2012)* pp 84172I-84172I

[8] Xiao J, Xiuwen Y, Bin Z, Shuguang Z and Pan H 2010 The influence of the thermal environment on the stray light performances of infrared telescope systems *Conf. 5th International Symposium on Advanced Optical Manufacturing and Testing Technologies, International Society for Optics and Photonics (Dalian, China, 26 April 2010)* pp 76540V-76540V

[9] Yan Z. 2010 Study of Stray Light Suppression by Cold Shield in Dewar *Infrared* **7** 002

[10] Zhiqiang H, Tingwen X 2006 Principle and realization of baffle and vane's programmable design *Opto-Electronic Engineering* **33** 135-142

[11] Smith W J. 1966 *Modern optical engineering* (Tata McGraw-Hill Education) pp 148-150

Plasmonic Fano resonances in compositional heterogenous Al-Au nanorod dimers

Botao Wu, Yingxian Xue, Qiang Ma, Chengjie Ding, Youying Rong, Yan Liu, Lingxiao Chen, E Wu and Heping Zeng

State Key Laboratory of Precision Spectroscopy, East China Normal University, Shanghai, 200062, China

E-mail: btwu@phy.ecnu.edu.cn

Abstract. We have investigated theoretically the plasmon resonance coupling in compositional heterogenous Al-Au nanorod dimers organized in a close proximity by end-to-end. It has been proved that the destructive interference between the bright dipole mode from Al nanorod and the dark quadrupole mode from Au nanorod nearby results in the appearance of apparent Fano resonance in the extinction spectra. The Fano resonance response on the structural dimension modifications in the proposed nanorod dimers have been estimated and determined. The Al-Au heterogeneous nanorod dimer shows a high sensitivity to the surrounding environment with a local surface plasmon resonance figure of merit of 7.6, which enables its promising applications in plasmonic sensing and detection.

1. Introduction

The unique optical properties of noble metal nanostructures because of the collective oscillations of free electrons on metal surface when coupled with light have attracted tremendous interest due to their ability to confine and manipulate electromagnetic waves beyond the diffraction limit and to tailor light-matter interactions down to nanometer scale, and thus have found a rich variety of fundamental techniques and applications in plasmonic photovoltaic cells,[1] surface-plasmon enhanced spectroscopies,[2-6] photochemistry and photocatalysis,[7,8] sensing,[9] and quantum optics.[10-12] The plasmonic properties of a metal nanostructure are closely related to its composition, shape, size, and the surrounding environment. When two or more metal nanostructures are placed closely, their individual plasmon modes will be coupled strongly by near-field interactions and produce hybridized plasmon modes.[13,14] Plasmon hybridization of complex metal nanostructures can induce extremely strong localized near field enhancement around the junctions between adjacent nanostructures, and enables applications in surface enhanced spectroscopies,[2-4] sensing,[9] and plasmonic rules.[15]

Plasmonic Fano resonance, originating from the interference between spectrally overlapping broad super-radiant bright and narrow sub-radiant dark plasmonic modes, is a unique consequence of near-field coupling in complex metal nanostructures. Fano resonance has been investigated in various plasmonic nanostructures formed by the same metal, including nanoparticle aggregates,[16,17] metal dimers,[18] ring/disk cavities,[19] metallic core-shell nanostructures.[20] Fano resonance has also been reported to a lesser extent in compositional heterogenous metal nanostructures, theoretically and experimentally.[21-24] Due to its narrow lineshape and high sensitivity to environmental media, plasmonic Fano resonance shows potential applications in surface enhanced Raman scattering[17] and plasmonic biosensing.[19]

In this paper, we explored theoretically the optical properties of a new heterogeneous metal dimers composed of an aluminum (Al) and a gold (Au) nanorod. As shown in Fig. 1(a), the Al and Au nanorods are arranged by end-to-end with a small gap in between. Recently, Al nanorods have been reported to exhibit highly tunable plasmonic resonances from the deep ultraviolet to the visible wavelength region.[25,26] Here, the short Al nanorod supports a broad bright mode while the long Au nanorod supports a narrow dark mode. The interaction between them induces an apparent Fano dip in the extinction spectra of Al-Au nanorod dimers. The Fano resonance can be tuned by changing the Al or Au nanorod length and spatial separation in between, and shows a high sensitivity to the surrounding environment.

2. Computation methods

The numerical simulations were performed by the finite difference time domain (FDTD) Solutions software (Version 8.5, Lumerical Solution, Inc. Canada). The dielectric constants of aluminum and gold are taken from Ref. 27. The schematic geometry of Al-Au nanorod dimer is shown in Fig. 1(a). The Al and Au nanorods are coaxially arranged with the lengths of L_1 and L_2, respectively and the diameter $2R$. The gap between the two nanorods is d. The refractive index of the surrounding matrix is set to be 1. A plane wave total field-scattered field source ranging from 200 to 900 nm is utilized as the incident light beam with linear polarization direction along the longitudinal axis of the nanorod dimer shown as x direction in Fig. 1(a). A three-dimensional nonuniform meshing is used, and a grid size of 0.5 nm is chosen for the inside and immediate vicinity of Al-Au nanorod dimers. We use perfectly matched layer absorption boundary conditions as well as symmetric boundary conditions to reduce the memory requirement and computational time. All the numerical results have been after prior convergence testing.

3. Results and discussion

Firstly the bright and dark modes supported by single Al and Au nanorods are characterized and discussed. Fig. 1(b) presents the extinction spectrum of one single Al nanorod with the length $L_1 = 187$ nm and the diameter $2R = 40$ nm. A dipole plasmonic resonance peak around 630 nm is observed, and serves as a bright mode in Al-Au nanorod dimers. Meanwhile, although the dark mode has a weak coupling with the plane wave, it can be excited by a point source.[28] Fig. 1(c) exhibits the nonradiative enhancement of one single Au nanorod excited by a dipole source as well as its extinction spectrum illuminated by the normal plane wave. The diameter $2R$ of Au nanorod is also 40 nm and its length L_2 is 200 nm. The dipole source is placed along the longitudinal axis of Au nanorod with a gap distance d = 10 nm from one end of the nanorod and its polarization is also along the longitudinal axis of the nanorod, as shown in the inset schematic of Fig. 1(c). It is seen that there is a resonance peak around 635 nm excited by the dipole source, but it cannot be excited by the normal plane wave, since it is a quadrupole plasmon resonance and can serve as a dark mode. Obviously, if the dipole source is replaced by the Al nanorod the dark mode in the Au nanorod will be excited when illuminated by the plane wave, and concomitantly Fano resonance will appear due to their good spectral overlapping.

Fig. 2(a) shows the extinction, scattering and absorption spectra of Al-Au heterogenous nanorod dimer consisting of the above mentioned Al and Au nanorods with $d = 10$ nm. Clearly, a dip at 643 nm appears in the extinction spectrum. Actually the dip is induced by Fano interference. According to Ref. 14, when the dimer is illuminated at the frequencies resonating with both bright and dark modes, the bright mode of Al nanorod will be excited by two pathways: $|I\rangle \rightarrow |B\rangle$ and $|I\rangle \rightarrow |B\rangle \rightarrow |D\rangle \rightarrow |B\rangle$, where $|I\rangle$, $|B\rangle$ and $|D\rangle$ are excitation source, bright mode and dark mode, respectively. Fano-like interference occurs when the cumulative phase shift from $|B\rangle \rightarrow |D\rangle \rightarrow |B\rangle$ is π so that the two pathways interfere destructively, canceling the polarization of the bright mode and resulting in a narrow Fano dip in the extinction spectrum as shown in Fig. 2(a). Simultaneously, the dipole and quadrupole plasmon modes supported by the short Al and long Au nanorods hybridize in a constructive way to form two collective resonant modes: a high-energy antibonding mode at 592 nm and a low-energy bonding mode at 696 nm. The electric field distributions of the dimer with the separation d = 10 nm at the wavelength of 592, 643 and 696 nm are shown in Fig. 2(b)-(d). The electric field enhancement of the bright mode Al nanorod at the Fano dip is greatly depressed due to the cancelation of its polarization in the Fano resonance (see Fig. 2c).

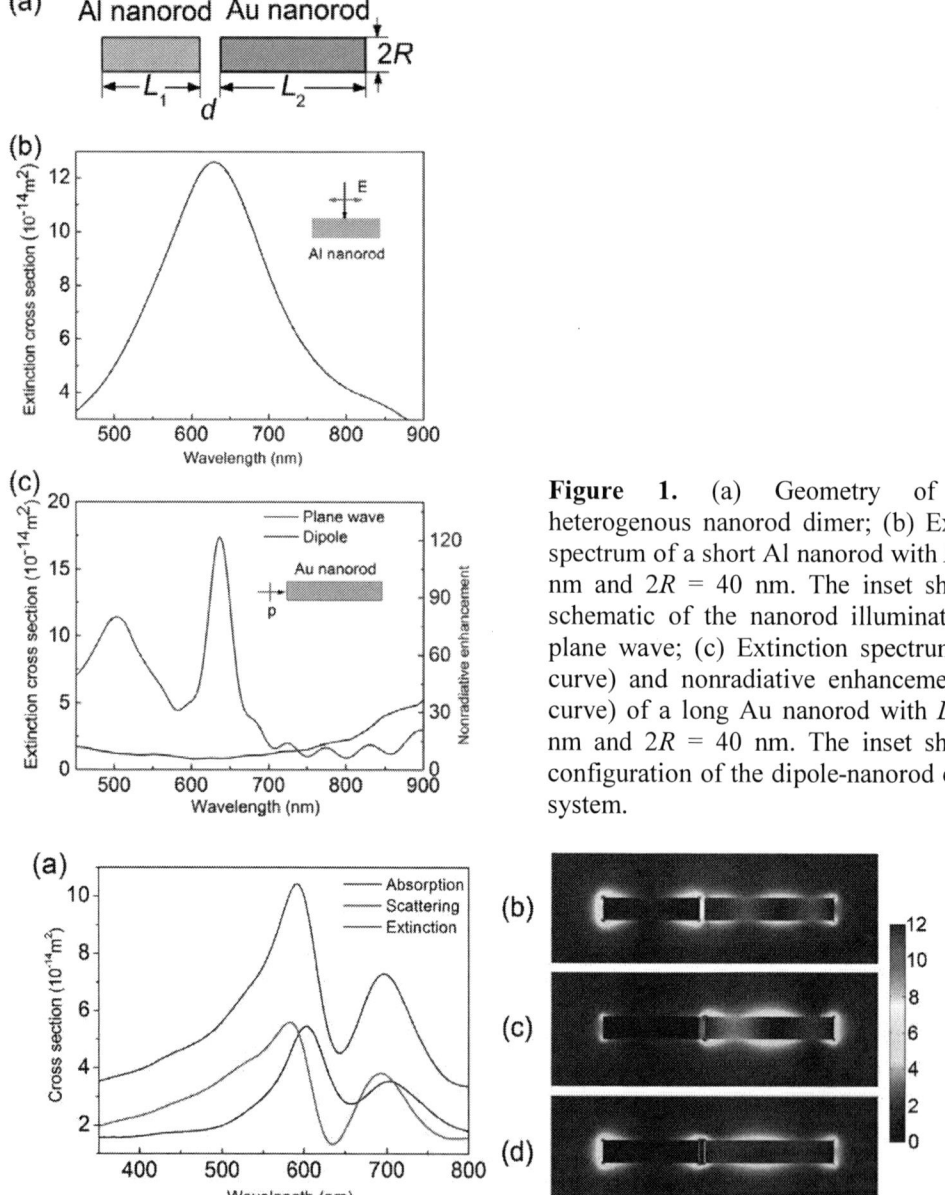

Figure 1. (a) Geometry of Al-Au heterogenous nanorod dimer; (b) Extinction spectrum of a short Al nanorod with L_1 = 187 nm and $2R$ = 40 nm. The inset shows the schematic of the nanorod illuminated by a plane wave; (c) Extinction spectrum (black curve) and nonradiative enhancement (blue curve) of a long Au nanorod with L_2 = 250 nm and $2R$ = 40 nm. The inset shows the configuration of the dipole-nanorod coupling system.

Figure 2. (a) Extinction, absorption and scattering spectra of Al-Au nanorod dimer with L1 = 187 nm, L2 = 250 nm, and 2R = 40 nm; Electric field distributions across the central cross section of the Al-Au nanorod dimer at (b) λ = 592 nm; (c) 643 nm and (d) 696 nm.

As shown in Fig. 3, the Fano resonance characteristics in Al-Au nanorod dimer can be tuned by modulating the gap distance between the two nanorods. Decreasing the separation between the two nanorods enhances the coupling strength between the bright and dark modes, and concomitantly the Fano dip becomes deeper.[28] The deeper Fano dip may be useful in the application of sensing and plasmon induced transparency. In addition, the change of plasmon coupling strength shows more apparent effect on the low-energy peak (λ = 696 nm) than that on the high-energy one (λ = 592 nm). With the gap d increasing from 10 to 30 nm, the low-energy peak shows a blue shift of about 27 nm (from 696 to 669 nm), whereas the high-energy peak shows a red shift of about 7 nm (from 592 to 599 nm). In the case of Fano dip, it only shifts from 643 to 640 nm.

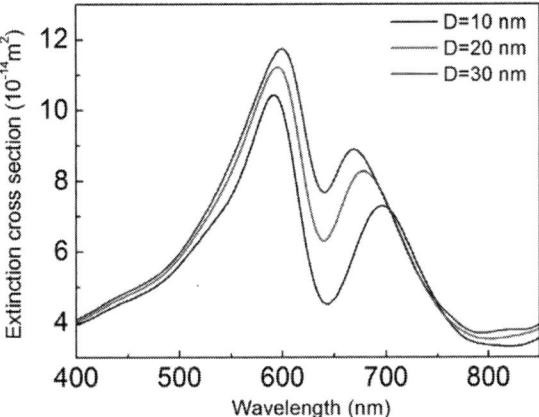

Figure 3. Extinction spectra of Al-Au nanorod dimers with gap distances d = 10, 20 and 30 nm, respectively. The lengths of Al and Au nanorods are 187 and 250 nm, respectively, and the diameter of each rod is 40 nm.

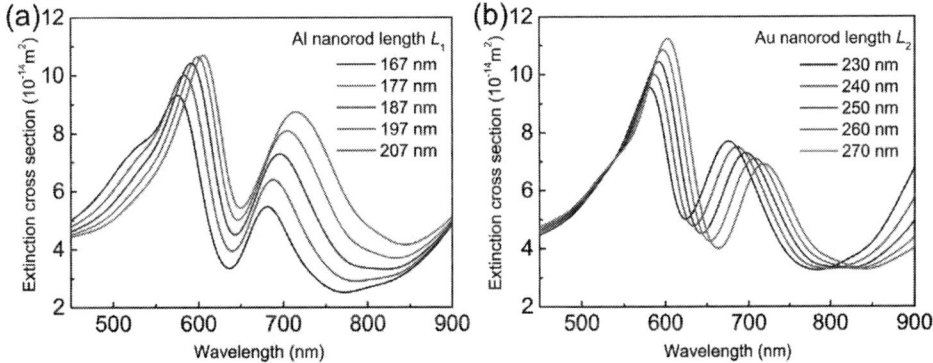

Figure 4. (a) Extinction spectra of Al-Au nanorod dimers with different Al nanorod length L1, and other parameters are fixed at L2 = 250 nm and d = 10 nm; (b) Extinction spectra of Al-Au nanorod dimers with different Au nanorod length L2, and other parameters are fixed at L1 = 187 nm and d = 10 nm.

The Fano resonance can also be tuned by adjusting the resonant frequencies of the bright (Al nanorod) or dark (Ag nanorod) mode, as shown in Fig. 4. In Fig. 4(a), as the bright mode wavelength redshift by increasing the length of the short Al nanorod from 167 to 207 nm with a fixed length 250 nm of Au nanorod, the Fano dip shows a slight redshift and becomes shallower. When the dark mode wavelength redshift by increasing the length of the long Au nanorod from 230 to 270 nm with a fixed short Al nanorod length at 187 nm, the Fano dip still shows a red shift but becomes deeper (see Fig. 4b).

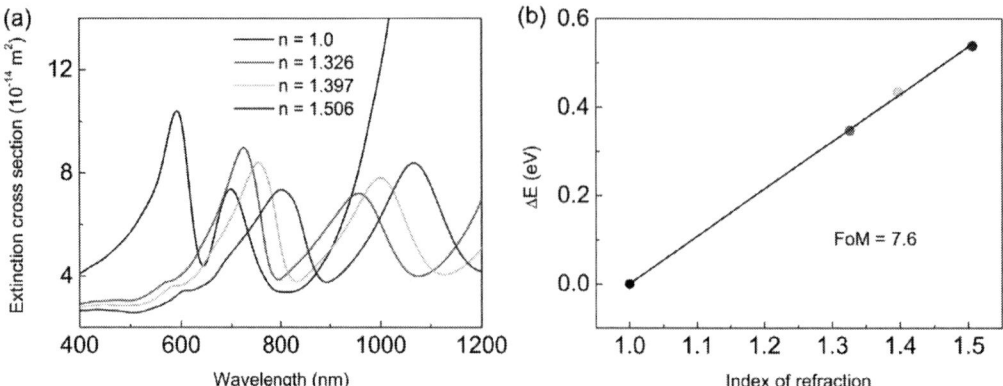

Figure 5. LSPR sensing in Al-Au nanorod dimer. (a) Extinction spectra of Al-Au nanorod dimer with the environment refractive index n of 1.0, 1.326, 1.397 and 1.506, respectively. (b) Linear plot of Fano dip shifts vs refractive index of the surrounding media.

One of the very interests in Fano resonance in plasmonic systems is its potential as effective localized surface plasmon resonance (LSPR) sensors, which stems from its inherent sensitivity to the surrounding environment. The efficiency of LSPR sensors is typically evaluated by their figure of merit (FoM), defined as the ratio of the plasmon energy shift per refractive index unit change in the surrounding medium, divided by the width of the spectral peak.[16] Here, we choose the Al-Au nanorod dimer with the same dimensional parameters as those in Fig. 2(a) to estimate its LSPR sensitivity to the refractive index change of the surrounding medium numerically. The resulting extinction spectra for the Al-Au nanorod dimer embedded in various media: ethanol ($n = 1.326$), butanol ($n = 1.397$) and index matching oil ($n = 1.506$, Cargille immersion oil) is presented in Fig. 5(a). The figure shows a pronounced redshift of the Fano resonance with the increasing surrounding refractive index. In air, the Fano resonance appears at 643 nm, and in immersion oil with a refractive index $n = 1.506$, it shifts to 892 nm. To determine the LSPR sensitivity, a slope of about 1.07 is obtained by a linear fit for the energy shift of the Fano dip as a function of the refractive index of the surrounding media (see Fig. 5b), and divided by the Fano line width (0.147 eV). The resulting FoM is 7.6, which is larger than those of isolated nanoparticles such as nanocubes (FoM = 5.4)[29] and nanoclusters (FoM = 5.7)[16], and comparable with that of noncentric ring/disk nanocavity (FoM = 8)[30]. Comparing with homogenous plasmonic nanostructures, the obstacle in experiment is the fabrication of heterogenous Al-Au nanorod dimers. One possible method can be realized by means of two-step electron-beam lithography and a combination of etch-down and lift-off with well-controlled spatial alignment reported in the recent work[31]. The proposed heterogenous nanorod cluster may find potential application in high sensitive sensors.

4. Conclusions

In this work, we investigated the hybridization of plasmonic modes between closely spaced Al-Au heterogeneous nanorod dimers arranged in a close proximity by end-to-end. Optical properties of Al-Au nanorod dimers are studied by FDTD simulation method. A pronounced Fano dip in the extinction spectra is observed, which strongly depends on both the geometry parameters of the complex nanostructure. The LSPR sensitivity of the heterogeneous metallic complex nanostructures is also checked and a FoM of 7.6 is obtained, which may find applications in biological sensing and molecule detection based on the coherent plasmonic coupling.

Acknowledgements

This work was funded in part by the National Nature Science Fund (11104079) and the National Key Scientific Instrument Project (2012YQ150092).

References

[1] Atwater H A and Polman A 2010 *Nat. Mater.* **9** 205.
[2] Lal S, Grady N K, Kundu J, Levin C S, Lassiter J B and Halas N J 2008 *Chem. Soc. Rev.* **37** 898.
[3] Kinkhabwala A, Yu Z, Fan S, Avlasevich Y, Muellen K and Moerner W E 2009 *Nat. Photonics* **3** 654.
[4] Wu E, Chi Y, Wu B, Xia K, Yokota Y, Ueno K, Misawa H and Zeng H 2011 *J. Lumin.* **131** 1971.
[5] Song M, Chen G, Liu Y, Wu E, Wu B and Zeng H 2012 *Opt. Express* **20** 22290.
[6] Song M, Wu B, Chen G, Liu Y, Ci X, Wu E and Zeng H 2014 *J. Phys. Chem.* C **118** 8514.
[7] Wu B, Ueno K, Yokota Y, Sun K, Zeng H and Misawa H 2012 *J. Phys. Chem. Lett.* **3** 1443.
[8] Zhong Y, Ueno K, Mori Y, Shi X, Oshikiri T, Murakoshi K, Inoue H and Misawa H 2014 *Angew. Chem. Int. Ed.* **53** 10350.
[9] Liu N, Tang M, Hentschel M, Giessen H and Alivisatos A P 2011 *Nat. Mater.* **10** 631.
[10] Akimov A V, Mukherjee A, Yu C L, Chang D E, Zibrov A S, Hemmer P R, Park H and Lukin M D 2009 *Nature* **450** 402.
[11] Chi Y, Chen G, Jelezko F, Wu E and Zeng H 2011 *IEEE Photon. Technol. Lett.* **23** 374.
[12] Chen G, Liu Y, Song M, Wu B, Wu E and Zeng H 2013 *IEEE J. Sel. Top. Quantum Electron* **19** 4602404.
[13] Prodan E, Radloff C, Halas N J and Nordlander P 2003 *Science* **302** 419.
[14] Fan J A, Wu C, Bao K, Bao J, Bardhan R, Halas N J, Manoharan V N, Nordlander P, Shvets G and Capasso F 2010 *Science* **328** 1135.
[15] Jain P K, Huang W Y and El-Sayed M A 2007 *Nano Lett.* **7** 2080.
[16] Lassiter J B, Sobhani H, Fan J A, Kundu J, Capasso F, Nordlander P and Halas N J 2010 *Nano Lett.* **10** 3184.
[17] Ye J, Wen F, Sobhani H, Lassiter J B, Dorpe P V, Nordlander P and Halas N J 2012 *Nano Lett.* **12** 1660.
[18] Yang Z, Hao Z, Lin H and Wang Q 2014 *Nanoscale* **6** 4985.
[19] Cetin A E and Altug H 2012 *ACS Nano* **6** 9989.
[20] Chen H, Shao L, Man Y, Zhao C, Wang J and Yang B 2012 *Small* **8** 1503.
[21] Bachelier G, Russier-Antoine I, Benichou E, Jonin C, Fatti N D, Valle'e F and Brevet P 2008 *Phys. Rev. Lett.* **101** 197401.
[22] Yang Z, Zhang Z, Zhang W, Hao Z and Wang Q 2010 *Appl. Phys. Lett.* **96** 131113.
[23] Peña-Rodríguez O, Pal U, Campoy-Quiles M, Rodríguez-Fernández L, Garriga M and Alonso M I 2011 *J. Phys. Chem.* C **115** 6410.
[24] Wu D, Jiang S and Liu X 2011 *J. Phys. Chem.* C **115** 23797.
[25] Knight M W, Liu L, Wang Y, Brown L, Mukherjee S, King N S, Everitt H O, Nordlander P and Halas N J 2012 *Nano Lett.* **12** 6000.
[26] Schwab P M, Moosmann C, Wissert M D, Schmidt E W, Ilin K S, Siegel M, Lemmer U and Eisler H 2013 *Nano Lett.* **13** 1535.
[27] Palik E D 1985 *Handbook of Optical Constants of Solids* (Academic).
[28] Yang Z, Zhang Z, Zhang L, Li Q, Hao Z and Wang Q 2011 *Opt. Lett.* **36** 1542.
[29] Sherry L J, Chang S H, Schatz G C, Duyne R P V, Wiley B J and Xia Y 2005 *Nano Lett.* **101** 2034.
[30] Hao F, Sonnefraud Y, Dorpe P V, Maier S A, Halas N J and Nordlander P 2008 Nano Lett. **8** 3983.
[31] Aouani H, Rahmani M, Navarro-Cía M and Maier S A 2014 *Nat. Nanotech.* **9** 290.

High Accuracy Tiny Crack Detection in Metal by Low Frequency Electromagnetic Technique

Weimin Lou[1], Changyu Shen[1], Fengying Shentu[1], Guanghai Li[2], Yu Chang[2], and Xinyuan Lu[2]

[1]Institute of Optoelectronic Technology, China Jiliang University, Hangzhou, China
[2]China Special Equipment Inspection and Research Institute, Beijing, China

E-mail: 1100402224cjlu@sina.com

Abstract. A low frequency testing technology based on eddy current technique is proposed for detecting defects in some special equipment surface. A two-dimension model is built to simulate the distribution of low frequency (10 Hz) magnetic flux density nearby the surface of a metal plate. The influence of lift-off effect, coil diameter, crack shape on the measurement are discussed. And the crack measurement sensitivity of 0.6μm was obtained.

1. Introduction

Nondestructive testing technology [1] has been widely applied in industry for detecting defects in some special equipment. Compared to other detecting technologies, low frequency electromagnetic technique [2] has many advantages. For example, the deeper skin depth can be obtained, checkout equipment is convenient to take and the inspection result can be obtained quickly. The low frequency electromagnetic technique derived from eddy current technique [3] is based on the theory of Faraday's law of electromagnetic induction. In this paper, a two-dimension model is built in COMSOL Multiphysics. Some interesting results are obtained by the finite element analysis.

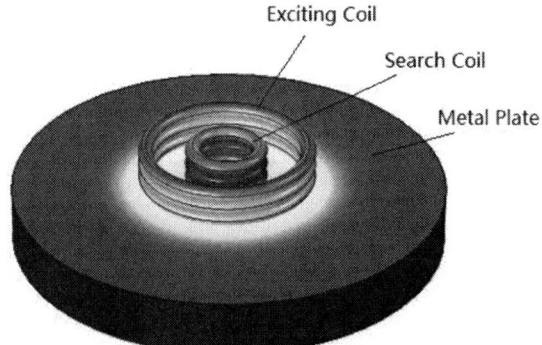

Figure 1. Eddy current technique

The principle of eddy current technique is illustrated in Figure1. Alternating electromagnetic field is aroused after alternating current is applied on the exciting coil. Since the metal plate is the conductive material, some eddy current appears and mainly distributes in the plate surface due to the skin effect. When there is a crack nearby the plate surface, the flow direction of eddy current will change and its induced magnetic field will also change. Hence, we can know if defect exists nearby the plate surface by putting a detector near the sample plate.

Content from this work may be used under the terms of the Creative Commons Attribution 3.0 licence. Any further distribution of this work must maintain attribution to the author(s) and the title of the work, journal citation and DOI.
Published under licence by IOP Publishing Ltd

2. Simulation

With the help of COMSOL Multiphysics, we use the finite element analysis method to simulate a series of experiments to know the properties of the low frequency testing technology. And our simulation is totally under a two-dimension model. Since the experiment is carried out by the software and simulation results can be easily obtained, the search coil needn't be set. The simulation parameters are shown in Table1. The simulation model is illustrated in Figure 2.

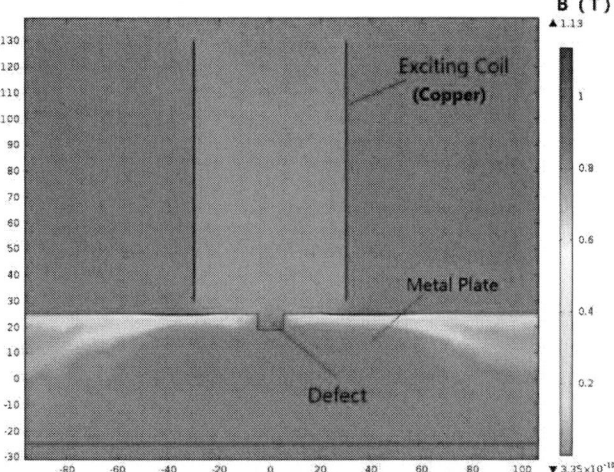

Figure 2. Two-dimension model

The exciting coil is made of copper and is put near the sample plate. There is a rectangular groove nearby the plate surface. By changing the coil diameter, lift-off value, defect size, defect shape and so on, we can find some interesting results.

Table 1. Parameters

Parameter	Wire diameter	Turns	Current	Frequency	σ (plate)	μ_r (plate)
Value	0.2mm	500	10A	10Hz	$5.727 \times 10^6 \, S/m$	2000

3. Results and Discussion

As shown in Figure 3, the lift-off value changed, the distribution of magnetic flux density nearby plate surface is different. The curve dip means there is a crack. The depth of curve dip decreases with the increase of lift-off value. From Figure 4, we can know that the curve span increases with the increase of coil diameter.

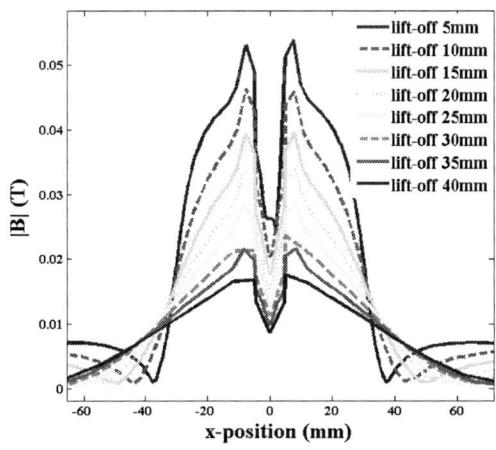

Figure 3. Effect of the coil lift-off value

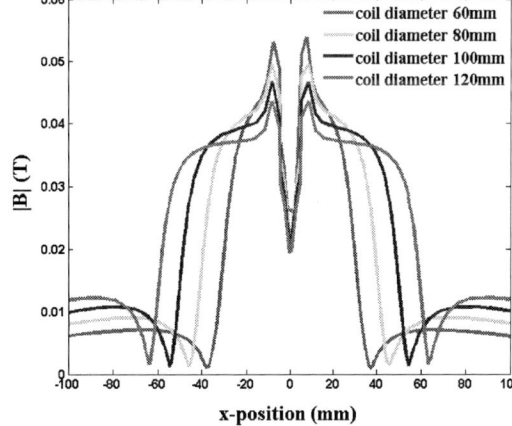

Figure 4. Effect of the coil diameter

The effects of rectangular groove, V-groove, dovetail groove and combination groove (rectangle+V) are illustrated in Figure 5 and the accuracy of measurement is shown in Figure 6.

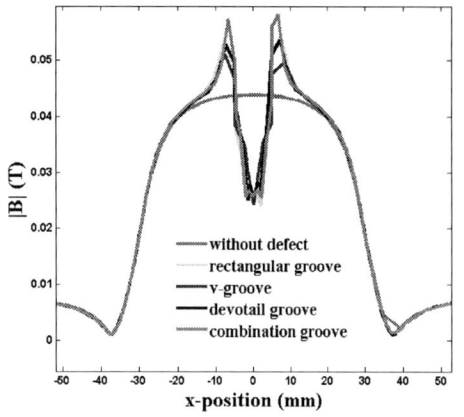

Figure 5. Effect of the defect shape **Figure 6.** Accuracy of measurement

The sequence of magnetic flux density from strong to week is as follows: combination groove, rectangular groove, dovetail groove, V-groove. The minimum fractional error of 0.03% (0.6μm) was obtained as the separation distance is rightly equal to the crack width.

4. Conclusion
On the condition of our given parameters, we gain the minimum accuracy of measurement when crack spacing is rightly equal to the crack width. For the future work, we will develop an intelligent algorithm to do the optimum work under the different conditions. And also, experiments will be done to verify our simulation results.

5. References
[1]   Y. Gotoh and N.Takahashi. 2007. *TRANSACTIONS ON MAGNETICS*. **43** 1733-1736
[2]   H.Yamada and T.Hasegawa. 2007. *NDT&E Int*. **41** 108–111
[3]   A. Egorov and V. Polyakov. 2015. *Defence Technology*. **11** 99-103

Acknowledgments
This work was supported by Special Fund for Quality Inspection Research in the public Interest of China (Grant No. 201510066), and the National Natural Science Foundation of China (Grant No.61405185).

An optical liquid level sensor based on core-offset fusion splicing method using polarization-maintaining fiber

Weimin Lou, Debao Chen, Changyu Shen, Yanfang Lu, Huanan Liu and Jian Wei

Institute of Optoelectronic Technology, China Jiliang University, Hangzhou, China

E-mail: 1100402224cjlu@sina.com

Abstract. A simple liquid level sensor using a small piece of hydrofluoric acid (HF) etched polarization maintaining fiber (PMF), with SMF-PMF-SMF fiber structure based on Mach-Zehnder interference (MZI) mechanism is proposed. The core-offset fusion splicing method induced cladding modes interfere with the core mode. Moreover, the changing liquid level would influence the optical path difference of the MZI since the effective refractive indices of the air and the liquid is different. Both the variations of the wavelength shifts and power intensity attenuation corresponding to the liquid level can be obtained with a sensitivity of -0.4956nm/mm and 0.2204dB/mm, respectively.

1. Introduction

In recent year, more and more liquid level sensors based on fiber sensing principles occur with the developments of optical fiber sensing technologies. Compared to the traditional electric-based liquid level sensors, the optical fiber liquid level sensors benefit from immune to electromagnetic interference, easy to fabricate, resistance to erosion, high sensitivity and capability of remote sensing, adapt to work in harsh environments, and thus there are lots of considerable meanings to research the fiber-optic liquid level sensors[1,2]. For example, Sarfraz Khaliq proposed a fiber-optic liquid level sensor based on a long period grating (LPG) in 2001[3]. Tuan Guo reported a fiber Bragg grating (FBG) liquid level sensor based on a bending cantilever beam in 2005[4].These sensors possess advantages such as absolute response parameter and high sensitivities. However, they are limited to large temperature cross sensitivities and complex fabrication processes which require the expensive ultraviolet light laser, phase masks, etc. In the present study, we propose a brief approach, employing a small piece of hydrofluoric acid (HF) etched polarization maintaining fiber (PMF), with SMF-PMF-SMF fiber structure based on MZI for the sensor application of liquid level. The MZI is formed by inserting a PMF with a length of 25 mm between two single mode fibers (SMF-28). The sandwich structure is fabricated by arc discharge fusion splicing with a small core-offset of 4.5, so it is compact and robust. Since the core of the fusion splicing cross section is offset, the core mode and cladding modes of the PMF could be excited at the same time in the first fused splicing cross section and interference would occur in the second fusion splicing cross section. By detecting the intensity variations and wavelength shifts of the selected resonance peaks, the liquid level variations information can be obtained. It is also notable that the fabrication of the sensor is simple and cost effective, including only the fusion splicing and a brief HF acid etching processing. Experimental results indicate that the sensitivity of the sensor reaches up to -0.4956nm/mm and 0.2204dB/mm, respectively.

Content from this work may be used under the terms of the Creative Commons Attribution 3.0 licence. Any further distribution of this work must maintain attribution to the author(s) and the title of the work, journal citation and DOI.

Published under licence by IOP Publishing Ltd

2. Experiments

The schematic diagram of the proposed liquid level sensor is shown in Fig.1. A gain-flattened Erbium-doped fiber amplified spontaneous emission (ASE) source with wavelength range of 1450 to 1650nm is used as the light source. And the polarization state of the output light from the ASE source is a non-polarized light. Thus a polarization controller (PC) is used to adjust the polarization states of the input light in order to obtain a high fringe visibility. The transmission spectrum is detected with an optical spectrum analyser (OSA, AQ6370, Japan). The maximum resolution of the OSA is 20 pm. A length of 25 mm PMF(PANDA, 1017-C, YOFC the birefringence index of the PMF is 7.7024×10-4.) is spliced between two conventional SMFs with mismatched core-offset fusion splicing to construct an all-fiber modal interferometer which we call in-fiber MZI. Both ends of the PMF are mismatch fusion spliced (using a commercial fusion splicer (Fujikura FSM-40S)) with the lead-in SMF1 and lead-out SMF2, respectively. The core offset size is about 4.5. The core and cladding diameters of the used SMF are 9 and 125, respectively. The PMF inserted between the two SMFs has the same core and cladding diameters as the SMF. Although the processing of the sensor head is only using the splicing method, it still needs carefully cleaving and fusion splicing procedures. The splicing loss could reduce the fringe visibility and transmission loss of the interference spectrum. After finished the fabrication of sensor head, we slowly etched the cladding of the PMF in 4% HF acid for~2h.

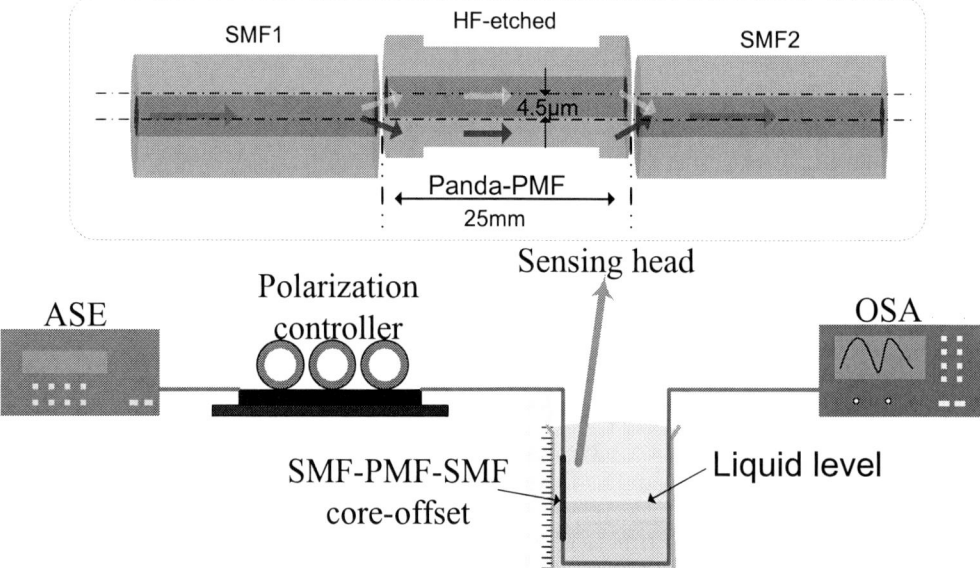

Fig 1. The schematic diagram of the proposed sensor.

3. Results and Discussion

It's well known that only one exited cladding mode is dominant in the interference between the core mode and the cladding modes by using the fast Fourier transform method [5-10]. The power intensity is mainly distributed in the core mode and a strong cladding mode. Other excited higher order cladding modes are weak. The main interference pattern is mainly formed by the interference of the dominant strong cladding mode with the core mode.

Fig.2 expresses the interference patterns of the proposed in-fiber MZI sensor corresponding to the liquid level variations with a liquid refractive index of 1.333. The dip wavelength has a blue shift of about 10nm as the liquid level increasing from 0 mm to 20 mm and the fringe visibility of the interference patterns is decreased gradually. The wavelength shifts and intensity attenuation of the resonant dip around 1536nm are researched with the liquid level increasing shown as the inset picture in Fig.2. The different colour spectrums represent the different liquid levels. The interference spectrum variations show a considerable dependence on the changing liquid level. Wavelength blue shift and intensity attenuation occurs with the increasing of the liquid level. Finally, the sensitivity of the fiber-optic liquid level sensor is calculated as -0.4956nm/mm and 0.2204dB/mm with a linear fitting method shown in Fig.3. The high sensitivity and correlation coefficient contribute to the great potential as a liquid level sensor in many applications.

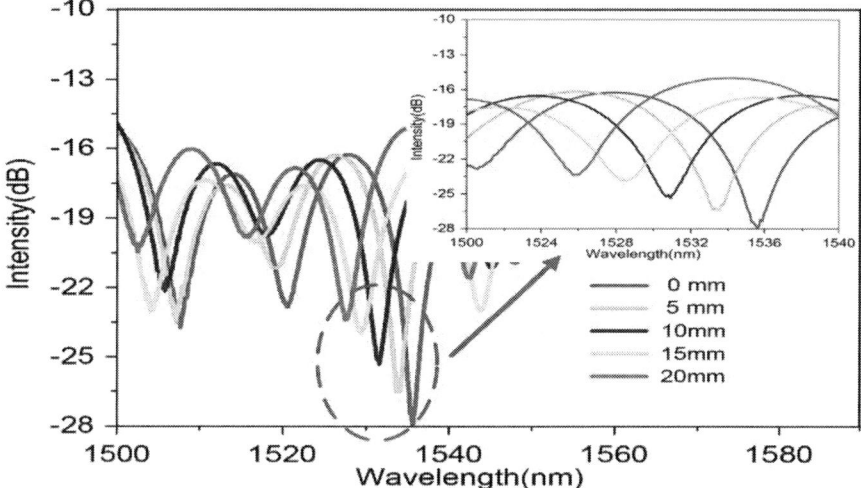

Fig 2. Transmission spectrums of the proposed liquid level sensor

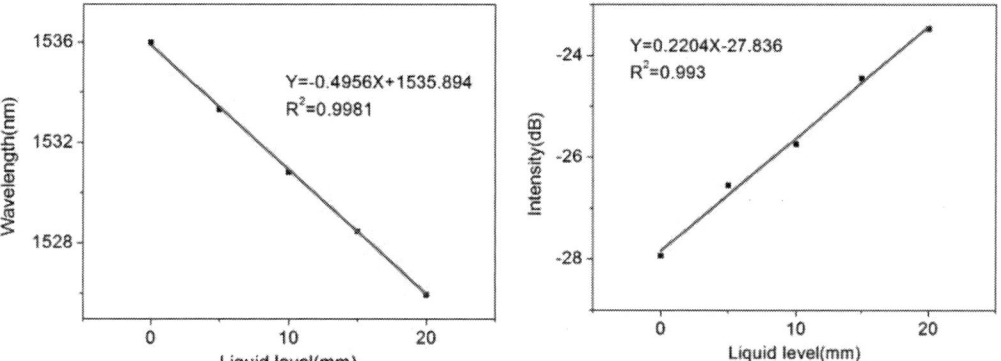

Fig 3. (a) Wavelength shifts **Fig 3.** (b) Power intensity attenuation

4. Conclusion

In conclusion, a compact polarization-dependent in-fiber MZI based on core-offset fusion splicing is proposed and demonstrated experimentally and theoretically for the application of liquid level measuring. The MZI is fabricated by core-offset fusion splicing one section of PMF between two SMFs. The core mode and the dominant cladding modes resulted in interference in the transmission spectrum. Therefore, the liquid level variations on the MZI part can be measured by detecting the wavelength shifts and power intensity attenuation which depend on the ratio of the length of PMF immersed in the liquid and air. Experiment results show that high sensitivity of -0.4956nm/mm and 0.2204dB/mm are obtained within the measurement range of 0 to 20mm. Compared with other liquid level sensors, the alternative simple fabrication process and high sensitivity shows the proposed sensor has a promising for industrial liquid level measurement applications.

5. References

[1] Chih-Wei Lai, Yu-Lung Lo, Jiahn-Piring Yur 2012 *J. Sensors Journal.* **12** 827-831.

[2] Kyung-Rak Sohn 2010 *J. Sensors and Actuators A: Physical.* **158** 193-197.

[3] Sarfraz Khaliq, Stephen W. James 2001 *J. Optics Letters.* **26** 1224-1226.

[4] Tuan Guo, Qida Zhao, Qingying Dou 2005 *J. Photonics Technology Letters.* **17** 2400-2402.

[5] Changyu Shen, Chuan Zhong, Xinyong Dong 2012 *J. Optics Express.* **20** 15407-15416.

[6] Bobo Gu, Wenliang Qi, Yanyan Zhou 2014 *J. Optics Express.* **22** 11834-11839.

[7] Xiaodong Wen, Tigang Ning, Chao Li, Zexin Kang 2014 *J. Applied Optics.* **53**(2014)71-75.

[8] HuapingGong, Haifeng Song, Xinyong Dong 2014 *J. Sensors and Actuators A: Physical.* 204-207

[9] Lecheng Li, Li Xia, Zhenhai Xie 2012 *J. Optics Express.* **20** 11109-11120.

[10] Chuan Zhong, Changyu Shen, Xinyong Dong 2012 *J. Journal of the Optical Society of America B.* **29** 1136-1140.

Acknowledgments

This work was supported by Special Fund for Quality Inspection Research in the public Interest of China (Grant No. 201510066), and the National Natural Science Foundation of China (Grant No.61405185).

Fast Perturbation Monte Carlo simulation for heterogeneous medium and its utilization in functional near-infrared spectroscopy

Y M Song[1], J W Li[1] and F H Cai[1,2,3]

[1]State Key Laboratory of Modern Optical Instrumentations, Centre for Optical and Electromagnetic Research, Zhejiang University, 310058 ,Hangzhou, China
[2]Suzhou WilHealth Information Technology Co. Ltd, 215500 ,Suhou, China

E-mail: caifuhong@zju.edu.cn

Abstract. In near-infrared spectroscopy, fiber optic probe is usually applied to incident light into the bio-sample and detect the spatial and temporal resolved optical signal re-emitted from the turbid medium. In this point-source-point-detector measurement system, seed Perturbation Monte Carlo (Pmc) method is an effective model to perform the forward simulation. In our study, the integration of parallel computing with graphics processing units(GPU) into the existing seed Pmc method substantially accelerate the speed of the original simulation. The GPU based seed Pmc provide an excellent solution for the application of fiber optic probe in both homogeneous a heterogeneous turbid medium.

1. Introduction

The forward model for simulating light migration in turbid sample is mainly based on the radiative transport equation (RTE) [1]. In practice, when $\mu_s \gg \mu_a$ and the source is far away from the detectors, diffusion approximation can be introduced into RTE to derive the diffuse reflectance or transmittance more effective [2]. However, in fiber optic probe to interrogate bio-sample, in order to manufacture a small probe, the incident fiber is assembled side by side next to the detection fiber. [3]Furthermore, for some special fiber optic probe, it is hard to concern the complex boundary condition between the probe and the tissue sample. Monte Carlo simulation offers a flexible method based on tracing photon emitted from fiber and propagating in turbid medium. Therefore in MC simulation, the fiber optic probe can be designed in arbitrary geometric structure [3] and placed on arbitrary position of the turbid medium. But this method is statistical in nature and it must trace a large number of photons to reduce the statistical error. Alex et.al. have introduce GPU to accelerate MC simulation and gain 1000 speeded-up compared with a traditional CPU based program. Ref [5] indicates that for a point-source-point-detector measurement system (i.e. fiber optic probe), the vast majority of photons do not contribute to the detection result because the detection rate is limited by the diameter of fiber core. Hence, a new sampling method (i.e. seed Perturbation Monte Carlo (Pmc)) can be introduced to improve the efficient of MC simulation for fiber probe system. The seed Pmc records the detected photon and abandon the other photons during its first step. Once the valuable seeds for the detected photons are recorded, they can be used in the second-step of seed Pmc simulation to effectively

---

[3] To whom any correspondence should be addressed.

Content from this work may be used under the terms of the Creative Commons Attribution 3.0 licence. Any further distribution of this work must maintain attribution to the author(s) and the title of the work, journal citation and DOI.
Published under licence by IOP Publishing Ltd

calculate the diffusive reflectance or transmittance. The less the ratio of the detected photons to the total simulation photons is, the more effective the seed Pmc model will be. However, the photon trajectory is determined by the scattering coefficient, in order to perform the seed Pmc within large range of scattering coefficients, it needs to record the valuable seeds by running the first-step seed Pmc with the corresponding scattering coefficients. It's worth to mention that the absorption coefficient can be varied arbitrarily in the second-step.

In our previous work [6], we have applied the seed Pmc model in GPU program and verify that the seed Pmc model has the ability to calculate the diffuse transmittance with various scattering coefficient in point-source-planar-detector system. In the present study, the similar simulation program will be used in the point-source-point-detector measurement system. The traditional MC and seed Pmc are all performed by using GPU, and the seed Pmc is at least 1000 factor faster than MC. Last, the brain model will be utilized in the seed Pmc method. The result demonstrates that the seed Pmc can also generate the Time domain reflectance more effectively.

2. Two-step seed Pmc to solve RTE with arbitrary scattering and absorption coefficients

Fiber optic probe provides a flexible solution for an optical platform between the spectroscopic device and the sample to be interrogated in situ. The illumination and detection fibers are usually separate to improve the system SNR [3]. To minimize the size of the probe, the separation of the illumination and detection fiber must be small enough. In this case, the diffusion equation is failed to calculate the reflectance result. In our study, the seed Pmc is consider to simulate the spatial resolved reflectance of fiber optic probe with fibers cling to each other. The seed Pmc can be described as two-step MC simulation, as shown in Fig.1. The main procedure of simulation of photon in the heterogeneous medium is according to Ref [7]. The voxel size is 1 mm. In the first-step Pmc, a GPU (NVIDIA® GeForce™ GT 240M) is used to perform the MC for fiber-based detection and record the valuable seeds in the computer disk. Here, the valuable seeds can 'move' the photons from the source fiber into the turbid sample and backscatter into the detection fiber. The valuable seeds can only work for a given scattering coefficient. Hence the first step will be run for 100 times for scattering coefficient varied from 1cm⁻¹ to 100cm⁻¹ to derive and store the corresponding valuable seeds. Note that, the medium is heterogeneous here, the scattering coefficient for all voxel is set as the same value and the absorption coefficient can be set as arbitrary value for each voxel. Then, the valuable seeds can be applied in the second-step Pmc to generate the diffusive reflectance with a large range of scattering coefficient (1cm⁻¹ to 100cm⁻¹, the value of scattering coefficient must be integer between 1-100) and arbitrary absorption coefficient. The first-step Pmc is just performed to record the valuable seeds and the second-step Pmc is the mainly procedure to calculate the reflectance for various scattering and absorption coefficients, therefore in the following only the running time of second-step Pmc is recorded.

AOM2015   IOP Publishing
Journal of Physics: Conference Series **680** (2016) 012019   doi:10.1088/1742-6596/680/1/012019

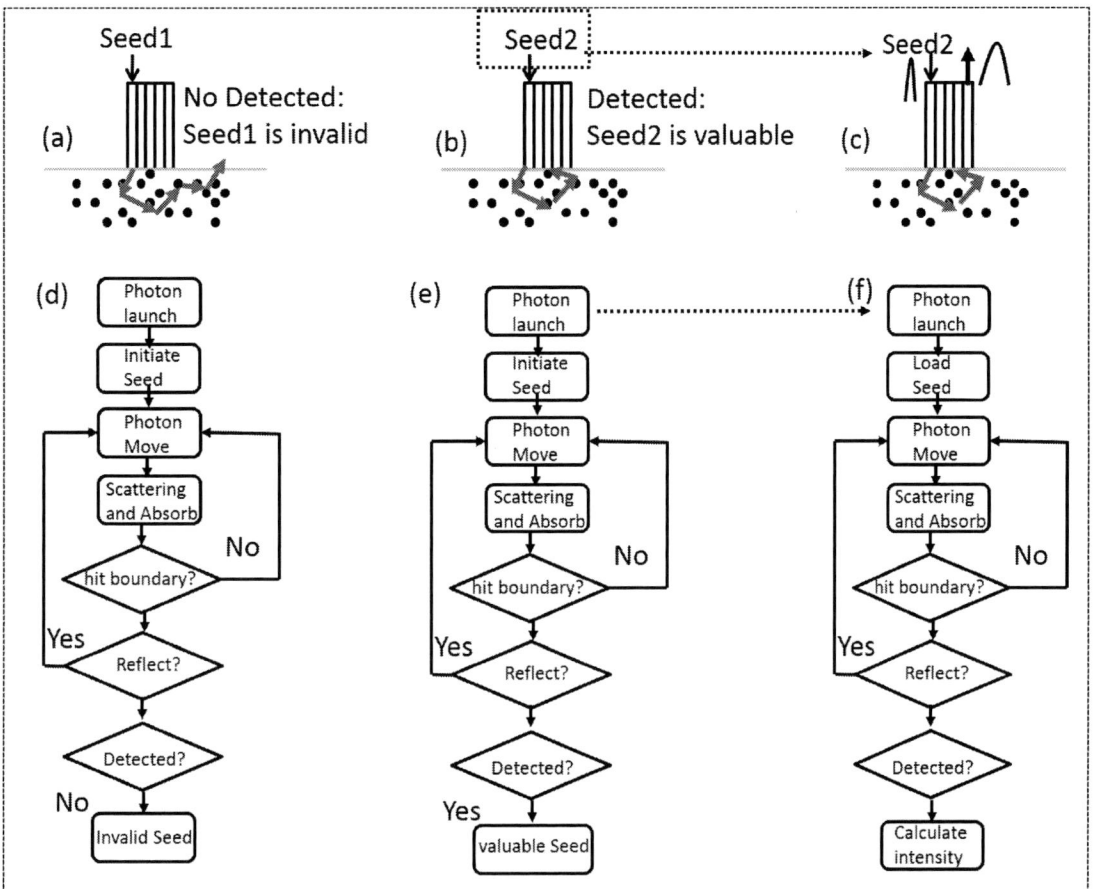

Figure 1. The scheme of the GPU based two-step PMC for fiber probe. (a) and (d) the processing procedure in the first step PMC for Non-Detected photon; (b) and (e) the processing procedure in the first step PMC for Detected photon; (c) and (f) in the second step PMC, using the valuable seed to perform the simulation.

3. Simulation of spatial resolved reflectance by fiber optic probe

A simple fiber reflectance probe is simulated in this study to demonstrate the ability of the seed Pmc. As shown in the inset of Fig.2(b), five-fibers are placed on the tissue to illuminate the sample and pick up reflectance light. This fiber probe has been used widely in tissue detection to obtain the scattering and absorption coefficients of turbid medium according to the reflectance. [3] The turbid medium is set as homogeneous turbid sample in this case. The fibers in the probe have 200 μm core and 300 μm cladding. The numerical aperture is 0.22. In order to overcome the statistical error, the stop criterion for MC and seed Pmc is that the fibers 2-4 pick up to 40000 reflected photons during the simulation. Fig.2(a) illustrates the reflectance of fibers 3-5 according to different scattering coefficients. It's worth noting that the sampling method in the seed Pmc is based on the valuable seeds (i.e. all simulated photons will be detected by the detected fibers in the second-step Pmc.), so the absolute reflectance of the fiber optical probe cannot be derived. In Fig2.(a), F3-F5 are the reflectance of fibers 3-5 which are normalized by reflectance of fiber 2. The absorption coefficient and the anisotropy factor are set as 0.01 mm^{-1} and 0.8 respectively. The relative error between PMC and traditional MC is shown in the inset of Fig.2(a). The speedup factors are illustrated in Fig.2(b). To detect 40000 photons by fibers 2-5, it up to takes 0.06 second to finish the second-step Pmc.

AOM2015
Journal of Physics: Conference Series **680** (2016) 012019

IOP Publishing
doi:10.1088/1742-6596/680/1/012019

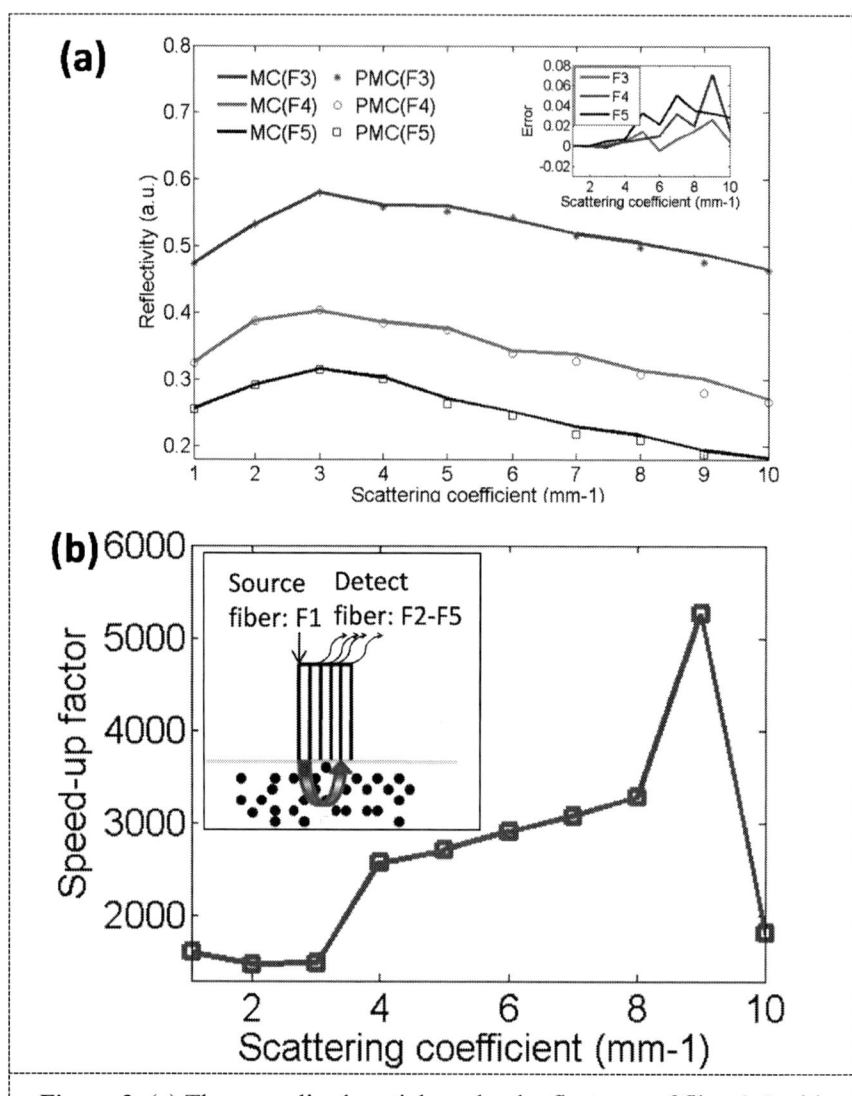

Figure 2. (a) The normalized spatial resolved reflectance of fiber 3-5 with different scattering coefficients. (b) The speed-up factor compared with seed Pmc and traditional MC simulation.

4. Simulation of temporal resolved reflectance in functional near infrared spectroscopy

Furthermore, the seed Pmc can be applied in heterogeneous brain medium. Herein the voxle based anatomical model of human head [8] is used. The input light wavelength is 632 nm and the optical coefficients of the skin, skull, white matter, grey matter and CSF are obtained from [8]. In functional near infrared spectroscopy (fNIRS) [9], fibers are usually exploited as transceiver to monitor the blood concentration of cerebral cortex. A schematic diagram is shown in Fig.3(a). One fiber is placed on the top of the head to incident a 632 nm laser into the head and another fiber, which is 1 cm apart from the first one, is utilized to measure reflectance light in time-domain. Two-dimensional representation of the distribution of optical intensity is shown in Fig.3(a), which illustrates that the 632 nm laser can penetrate the skull and arrive the brain region. Besides, seed Pmc is capable of calculating the temporal resolved reflectance in the head model. In the temporal resolved seed Pmc, the length of the photon's trajectory is recorded during the simulation and contributes to different time frame. The seed Pmc proves to be about 3000 factor faster than traditional MC. As mentioned before, the use of seed

Pmc simulation is more accurate and convenient than diffusion equation, which is mainly used at present, hence seed Pmc holds great potential to become a better forward model to solve the change of the scattering and absorption coefficients due to the brain activity [9].

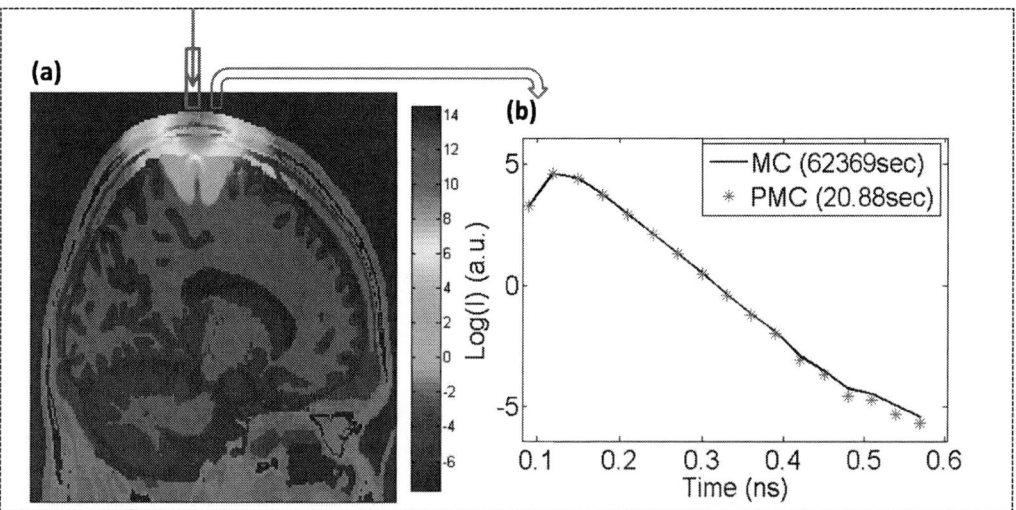

Figure 3. (a) The distribution of optical intensity inside a human head. A fiber is used to input the light into the head, and another fiber, which is 1 cm apart from the source fiber, is used to take the temporal resolved signal, as shown in (b). The perform time for derive the temporal resolved signal by MC and seed Pmc is 62369 second and 20.88 second respectively.

5. Conclusions

All of the seed Pmc simulation is performed on a laptop computer with a modest GPU (GT 240M). Even so, the seed Pmc behaves much better than the traditional MC model. In fiber optical probe (the inset of Fig.2(b)), it only takes about 0.06 second to generate the simulation result. Also, the seed Pmc can handle an arbitrarily complex medium, such as the human head and also gains a remarkable speedup compared with MC model. As mentioned above, the seed Pmc can deal with a sample with complex boundary condition as the traditional MC model does. By using an advanced GPU or Scalable Link Interface(SLI) to combine several modest GPU, this seed Pmc model has the realistic ability, rather than the potential, to realize real-time forward simulation for light migration in turbid medium with a point-source-point-detector system.

References

[1]   Wang L, Jacques S L and Zheng L 1995 *Comput. Methods Programs Biomed.* **47** 131-146
[2]   Contini D, Martelli F and Zaccanti G 1997 *Appl. Opt.* **36** 4587-4599
[3]   Utzinger U, Richards-Kortum R R 2003 *J. Biomed. Opt.* **8** 121-147
[4]   Alerstam E, Svensson T and Andersson-Engels S 2008 *J. Biomed. Opt.* **13** 060504-060504-3
[5]   Sassaroli A 2011 *Opt. Lett.* **36** 2095-2097
[6]   Cai F and He S 2012 *J. Biomed. Opt.* **17** 0405021-0405023
[7]   Boas D Culver J Stott J and Dunn A 2002 *Opt. Express* **10** 159-170
[8]   Fang Q and Boas D A 2009 *Opt. Express* **17** 20178-20190
[9]   Ferrari M and Quaresima V 2012 *NeuroImage* **63** 921-935

Enhanced Second Harmonic Generation in Au/Al₂O₃/Au absorber

Fenglun Huang, Songang Bai, Qiang Li, Yurui Qu and Qiu Min

State Key Laboratory of Modern Optical Instrumentation, College of Optical Science and Engineering, Zhejiang University, Hangzhou, 310027, China

E-mail: qiangli@zju.edu.cn

Abstract. A kind of metal-insulator-metal (MIM) metamaterial absorber for generating second harmonic signal is investigated. The absorbers exhibit high absorption efficiency at the dip and notably enhance the generated second harmonic signal by a factor of over 30, in contrast to an Au/alumina double-layer without Au disk on the top. This study demonstrates the potential of metamaterial absorber for nonlinear photonics.

1. Introduction

Second harmonic generation (SHG) is a typical nonlinear process, which occurs when the light of suitable wavelength interacts with some nonlinear materials[1,2]. SHG originates from both the surface and the bulk of various nonlinear materials. Unlike other nonlinear processes, the SHG relating to electric-dipole is forbidden in the bulk of a centrosymmetric material, with only negligible electric quadrupole and magnetic dipole predominant[3]. Due to the break of the inversion symmetry at the surface of this kind of material, SHG occurs with a low efficiency[4,5]. The metamaterial can confine light, enhance local field and avoid involvement of phase matching, making surface plasmon resonance from metal nanostructures a good candidate for nonlinear effect[6-9].

Numerous nano-scale metallic structures have been proposed and studied both theoretically and experimentally over the past several years, such as simple geometric hole-type[10-13], metal particles and convex geometry[14-16] structures on isolator substrates, metal film with bull's eye patterns[17], bowtie antennas[18], core-shell on nonlinear crystal substrate[19], special hybrid structures[20] and so on. However, few studies have paid close attention to a special MIM metamaterial absorber structure for exciting SHG with ultrahigh absorption peak.

Here we demonstrate an Au/Al2O3/Au three-layer absorber on silica substrates, as illustrated in Fig. 1a. Figure 1b shows one of its unit cell, where t, d and h denote the thickness of three layers, r represents the radius of the gold nanoparticle on the top, and p is period of the array.

Content from this work may be used under the terms of the Creative Commons Attribution 3.0 licence. Any further distribution of this work must maintain attribution to the author(s) and the title of the work, journal citation and DOI.

Published under licence by IOP Publishing Ltd

AOM2015   IOP Publishing
Journal of Physics: Conference Series **680** (2016) 012020   doi:10.1088/1742-6596/680/1/012020

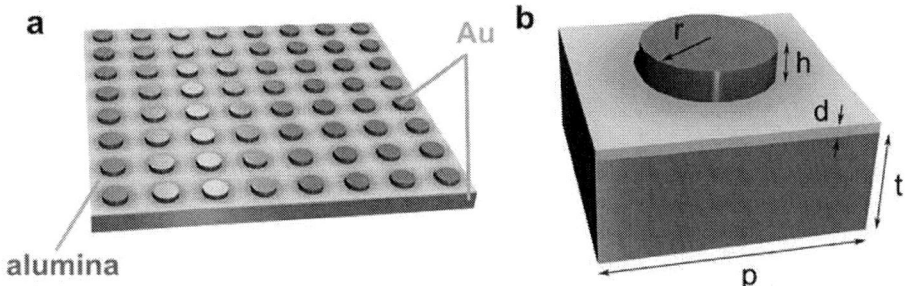

Figure 1. (a) Geometry of MIM three-layer absorber. (b) Unit cell of absorber. Here, t, d, h represent thickness of three layers. p is the period of array and r denotes radius of Au disk on the top.

2. Numerical simulation

We perform numerical calculations of nonlinear optical properties of the absorbers at normal incidence with finite difference time domain (FDTD) method. By changing structural parameters, influence of different thicknesses, radii and periods on the nonlinearity is studied. Here, thickness of the top Au disk (h) is fixed at 30nm, and the thickness of bottom Au (t) is set to be thick enough to make sure that no light is transmitted.

The results of the simulation are presented in Fig. 2. At normal incidence, the fundamental resonances at reflection spectra for different radii of Au disks are shown in Fig. 2a. The period p and

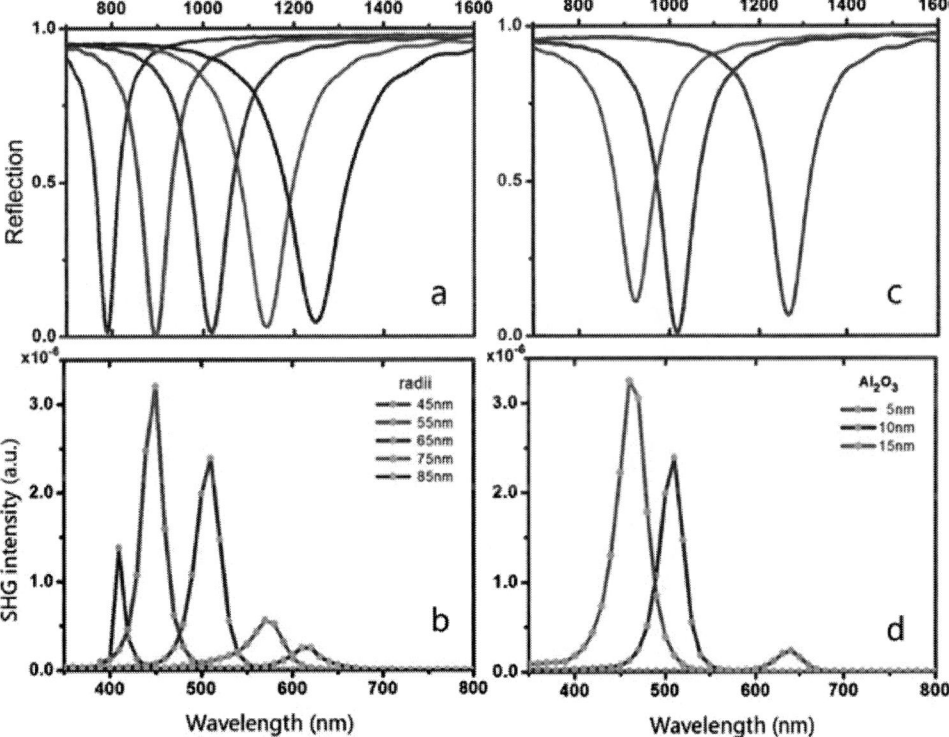

Figure 2. Simulated reflection spectra (a) and SHG intensity spectra (b) of the absorber with different Au disk radii (r), where p=250nm and d=10nm. Simulated reflection spectra (c) and SHG intensity spectra (d) of the absorber with different alumina thickness (d), where p=250nm and r=65nm.

thickness d are set to be 250 nm and 10 nm, respectively. At resonant dip, the reflection is close to zero, indicating a high absorption. The resonance shifts to longer wavelength and becomes slightly weaker as the radius of Au disk increases. Figure 2b shows the SHG intensity spectra of absorbers with different radii r, where the peaks of each spectrum clearly correspond to the resonance dips. Specifically, the peaks with radii of 55nm and 65nm are much higher than others.

Figure 2c and 2d shows the reflection and SHG intensity spectra of absorbers with different alumina thickness (d), where r=65nm and p=250nm. Resonance shifts towards shorter wavelength as the alumina becomes thicker. SHG peaks are in agreement with the resonance of fundamental wavelength. The maximum SHG is obtained with the thickness of 15nm.

3. Experiment

In our experiment, we fabricate the three-layer MIM structure. The substrate is firstly coated with 100-nm-thickness Au film. Then a 10-nm-thickness alumina is evaporated. After that, Au disks with different radii (with a period of 250nm) are patterned by electron-beam lithography. Figure 3a shows the SEM image of a representative absorber with the radius of Au disk at 65nm. The laser pulse for SHG is generated by a Mira HP-D femtosecond Ti:Sapphire oscillator, successively tuned by a Mira OPO. The power arriving at the sample is attenuated down to 28mW. The SHG signal is collected and measured by a PI TriVista spectrometer in the reflection direction.

Figure 3b shows the measured and simulated reflection spectra of the sample with r=65nm. The measured spectrum has a resonant reflection of 30% around 1000nm, while the simulation indicates nearly no reflection at resonance. We experimentally measure the SHG signals reflected by the sample with the incident light near 1000nm where the resonance occurs. The measured and simulated results have been displayed in Fig. 4. The measured SHG signals (dark cyan) agrees well with the simulation except that it is shifted by around 10 nm, which originates from the shift in the fundamental wavelength in both simulation and experiment. The pink hollow circles in figure 4 are calculated SHG intensity spectrum excited by different fundamental wavelengths from 960nm to 1060nm, which illustrates that the strongest SHG can be obtained at resonant fundamental wavelength. We also compare the measured SHG signal from the absorber with that from Au/Al_2O_3 double-layered film without Au disk on the top in experiments, the SHG generated from the absorber is enhanced by a factor of over 30.

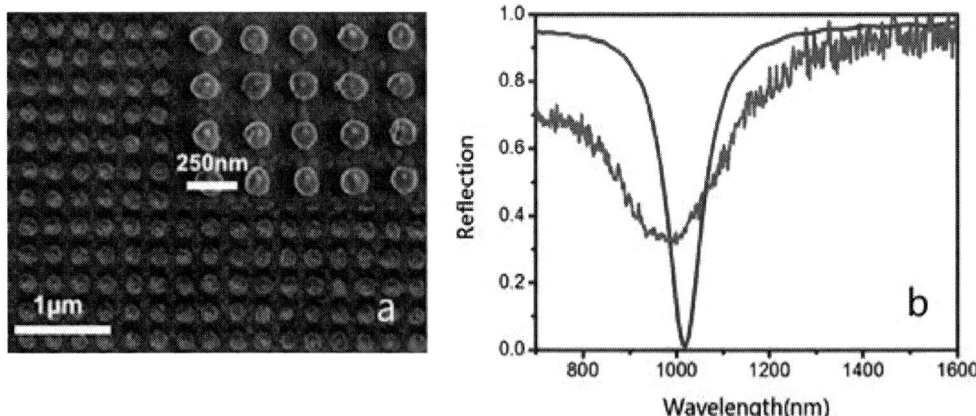

Figure 3. (a) SEM image of fabricated metamaterial absorber sample. (b) Measured (dark cyan) and simulated (purple) reflection spectra of sample with p=250nm, d=10nm and r=65nm.

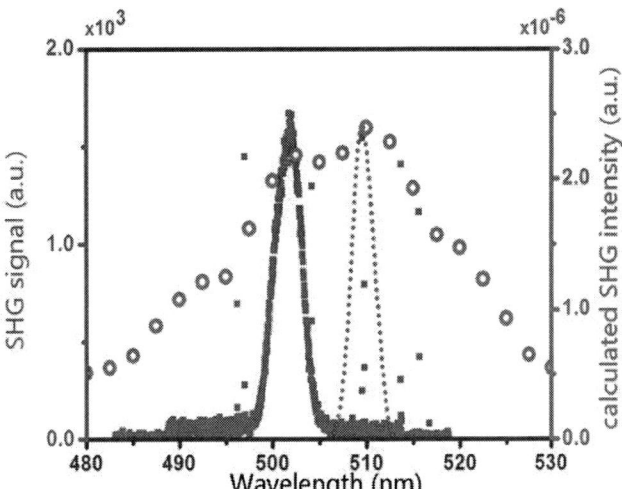

Figure 4. Measured SHG signal (dark cyan) and simulated SHG intensity (pink dot line) spectra of the sample with p=250nm, d=10nm at r=65nm, at their respective fundamental resonant wavelength of incidence. Simulated SHG intensity spectrum (pink hollow circle) of the sample with different fundamental wavelengths from 960nm to 1060nm.

4. Conclusion

We have demonstrated an $Au/Al_2O_3/Au$ metamaterial absorber can be used for SHG enhancement. The absorbers exhibit high absorption efficiency at the dip and notably enhance the generated second harmonic signal by a factor of over 30, in contrast to an Au/alumina double-layer without Au disk on the top. This study demonstrates the potential of metamaterial absorber for nonlinear photonics.

Acknowledgment

This work is supported by the National Natural Science Foundation of China (Grant Nos. 61235007, 61425023, 61575177, 61205030 and 61275030).

References

[1] Boyd R W 2008 *Nonlinear Optics* (Amsterdam: Elsevier)
[2] Franken P A, Hill A E, Peters C W and Weinreich G 1961 *Phys. Rev. Lett.* **7** 118-119
[3] Bloembergen N, Chang R K, Jha S S and Lee C H 1968 *Phys. Rev.* **174** 813–822
[4] Sipe J E, So V C Y, Fukui M and Stegeman G I 1980 *Phys. Rev.* B **21** 4389–4402
[5] Sionnest P G and Shen Y R 1988 *Phys. Rev.* B **38** 7985–7989
[6] Scalora M, Vincenti M A, Ceglia D, Roppo V, Centini M, Akozbek N and Bloemer M J 2010 *Phys. Rev.* A **82** 043828
[7] Wang F X, Rodríguez F J, Albers W M, Ahorinta R, Sipe J E and Kauranen M 2009 *Phys. Rev.* B **80** 233402
[8] Van-Nieuwstadt J A H, Sandtke M, Harmsen R H, Segerink F B, Prangsma J C, Enoch S and Kuipers L 2006 *Phys. Rev. Lett.* **97** 146102
[9] Kim E, Wang F, Wu W, Yu Z and Shen Y R 2008 *Phys. Rev.* B **78** 113102
[10] Lesufflr A, Kumar L and Gordon R 2006 *Appl. Phys. Lett.* **88** 261104
[11] Wang B L, Wang R, Liu R J, Lu X H, Zhao J M and Li Z Y 2013 *Scientific Reports* **3** 2358
[12] Salomon A, Zielinski M, Kolkowski R, Zyss J and Prior Y 2013 *J. Phys. Chem.* C **117** 22377–22382
[13] Rodrigo S G, Laliena V and Moreno L M 2015 *J. Opt. Soc. Am.* B **32** 15-25
[14] Thyagarajan K, Butet J and Martin O J F 2013 *Nano Lett.* **13** 1847-1851

[15] Walsh G F and Negro L D 2013 *Nano Lett.* **13** 3111 −3117
[16] Czaplicki R, Makitalo J, Siikanen R, Husu H, Lehtolahti J, Kuittinen M and Kauranen M 2015 *Nano Lett.* **15** 530-534
[17] Nahata A, Linke R A, Ishi T and Ohashi K 2003 *Opt. Lett.* **28** 423–425
[18] Kim S, Jin J, Kim Y J, Park I Y, Kim Y and Kim S W 2008 *Nature* **453** 757–760
[19] Lehr D, Reinhold J, Thiele I, Hartung H, Dietrich K, Menzel C, Pertsch T, Kley E B and Tunnermann A 2015 *Nano Lett.* **15** 1025 −1030
[20] Grinblat G, Rahmani M, Cortes E, Caldarola M, Comedi D, Maier S A and Bragas A V 2014 *Nano Lett.* **14** 6660-6665

AOM2015
IOP Publishing
Journal of Physics: Conference Series **680** (2016) 012021
doi:10.1088/1742-6596/680/1/012021

The ordering alignment of gold nanorods in liquid crystals and its applications to polarization-sensitive SERS

Y L Wang[1,+], L Y Chen[1,+], Q K Liu[3], F H Cai[1,2,*], and J Qian[1]

1 State Key Laboratory of Modern Optical Instrumentations, Centre for Optical and Electromagnetic Research, Zhejiang University, Hangzhou 310058, Zhejiang Province, China

2 Suzhou WilHealth Information Technology Co. Ltd, Changshu 215500, Jiangsu Province, China

3 Department of Physics and Liquid Crystal Materials Research Center, University of Colorado, Boulder, Colorado 80309, United States

[*]E-mail: caifuhong@zju.edu.cn
[+] These authors contributed equally to this work.

Abstract. Gold nanorods (GNRs) were synthesized, coated with poly(ethylene glycol) (PEG) chains, and uniformly dispersed into the lyotropic nematic liquid crystal (LC) matrix by our proposed method. The GNRs-LC composites were found to exhibit good stability over days and have high density of GNRs. The extinction spectra of the composites were found to be polarization sensitive when the shearing force were applied, due to the alignment of GNRs driven by the LC molecules, which was also in accordance to the simulation results. A type of Raman reporter, 3,3'-diethylthiatricarbocyanine iodide (DTTC), was co-conjugated onto the GNRs with PEG molecules, and then incorporated into the LC matrix. Thus, its Raman signals could be enhanced by the localized surface plasmon resonance (LSPR) of the GNRs. These surface-enhanced Raman scattering (SERS) signals were found to be polarization sensitive when the shearing force was applied, due to the polarization sensitive enhancement of the local field of GNRs. The nanocomposites with tunable SPR peaks and SERS signals have potential applications in optoelectronic devices.

1. Introduction

Gold nanorods (GNRs), with unique optical properties and good chemical stability, have wide applications in sensing [1-2], imaging [3-4], surface-enhanced Raman scattering (SERS) [5-6], and so on. For a single GNR, its optical properties, like absorption, scattering, and local field, are sensitive to

Content from this work may be used under the terms of the Creative Commons Attribution 3.0 licence. Any further distribution of this work must maintain attribution to the author(s) and the title of the work, journal citation and DOI.
Published under licence by IOP Publishing Ltd

the polarization direction of the incident light, while the polarization sensitivity is lost for lots of GNRs when they are randomly dispersed. The polarization sensitivity of GNRs would be helpful in their applications in sensing, bioimaging, optoelectronic devices, and so on.

Many efforts have been paid to align GNRs with long-range orientational ordering, including Langmuir-Blodgett technique [7], electric field or magnetic field method [8], droplet evaporation method [9], and so on, while most of them could only align GNRs in a small area. By dispersing GNRs in poly(vinyl alcohol) (PVA) film and stretching them, Perez-Juste et al. obtained large scale of aligned GNRs [10], but the GNRs were fixed in the film and they couldn't be tuned by applied electrical field. Liquid crystal (LC), which could be aligned and tuned by the applied shearing forces, electrical field and magnetic field, is potentially capable to align GNRs with long-range orientational ordering.

In this paper, GNRs were synthesized and poly(ethylene glycol) (PEG) chains were conjugated to improve their stability in LC matrix. The GNRs were dispersed into lyotropic nematic LC matrix, and a much higher density of GNRs were found in the GNRs-LC composites than before. The GNRs-LC composites also had a better long-term stability than reported, and this was very beneficial for their applications in optoelectronic devices. The polarization dependence of the extinction spectra of the GNRs-LC composites were studied by a home-built polarization microscope system, and the results were found to be very different from the behavior of randomly dispersed GNRs. Discrete dipole approximation (DDA) simulation was applied to calculated the extinction spectra of a single GNR with different angles between the polarization direction of the incident light and the long axis of the GNR, and similarities were found between the simulated and experimental results, thus the GNRs should be aligned with long-range orientational ordering in the LC matrix.

Raman scattering has wide applications in sensing and imaging for its capability of providing molecular fingerprints of samples [11]. As the spontaneous Raman scattering is usually too weak to be detected, SERS is commonly used [12]. GNRs, with their novel localized surface plasmon resonance (LSPR) properties, are very effective SERS substrates [13-14], and GNRs-LC composites could be polarization sensitive SERS substrates. In this work, a type of Raman reporter, 3,3'-Diethylthiatricarbocyanine iodide (DTTC) was selected, and DTTC-GNRs-LC composites were prepared. The polarization dependence of SERS signals in DTTC-GNRs-LC composites were studied by a home-built polarization Raman microscope system, and it coincided with the calculated near field of a single GNR in LC matrix with different angles between the polarization direction of the incident light and the long axis of the GNR. Thus the SERS signals would be mainly enhanced by the near filed of the GNRs, and they could be modulated by varying the polarization direction of the incident light or the LC director n. The tunability of SERS signals in DTTC-GNRs-LC composites would be meaningful for their applications in sensing, imaging and related fields.

2. Methods and experiments

2.1. Materials and instruments

All chemicals in experiments were obtained from commercial suppliers and used without further purification. Thiolated poly(ethylene glycol) (SH-PEG) was purchased from JenKem Technology Co.,

Ltd. Other chemical reagents, which were not specially mentioned, were purchased from Sigma-Aldrich. Deionized (DI) water was used in all the experiments.

2.2. Synthesis of CTAB-GNRs and PEG-GNRs

The GNRs used in our experiments were synthesized according to the classical seed-mediated growth method with small modifications [15]. Firstly, the seed was prepared by pouring 600 μL of ice-cold sodium borohydride ($NaBH_4$) into the 10 mL of stirring deionized (DI) water solution containing 0.25 mM chloroauric acid ($HAuCl_4$) and 0.1 M cetyltrimethylammonium bromide (CTAB). It was kept stirring at 28℃ for 12 min until the growth of the seed had been completed. Then, the GNR growth solution was prepared by adding 90 μL of ascorbic acid (Vc) into 10 mL of water solution containing 0.5 mM $HAuCl_4$, 0.064 mM $AgNO_3$ and 0.1 M CTAB. After consequent addition of 12 μL seed solution, the GNRs started to grow. After 24h, the growth had been completed and the GNRs were purified by centrifugation and then redispersed in DI water. By adjusting the $AgNO_3$ concentration in the growth solution, different aspect ratios of GNRs can be obtained.

SH-PEG was used to replace CTAB on the GNRs as reported [4]. Briefly, 250 μL aqueous solution with 2mM 5 kDa SH-PEG was added into 5 ml 3.5 nM CTAB-GNRs solution, and the mixture was dialyzed in a 5 kDa cutoff cellulose membrane to replace the CTAB surfactants on GNRs for 16 h. After the dialysis, samples were purified with centrifugation to remove the excess PEG molecules.

2.3. Preparation and characterization of GNRs-LC composites

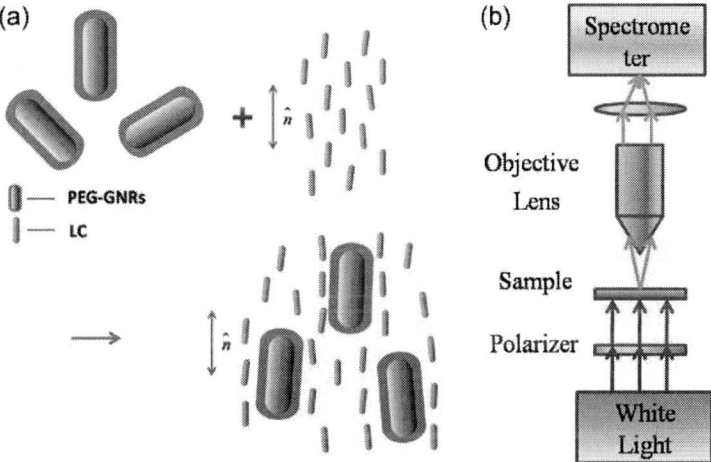

Figure 1. (a) The scheme for preparing GNRs-LC composites. (b) Set up for the polarization dependence measurement of the extinction spectra of GNRs-LC composites.

The surfactant-based lyotropic PEG-GNRs-LC composites in the nematic phase were prepared by mixing 37.5 wt% of sodium decyl sulfate (SDS), 5.5 wt% of 1-decanol and 57 wt% of aqueous suspension of PEG-GNRs from 10^{-8} M to 10^{-6} M together [16], as shown in Figure 1(a). The mixture was then centrifugated at 3000 rpm for 10 min and ultrasonicated for 30 min at room temperature. No visible aggregates were observed, indicating that GNRs were well-dispersed in the nematic phase.

Different amounts of GNRs were added to measure their solubility in LC matrix. The GNRs-LC composites were sandwiched between two glass slices separated by 0.02 mm gap using Mylar film spacers and uniformly aligned by shearing for further characterization. The CTAB-based lyotropic CTAB-GNRs-LC composites in the nematic phase was also prepared following the same procedures except that they were consisted of 25% CTAB and 75% of aqueous GNRs dispersion of 10^{-8} M.

The extinction spectra of these composites were measured by spectrometer (Ocean Optics, USB2000). The solubility of GNRs in LC matrix was obtained by recording the extinction spectra of GNRs-LC composites with varies GNRs concentrations. The long-term stability of GNRs-LC composites was characterized by measuring the extinction spectra at different time intervals.

The alignment of GNRs in LC matrix was verified by studying the polarization dependence of the extinction spectra of GNRs-LC composites with a home-built polarization microscope system, as shown in Figure 1(b). The white light from the halogen lamp was converted to be linearly polarized by a polarizer, while the polarization direction was modulated by its rotation angle. The sample was shined by the linearly polarized white light. The signal from the sample was collected by an objective lens (40×) and then to be directed into the spectrometer (Ocean Optics, USB2000).

DDA simulation was applied to calculated the extinction spectra of a single GNR with different angles between the polarization direction of the incident light and the long axis of the GNR [17]. The GNR was assumed to be a cylinder with hemispherical caps, with an effective radius of 20.3 nm and an aspect ratio of 1.25. The GNR was composed of gold and the surrounding media was LC with the refractive index of 1.39 [18]. The angle between the polarization direction of the incident light and the long axis of the GNR was set to be 0-90 degrees, with an interval of 10 degrees, and the corresponding extinction spectra of GNR were calculated.

2.4. Preparation and characterization of DTTC-GNRs-LC composites

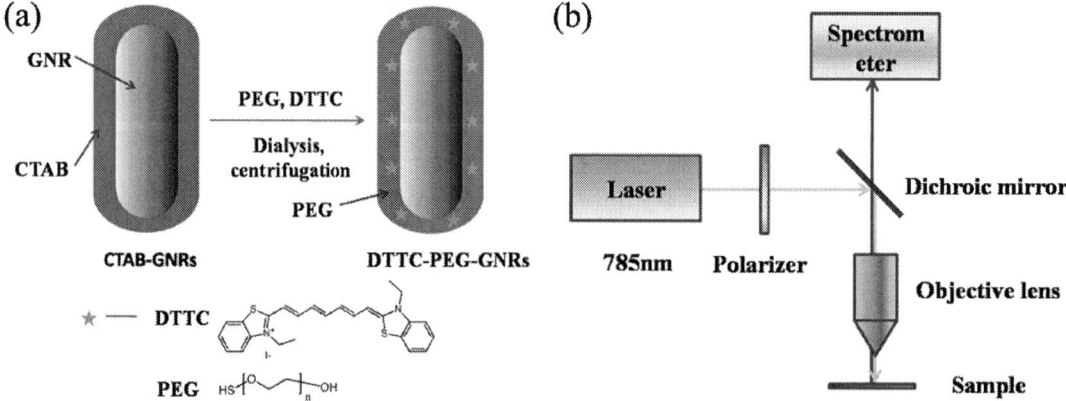

Figure 2. (a) The scheme for preparing DTTC-PEG-GNRs. (b) Set up for the polarization sensitive SERS detection.

DTTC was selected as the Raman reporter, and it was conjugated onto the GNRs as reported, as shown in Figure 2(a) [4]. Briefly, 100 μL ethanol solution with 1 mM DTTC and 250 μL aqueous solution with 2 mM 5 kDa SH-PEG were added into 5 mL 3.5 nM GNRs solution, and the mixture was dialyzed in a 5 kDa cutoff cellulose membrane for 16 h. After the dialysis, samples were purified with centrifugation to remove the excess reactants, and DTTC-PEG-GNRs were obtained. Then DTTC-GNRs-LC composites were prepared via the same routine as before, with DTTC-PEG-GNRs being added instead of PEG-GNRs.

The polarization dependence of the SERS signals from the DTTC-GNRs-LC composites was characterized by a home-built polarization Raman microscope system, as shown in Figure 2(b). The 785 nm laser beam from the source was converted to be linearly polarized by a polarizer. It was reflected by a Raman dichroic mirror (transmission: 802-900 nm, reflectance: 400-800 nm) and then focused onto the sample by an objective lens (40×). The Raman signal from the sample were collected with the same objective lens, filtered by the Raman dichroic mirror, and then directed into the Raman spectrometer (BWTEK). The output laser power in our experiments was 300 mW, and the Raman signal integration time was 10 s.

DDA simulation was applied for the near field calculations of a single GNR in LC matrix with different angles between the polarization direction of the incident light and the long axis of the GNR [19]. The structure of the GNR was as above, and the surrounding media was LC, with refractive index of 1.39. The incident light was 785 nm, propagating in the z direction, with electrical amplitude $|E|$ of 1. The angle between the polarization direction of the incident light and the long axis of the GNR was set to be 0, 30, 60, and 90 degrees. The near field electrical intensity $|E|^2$ of GNR at the plane $z = 0$ was calculated.

3. Results and discussion

3.1 The solubility of GNRs in LC matrix

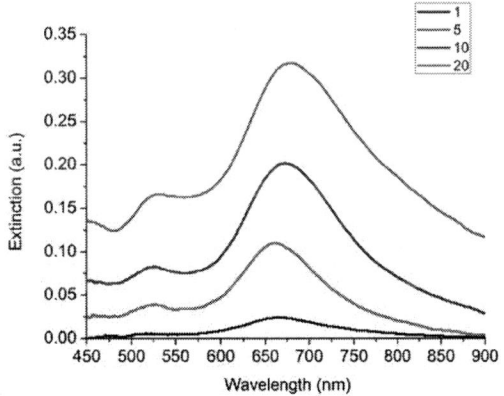

Figure 3. The solubility of PEG-GNRs-LC composites.

The maximum concentration of GNRs in our previous reported CTAB-GNRs-LC composites was about 10^{-8} M [20], and it was denoted as 1. Different amounts of PEG-GNRs were added into the LC matrix, and the extinction spectrums were shown in Figure 3. It was found that as many as 20 times of GNRs were found to be soluble in PEG-GNRs-LC composites, which was about 2×10^{-7} M. From the increasing extinction peak and fine spectrum shapes, the GNRs in LC matrix would have good dispersibility and stability. The high concentration of GNRs in PEG-GNRs-LC composites is very meaningful in manufacturing optical devices.

3.2 The long-term stability of GNRs in LC matrix

The long-term stability of the GNRs-LC composites was characterized by measuring the extinction spectra at different time intervals, as shown in Figure 4. For PEG-GNRs-LC composites, as shown in Figure 4(a), the extinction spectra kept stable until 108 h, with little decrease in the LSPR peak, indicating the good stability of PEG-GNRs in surfactant-based LC matrix. For CTAB-GNRs-LC composites, as shown in Figure 4(b), there was a sharp decrease in the LSPR peak at 52 h. At 110 h, the extinction spectrum had nearly been flat, which was an indication of the aggregation of GNRs in the composites. The improvement in long-term stability of PEG-GNRs-LC composites may come from the strong bond of thiol group with gold surface and the compact covering of polymer. The superiority in long-term stability of PEG-GNRs-LC composites makes them more appropriate in fabrication of nanoscale devices.

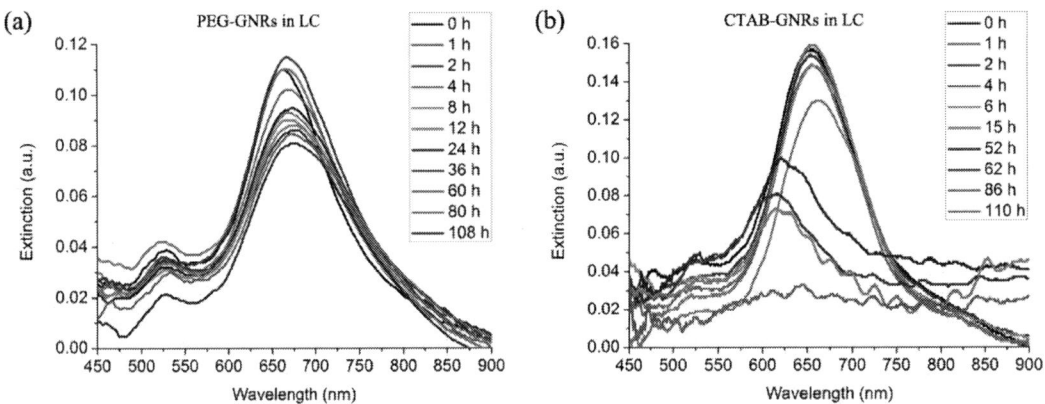

Figure 4. The long-term stability of PEG-GNRs-LC composites (a) and CTAB-GNRs-LC composites (b).

3.3 The polarization dependence of GNRs in LC matrix

The experimental extinction spectra of PEG-GNRs in LC matrix with different angles between the polarization direction of the incident light and the LC director n were shown in Figure 5(a). When the polarization direction of the incident light was parallel to LC director n, there was only one LSPR peak near 670 nm appeared. With increasing the angle between the polarization direction of the incident light and LC director n by rotating the polarizer, a new LSPR peak near 520 nm came up, while the LSPR peak near 670 nm fell down. When the polarization direction of the incident light was perpendicular to LC director n, the LSPR peak near 520 nm reached its maximum, while the LSPR peak near 670 nm vanished. This was quite different from the randomly dispersed GNRs in aqueous solutions, and it was very similar to the aligned GNRs, thus it would be an indication of the long-range orientational ordering alignment of GNRs in LC matrix.

AOM2015   IOP Publishing
Journal of Physics: Conference Series **680** (2016) 012021   doi:10.1088/1742-6596/680/1/012021

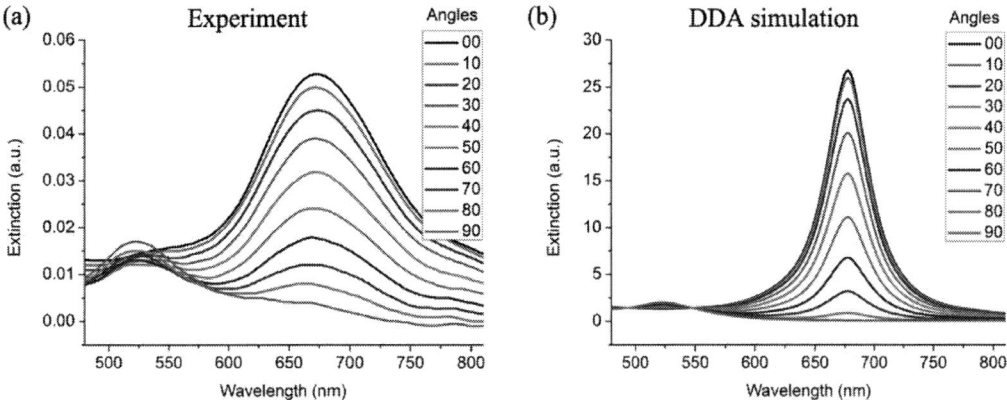

Figure 5. (a) The experimental extinction spectra of GNRs in LC matrix with different angles between the polarization direction of the incident light and the LC director **n**. (b) The calculated extinction spectra of GNRs in LC matrix with different angles between the polarization direction of the incident light and the long axis of the GNRs via DDA simulation.

The calculated extinction spectra of GNR in LC matrix with different angles between the polarization direction of the incident light and the long axis of the GNR via DDA simulation were shown in Figure 5(b). Compared with the experimental results on Figure 5(a), the same trend on the extinction spectra of GNR could be observed when varying the polarization angles. When the angle was 0 degree, the longitudinal LSPR peak reached its maximum. With increasing the angles, the longitudinal LSPR peak decreased and the transverse LSPR peak increased. When the angle was 90 degrees, the transverse LSPR peak reached its maximum. These changes would be another sign that the GNRs in LC matrix had been aligned with long-range orientational ordering.

3.4 The polarization dependence of SERS signals in DTTC-GNRs-LC composites

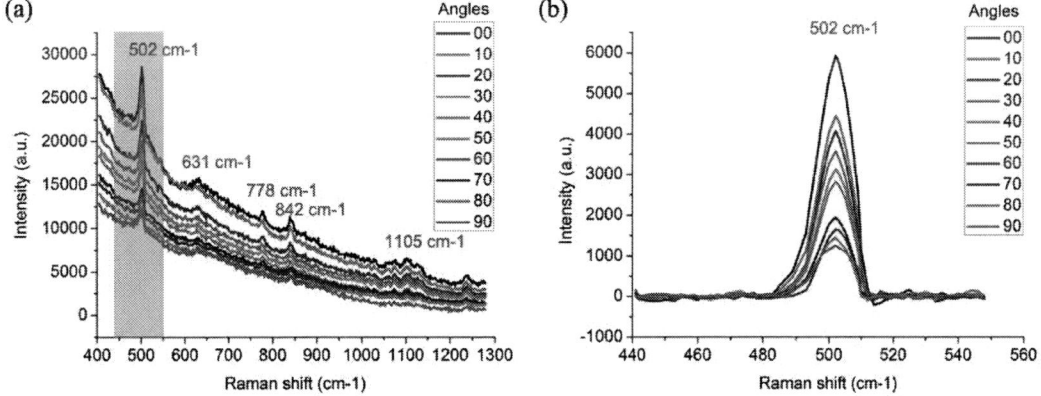

Figure 6. The SERS signals from DTTC-GNRs-LC composites with different angles between the polarization direction of the incident light and the LC director **n**. (a) with background, (b) background subtracted.

113

The SERS signals from DTTC-GNRs-LC composites with different angles between the polarization direction of the incident light and the LC director n was shown in Figure 6(a). The composites had several specific Raman bands such as 502, 631, 778, 842 and 1105 cm^{-1}, which coincided with those of DTTC molecules [21]. For clarity, the strongest Raman band at 502 cm^{-1} was picked out, and the background was subtracted, as shown in Figure 6 (b). When the polarization direction of the incident light was parallel to the LC director n, the Raman band at 502 cm^{-1} got its maximum value of 6000, indicating strong enhancement of Raman signals. When the angle between the polarization direction of the incident light and the LC director n increased, the Raman band at 502 cm^{-1} decreased. When the polarization direction of the incident light was orthogonal to the LC director n, the Raman band at 502 cm^{-1} got its minimum value of 1000. Thus the SERS signals from the DTTC-GNRs-LC composites were sensitive to the polarization direction of the incident light.

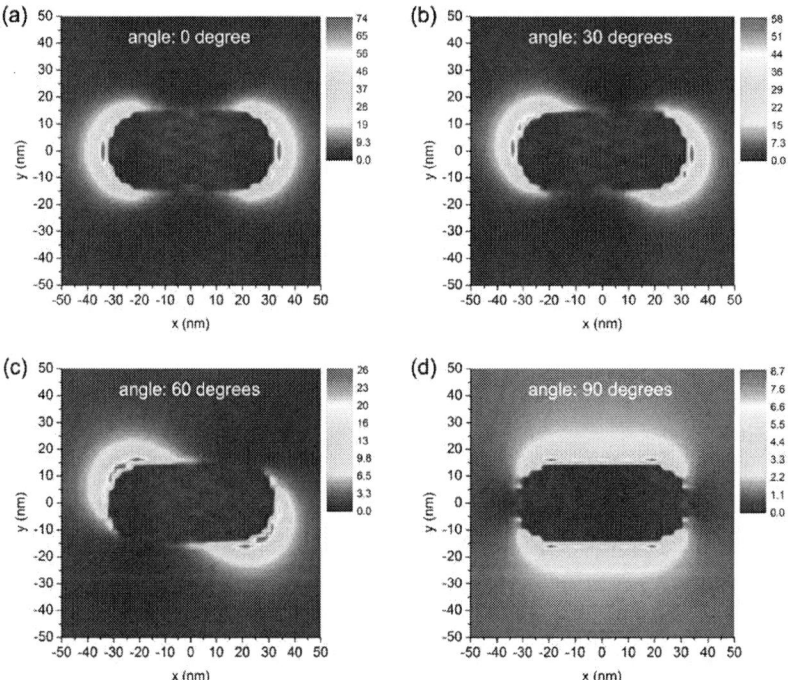

Figure 7. The calculated near filed electrical intensity $|E|^2$ of a single GNR in LC matrix with different angles between the polarization direction of the incident light and the long axis of the GNR. (a) The angle was 0 degree. (b) The angle was 30 degrees. (c) The angle was 60 degrees. (d) The angle was 90 degrees.

The calculated near filed electrical intensity $|E|^2$ of a single GNR in LC matrix with different angles between the polarization direction of the incident light and the long axis of the GNR was shown in Figure 7. When the angle was 0 degree, the near field electrical intensity of GNR reached its maximum. With increasing the angles, the enhancement of the near field electrical intensity of GNR decreased. When the angle was 90 degrees, the enhancement of the near field electrical intensity of

GNR reached its minimum. These coincided with the variations of the SERS signals from DTTC-GNRs-LC composites. Thus the polarization sensitivity of SERS signals of DTTC-GNRs-LC composites would be from the polarization sensitivity of the near field enhancement of aligned GNRs in LC matrix, and the SERS signals could be tuned by varying the polarization direction of the incident light.

4. Conclusions

In summary, GNRs were synthesized via the seed-mediated growth method, and they were conjugated with PEG chains to improve the stability. The PEG-GNRs were dispersed into lyotropic nematic LC matrix, with good solubility and long-term stability. The experimental extinction spectra of the GNRs-LC composites obtained from a home-built measuring system were found to be polarization sensitive, and the relationship was in accordance to the calculated results of a single GNR in LC matrix via DDA simulation, indicating that the GNRs had been ordering aligned. DTTC was selected as the Raman reporter, and DTTC-GNRs-LC composites were prepared. The SERS signals from DTTC-GNRs-LC composites were found to be polarization sensitive, and they were from the polarization sensitivity of the near field of GNRs. The alignment of GNRs and polarization sensitivity of SERS signals in GNRs-LC composites would be meaningful in their application in sensing, imaging, optoelectronic devices, and so on.

Acknowledgements

This work was supported by National Basic Research Program of China (973 Program: 2013CB834704), the National Natural Science Foundation of China (61275190 and 91233208), and the Program of Zhejiang Leading Team of Science and Technology Innovation (2010R50007).

References

[1] Huang X, Neretina S, and El-Sayed M A, "Gold Nanorods: From Synthesis and Properties to Biological and Biomedical Applications", 2009, *Adv. Mater.*, **21**(48), 4880-910.

[2] Li X, Jiang L, Zhan Q, Qian J, and He S, "Localized surface plasmon resonance (LSPR) of polyelectrolyte-functionalized gold-nanoparticles for bio-sensing", 2009, *Colloids Surf. A Physicochem. Eng. Asp.*, **332**(2-3), 172-9.

[3] Huang X H, El-Sayed I H, Qian W, and El-Sayed M A, "Cancer cell imaging and photothermal therapy in the near-infrared region by using gold nanorods", 2006, *J. Am. Chem. Soc.*, **128**(6), 2115-20.

[4] Qian J, Jiang L, Cai F H, Wang D, and He S L, "Fluorescence-surface enhanced Raman scattering co-functionalized gold nanorods as near-infrared probes for purely optical in vivo imaging", 2011, *Biomaterials*, **32**(6), 1601-10.

[5] Wang Z, Zong S, Yang J, Song C, Li J, and Cui Y, "One-step functionalized gold nanorods as intracellular probe with improved SERS performance and reduced cytotoxicity", 2010, *Biosens. Bioelectron.*, **26**(1), 241-7.

[6] Zhang Y, Qian J, Wang D, Wang Y L, and He S L, "Multifunctional Gold Nanorods with Ultrahigh Stability and Tunability for In Vivo Fluorescence Imaging, SERS Detection, and

Photodynamic Therapy", 2013, *Angew. Chem. Int. Edit.*, **52**(4), 1148-51.

[7]   Kim F, Kwan S, Akana J, and Yang P D, "Langmuir-Blodgett nanorod assembly", 2001, *J. Am. Chem. Soc.*, **123**(18), 4360-1.

[8]   van der Zande B M I, Koper G J M, and Lekkerkerker H N W, "Alignment of rod-shaped gold particles by electric fields", 1999, *J. Phys. Chem. B*, **103**(28), 5754-60.

[9]   Ming T, Kou X, Chen H, Wang T, Tam H-L, Cheah K-W, Chen J-Y, and Wang J, "Ordered Gold Nanostructure Assemblies Formed By Droplet Evaporation", 2008, *Angew. Chem. Int. Edit.*, **47**(50), 9685-90.

[10] Perez-Juste J, Rodriguez-Gonzalez B, Mulvaney P, and Liz-Marzan L M, "Optical control and patterning of gold-nanorod-poly(vinyl alcohol) nanocomposite films", 2005, *Adv. Funct. Mater.*, **15**(7), 1065-71.

[11] Nie S M and Emery S R, "Probing single molecules and single nanoparticles by surface-enhanced Raman scattering", 1997, *Science*, **275**(5303), 1102-6.

[12] Orendorff C J, Gearheart L, Jana N R, and Murphy C J, "Aspect ratio dependence on surface enhanced Raman scattering using silver and gold nanorod substrates", 2006, *PCCP*, **8**(1), 165-70.

[13] Nikoobakht B, Wang J P, and El-Sayed M A, "Surface-enhanced Raman scattering of molecules adsorbed on gold nanorods: off-surface plasmon resonance condition", 2002, *Chem. Phys. Lett.*, **366**(1-2), 17-23.

[14] Nikoobakht B and El-Sayed M A, "Surface-enhanced Raman scattering studies on aggregated gold nanorods", 2003, *J. Phys. Chem. A*, **107**(18), 3372-8.

[15] Nikoobakht B and El-Sayed M A, "Preparation and growth mechanism of gold nanorods (NRs) using seed-mediated growth method", 2003, *Chem. Mater.*, **15**(10), 1957-62.

[16] Liu Q, Qian J, Cai F, Smalyukh I I, and He S, "Switchable Polarization-Sensitive Surface Plasmon Resonance of Highly Stable Gold Nanorods-Liquid Crystals Composites", 2011, *Optoelectronic Materials and Devices Vi*, G.H. Duan.

[17] Draine B T and Flatau P J, "Discrete-dipole approximation for scattering calculations", 1994, *J. Opt. Soc. Am. A*, **11**(4), 1491-9.

[18] Johnson P B and Christy R W, "Optical constants of the noble metals", 1972, *Phys. Rev. B*, **6**(12), 4370-9.

[19] Flatau P J and Draine B T, "Fast near field calculations in the discrete dipole approximation for regular rectilinear grids", 2012, *Opt. Express*, **20**(2), 1247-52.

[20] Liu Q, Cui Y, Gardner D, Li X, He S, and Smalyukh I I, "Self-Alignment of Plasmonic Gold Nanorods in Reconfigurable Anisotropic Fluids for Tunable Bulk Metamaterial Applications", 2010, *Nano Lett.*, **10**(4), 1347-53.

[21] Jiang L, Qian J, Cai F H, and He S L, "Raman reporter-coated gold nanorods and their applications in multimodal optical imaging of cancer cells", 2011, *Anal. Bioanal. Chem.*, **400**(9), 2793-800.

Study of humidity on the structure and optical properties of cesium iodide thin film

Bo Liang[1], Shuang Liu[1], Lina Guo[1], Dejun Chen[1], Yong Liu[1], Zhiyong Zhong[2], Liufeng Xiong[3]

[1]University of Electronic Science and Technology of China, School of optoelectronic information, Chengdu, China

[2]University of Electronic Science and Technology of China, State Key Laboratory of Electronic Thin Films and Integrated Devices, Chengdu, China

[3]XL Innovative Technologies, Cleveland, OH 44111, USA

e-mail: lbstorm@126.com

Abstract. We report the influence of 70% 5% humid exposed air at room temperature on the structure of 500nm thick CsI film. By means of scanning electron microscopy (SEM) and X-ray diffusion (XRD) analysis, the surface morphology and polycrystalline structure were investigated. After 24 hour exposure to humid air, the grain size of CsI film has a drastic change, and so as to the preferential growth orientation. Infrared spectra showed a hydroxyl absorption peak after 3h exposure. And in long term exposure (24h), the absorption peak broadened with the band center shifting toward lower wavenumbers.

1. Instruction

With the high Quantum Efficiency (QE) in VUV/UV and X-ray energy range, cesium iodide (CsI) thin film are widely used as photoelectron converters in medical imaging, particle physics, position emission tomography and other various fields [1-2]. In recent years, CsI-based devices have been successfully applied in Ring Imaging Cherenkov experiments, such as A Large Ion Collider Experiment at the CERN LHC, High Acceptance Di-Electron Spectrometer at Gigascale Integration [3-6]. CsI thin film is sensitive to its surrounding. All of those experiments need CsI complete insulation from air. The problems of CsI thin film exposed to humidity air have been studied and discussed. Most of the studies concern the effects of the exposure to air on surface morphology and the quantum efficiency (QE) of CsI thin film. However, the grain-size mechanism of crystal column is still uncertain [7-12]. The theory of CsI recrystallization caused by water molecules [11] and the fusion of crystal surface caused by chemical reaction between water and CsI were put forward [12].

The aim of the present work is to examine the influence of humidity on the structure and optical properties of CsI thin film. In particular, the surface morphology and structure of CsI thin films, exposed to 70% 5% humidity air, were studied. Results of ultraviolet transmittance and infrared absorption rate affected by the water were also reported..

2. Experimental technique

The films were prepared by resistive heating evaporation in a high vacuum spherical chamber. CsI crystals were grinded into powder and placed into a molybdenum boat, under a shutter, inside the vacuum chamber. The quartz slides were used as substrates. After more than 6 hours soaked in chromic acid lotion, the substrates were ultrasonic vibration cleaned with acetone, alcohol and deionized water then dried. After creating a vacuum condition at 2.5 x 10-3 Pa (the order of 10-5 Torr), the shelf whirled at the rate of 10 r/min and the CsI crystals films of 3μm were deposited at a typical evaporation rate of 1-2nm/s. After the deposition completed, high-purity argon gas was filled into the vacuum chamber as a protective gas. And the films were transferred into a vacuum box immediately after the chamber opened.

Scanning electron microscopy (SEM, QUANTA FEG 450) and X-ray diffraction (XRD, PHILIPS HR-XRDScan) were used to investigate the surface morphology and inspect the structure of the films. The properties of transmittance and absorptance of the films were measured by UV Spectrophotometer (UV-1700) and Fourier Transform Infrared Spectroscopy.

3. Results

3.1. Surface morphology and structure analysis

In order to study the humidity impact, the surface of CsI film, scanned by SEM, are shown in Fig. 1. Fig. 1a shows the fresh film has a typical crystal columnar structure, with high grain density and small grain size. The SEM images of same deposition process samples with exposure to 70% 5% humid air for 3 h, 12 h and 24 h are shown in Fig. 1b, Fig. 1c and Fig. 1d respectively. As shown in the images, the grain density decreased obviously and the grain size increased rapidly. After being exposed to humid air for 3 h, the average grain size grew from 0.25μm to 0.68μm as is shown in Fig. 2. As can be seen in Fig. 2, the boundaries of the grains gradually become blurred due to the coalescence process.

It is considered the short-circuit diffusion played an important role in the process of the grains growing up. The spreading activation energy of atoms in the boundaries can be reduced by the diffusion of the attracted water molecules. That lends to the bending and moving of the grains boundaries, which causes the increasing of grain size [11].

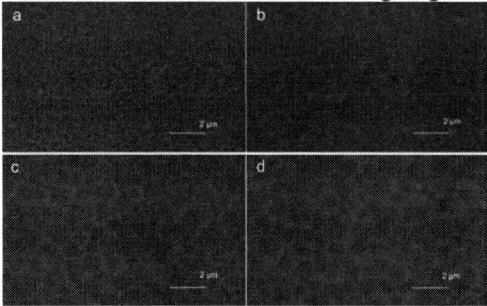 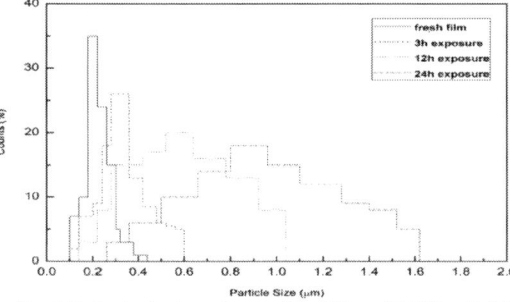

Fig. 1. Scanning electron microscopy (SEM) surface image of freshly evaporated. (b) 3h exposed. (c) 12h exposed. (d) 24h exposed .

Fig. 1. Grain size horizontal step of the $500nm$ CsI 500nm CsI film (a) thin film sample without exposure and exposed to humid air for 3h, 12h and 24h

The large difference of electronegativity between Cs and I makes CsI show a strong polarity. Therefore, the polar water molecules would be easily adsorbed on the surface of the grains. Moreover, the distortion and sparse of the grain boundaries tends to lower diffusion activation energy , but greater vacancy diffusion. Thus more and more water molecules are attracted. Its kinetics and presented morphology obey the Fick's law and the basic laws of thermodynamics.

The chemical reaction as: $CsI + H_2O = CsOH + HI \uparrow$ on the surface of grains is proved in Xie's work [12]. Considering the diffusion-water mainly being molecular state, the chemical equation rewritten as: $CsI + H_2O(gas) \rightleftharpoons CsOH + HI(gas)$, is more reasonable. Since the adsorption and chemical reaction happened with exothermic phenomenon, which lends to the increasing kinetic energy. And according to Fick' law [13] the surface self-diffusion coefficient D could be expressed as:

$$D = C_0 \exp(-\Delta E / kT)$$

Where ΔE is the diffusion activation energy, k is the Boltzmann constant, is the intrinsic factor unrelated to temperature. The larger kinetic energy and lower cause the more intense diffusion. Regarding to three particle junction system [14]

$$\frac{\gamma_{12}}{\sin\varphi_3} = \frac{\gamma_{23}}{\sin\varphi_1} = \frac{\gamma_{31}}{\sin\varphi_2}$$

Where $\varphi_1\, \varphi_2$, φ_3 is the junction angle; γ_{12}, γ_{23}, γ_{31} is the coefficient of the interfacial energy of the sector. When the junction angle is , the columnar has a shape with a hexagonal base and a straight boundary; otherwise the junction system is unstable and that makes the boundary bend to

AOM2015   IOP Publishing
Journal of Physics: Conference Series **680** (2016) 012022   doi:10.1088/1742-6596/680/1/012022

small-angle direction. In this case, the bending boundary moves along the curvature center, with interface decreased, to reduce the system free energy. According to the Young-Laplace equation, the movement of the boundary is driven by the unequal pressure of the different sides [15].The smaller grains were consumed to promote the bigger-around to grow. The larger grains annexed the smaller by the concave boundaries merging the convex, as shown in Fig. 3.

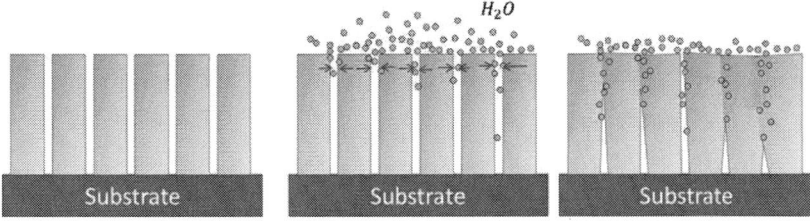

Fig. 3. A model for CsI grain growth boundary change

Fig. 4 shows the altered XRD spectra among the fresh, 3h, 12h and 24h exposed films. The spectra clearly shows that the diffraction intensity of (110), (211) and (220) crystal planes is enhanced with the exposure time, while the (200) is weakened. That mainly because the recrystallization process made by water molecules lends the grains to grow along the (110) and (211) direction.

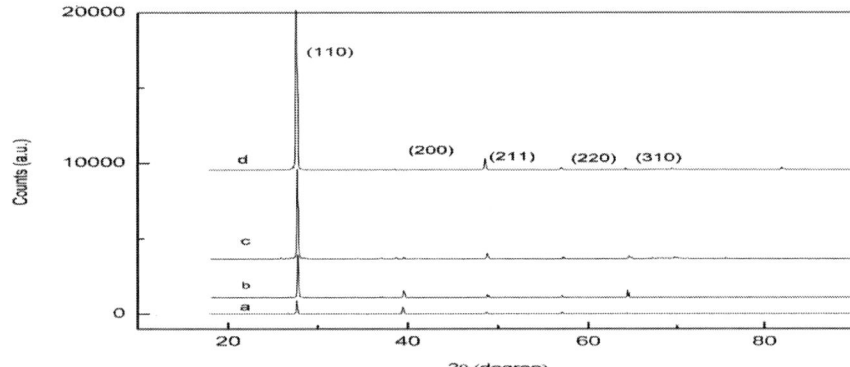

Fig. 4. XRD spectra of CsI film (a) without exposure (b) 3h exposed (c) 12h exposed (d) 24h expose

3.2. Optical properties analysis.

As a function of wavelength, Fig. 5a and Fig. 5b show the trends of the transmittance percentage T and the reflectance percentage R in the range of 190-500 nm, respectively, with two curves for 500μm thick CsI films without exposure and with 24h exposed to 70% 5% humid air.

The measurements for handling substrates were done to calibrate the figures. Thus the T and R curves are only referred to the CsI film.

The 24h exposed film shows a lower but smoother R curve. It is obviously to see the transmittance spectra relative to 24h exposure, which the average transmittance in the UV spectral range is about 50% is lower than the unexposed one while that is 78%. After exposure in the humid air, a white substance called milky effect is formed on the surface of the film [16]. That mainly due to the chemical reaction with water molecular mentioned in (1), do favor to cause the reflectance and transmittance reduction. In particular, considering the recrystallization crystalline structure introduced optical absorption centers lends to an increased absorbance in the UV range [17].

119

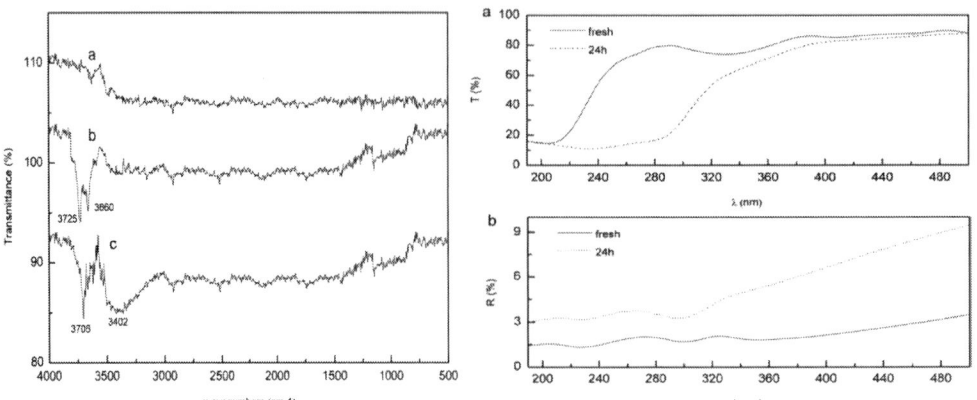

Fig. 5. The transmittance and reflectance of 500nm fresh And 24h exposed to humid air CsI film as function of wavelength.

Fig. 6. The infrared spectra of (a)500nm fresh, (b)3-h and (c)24-h exposed to humid air CsI film.

3.3. Infrared absorptance analysis

Absorption spectra of CsI films without exposure, 3h exposed and 24h exposed to 70%+5% humid air are shown in Fig.6 (10% paned up). Spectra shown in Fig.6 of obtained sequentially for increasing time depicts the apparent absorption of water molecules in corresponding coverages of 3760cm-1-2840cm-1. The OH (free) stretching vibration band centers are at 3652 and 3756cm-1, while that of liquid water is at 3400cm-1 [18]. It can be concluded the properties of water molecules on the surface of the grains are changed by the film. The affection by cation and anion of the film and the chemical reaction between CsI and water fracture the hydrogen bonding. Ion-dipole bond affects the hydroxyl vibration. Thus the spectra are more likely to exhibit the OH (free) absorption peak. With exposure time increasing, the absorption peak broadens with the band center shifting toward lower wavenumbers. That suggests the hydrogen bonding network in the film is more closed to liquidlike. Thus the absorption spectrum results in a state of liquid water.

4. Conclusion

In this paper, scanning electron microscope (SEM) and X-ray diffraction (XRD) were used to study the surface morphology and structure of 500nm freshly evaporated and humid-air (70%+5%) exposed CsI film. Adsorbed water diffusion between the particles boundary reduce the film tensile stress. The particles boundaries bent and moved showed a significant change in the grain size and preferential growth orientation. The average grain size grew from 0.25μm to 0.68μm. The affected transmittance and reflectance of the films in the band of 190-500nm due to the change, the 24h exposed film showed a lower but smoother R curve and also a reduce T spectrum. Infrared spectra showed a hydroxyl absorption peak after 3h exposure. And in long term exposure (24h) the absorption peak broadened with the band center shifting toward lower wavenumbers.

Acknowledgments

This work was supported by 973 Program of China (2012CB315701), the National Natural Science foundation of China (61177035), Sichuan Provincial International Cooperation Project (2013HH0002), and Sichuan Provincial Science and Technology Support Project (12ZC0245, 2012GZ0051). The authors would like to thank CCD Research Center of China Electronics.

References

[1]. Gratta, G., Newman, H., &Zhu, R. Y. (1994). Annu. Rev. Nucl. Part. S., 44(1), 453-500.
[2]. Di Mauro, A., Nappi, E., Posa, F., Breskin, A., Buzulutskov, A., Chechik, R., &Piuz, F. (1996). Nucl. Instrum. Meth. A., 371(1), 137-142.
[3]. Piuz, F. (1996). Nucl. Instrum. Meth. A., 371(1), 96-115.

[4]. Nagarkar, V. V., Gupta, T. K., Miller, S. R., Klugerman, Y., Squillante, M. R., & Entine, G. (1998). IEEE. T. Nucl. Sci., 45(3), 492-496.

[5]. Aprile, E., Arisaka, K., Arneodo, F., Askin, A., Baudis, L., & Behrens, A. (2010). Phys. Rev. Lett., 105(13), 131302.

[6]. Trefilova, L., Grinyov, B., Kovaleva, L., Kosinov, N., & Shpylynska, O. (2005). Phys. Status. Solidi.(c), 2(1), 101-104..

[7]. Nitti, M. A., Senesi, G. S., Liotino, A., Nappi, E., Valentini, A., & Singh, B. K. (2004). Nucl. Instrum. Meth. A., 523(3), 323-333.

[8]. Singh, B. K., Nitti, M. A., Valentini, A., Nappi, E., Coluzza, C., Di Santo, G., & Zanoni, R. (2007). Nucl. Instrum. Meth. A, 581(3), 651-655.

[9]. Fejdi, P., & Holocsy, A. (2001). Mater. Struct., 8(1), 22.

[10].Di Mauro, A., Martinengo, P., Piuz, F., Schyns, E., van Beelen, J., & Williams, T. D. (2001). Nucl. Instrum. Meth. A., 461(1), 584-586.

[11].Long F., Zhiwen Y., & Tao C. (2014). *Acta Phys. Sin. 63(14), 146801-146801.*

[12].Xie, Y., Zhang, A., Liu, Y., Liu, H., Hu, T., Zhou, L. & Lü, J. (2012). *Nucl. Instrum. Meth. A., 689, 79-86.*

[13].Xun C. et al. (2007). *Modern film materials and technology.* East China University of Science and Technology Press, China.

[14].Zhong Z. et al. (1993). *The surface and interface physics.* The University of Electronic Science and Technology Press, China.

[15].Plumer, M. et al. (2001). In: *The physics of ultra-high-density magnetic recording*, Van Ek, J., & Weller, D. (Eds.)., Chapter 2, Springer-Verlag: Berlin, 12.

[16].Nitti, M. A., Cioffi, N., Nappi, E., Singh, B. K., & Valentini, A. (2002). *Nucl. Instrum, Meth. A., 493(1), 16-24.*

[17].Foster, M. C., & Ewing, G. E. (2000). *J. Chem. Phys., 112(15), 6817-6826.*

Polarization-dependent refractive index fiber-optic sensor based on the core-offset with a taper

Youqing Wang, Changyu Shen, Weimin Lou and Fengying Shentu

Institute of Optoelectronic Technology, China Jiliang University, Hangzhou, 310018, China,

E-mail: wyq91123@sina.com

Abstract. A polarization-dependent refractive index sensor based on the in-fiber Mach-Zehnder interferometer which is constructed by core-offset fusion splicing a tapered polarization maintaining fiber (PMF) with a length of 20 mm between the core-offset and the taper is proposed. Due to the introduced high environment-sensitivity of the two orthogonal polarization modes in the PMF and the enhancing effect of the tapered PMF, The sensitivity of the fast axis and the slow axis up to ~-107.51 nm/RI and ~-74.54 nm/RI in the RI range between 1.333 and 1.381 is obtained, respectively. Such kinds of low-cost and highly sensitive fiber-optic RI sensors would find applications in chemical or biochemical sensing fields.

1. Introduction

Refractive index (RI) sensing based on the conventional optical fiber has been intensively studied in recent years, especially in chemical sensing field. The high-precision RI sensors are mainly depending on the Mach-Zehnder interferometer (MZI) technology, grating-based MZIs have a large range measurement and high RI sensitivity, but they require precise and expensive lasers to obtain the Fiber Bragg Grating (FBG) or Long-period Grating (LPG) [1, 2]. Recently, some special structures were proposed to fabricate the MZI-type RI fiber sensors, such as single mode (SM) –multimode (MM)– single mode (SM) fiber structure [3, 4], in-line two-tapered MZI [5, 6] and etc. All of these fiber-based structures are easily fabricated and much cheaper.

In this paper, an improved RI fiber sensor constructed by core-offset fusion splicing a tapered polarization maintaining fiber (PMF) with a length of 20 mm between the core-offset and the taper is reported. Unlike the normal taper, the proposed taper was made by two identical PMFs. Since a tapered fiber can enhance the evanescent field and the high environment-sensitivity of the two orthogonal polarization modes because of the birefringence of the PMF, a much higher accuracy of RI measurement is expected. The sensitivity of fast axis and slow axis up to ~-107.51 nm/RI and ~-74.54 nm/RI in the RI range between 1.333 and 1.381 is obtained, respectively. Compared with [5], the proposed RI sensor has larger measurement range, such kind low-cost and highly sensitive fiber-optic RI sensor would find applications in chemical or biochemical sensing fields.

2. Experiments

The experimental setup is shown in Fig.1. A broadband source (BBS) was utilized as the input light. And the polarization state of the output light from the BBS was adjusted by a polarized controller (PC) to obtain a high fringe visibility. The output spectrum of the proposed structure was detected by an optical spectrum analyzer (OSA: AQ6317B, Advantest, Japan) with wavelength resolution set to 0.2 nm.

The taper was made easily by using a special function of a commercial fusion splicer (Fujikura FSM-100P) with PMF1 and PMF 2 (PANDA, 1017-C, YOFC), the birefringence index of the PMF 1 and PMF 2 are 7.7024 ×104. The fast axis and slow axis of PMF1 and PMF2 are alignment with each other. And then, the end of the PMF 1 was mismatch fusion spliced with SMF (SMF-28), the core-offset size is about 4.5 μ m. The core and cladding diameters of the used SMF are 9 μ m and 125 μ m, respectively. The applied PMFs have the same core and cladding diameters as the SMF. The length between the core-offset and the taper region is ~20 mm. Both sides of the sensing head were fixed and then immersed in the solution (sucrose solutions, ranging from 1.333 to 1.381).

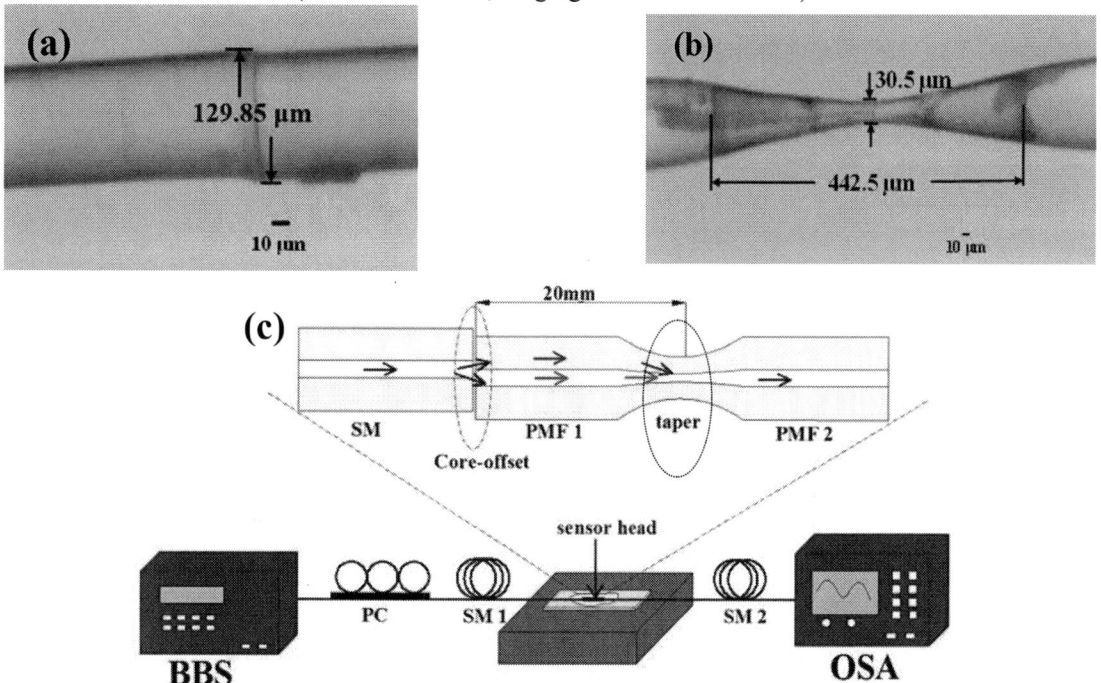

Fig.1. Schematic diagram of the proposed RI sensor. (a) Microscopic image of the core-offset, (b) Microscopic image of the taper, (c) the partially enlarged drawing of the sensor head.

As shown in the inset of Fig. 1, When input light (black arrow) reaches the core-offset, core mode of the SMF is partly coupled into the cladding of the PMF to excite massive cladding modes (blue arrow), and the remaining light still propagates in the core as the fundamental core mode (red arrow). Then, the cladding modes are re-coupled into the fiber core by the taper to interfere with the fundamental core mode. Lengths of this kind MZI's two arms are exactly same, but the optical lengths are different because of the different effective indices of each optical length [7]. Theoretically, phase difference between the core mode and the excited cladding modes of the formed MZI can be expressed as [8, 9],

$$\Delta\Phi = 2\pi n_{eff}^{m} L/\lambda \tag{1}$$

Where n_{eff}^{m} is the effective refractive index difference between the core and the m_{th} cladding mode, λ is the wavelength of the input light, L is the length between PMF1 and PMF2. And the output intensity of the MZI can be described as,

$$I = I_1 + I_2 + 2\sqrt{I_1 I_2}\cos(\Delta\Phi) \tag{2}$$

Where I_1 and I_2 are the intensities of the light propagating along the fiber core and cladding, respectively.

Owing to the difference of the effective refractive index between the two orthogonal core modes and the excited cladding modes, the resonant dip wavelengths of the interference patterns corresponding to fast axis (λ_f) and slow axis (λ_s) polarization modes are different. By adjusting the PC at the appropriate state, initial interference patterns of the proposed MZI is recorded as shown in Fig.2, the resonant dip wavelength satisfies the equation of $\Delta\Phi = (2k+1)\pi$, k is natural number. According to [10], the resonant dip wavelength corresponding to fast axis (λ_f) and slow axis (λ_s) polarization modes can be described as,

$$\lambda_f = \frac{2(n_{eff,f}^{core} - n_{eff,f}^{clad})L}{2k+1} \tag{3}$$

$$\lambda_s = \frac{2(n_{eff,s}^{core} - n_{eff,s}^{clad})L}{2k+1} \tag{4}$$

Where $n_{eff,f}^{core}$ and $n_{eff,f}^{clad}$ are the effective refractive index of the core mode and excited cladding mode on the fast axis of the PMF, respectively. $n_{eff,s}^{core}$ and $n_{eff,s}^{clad}$ are the effective refractive index of the core mode and excited cladding mode on the slow axis of the PMF, respectively.

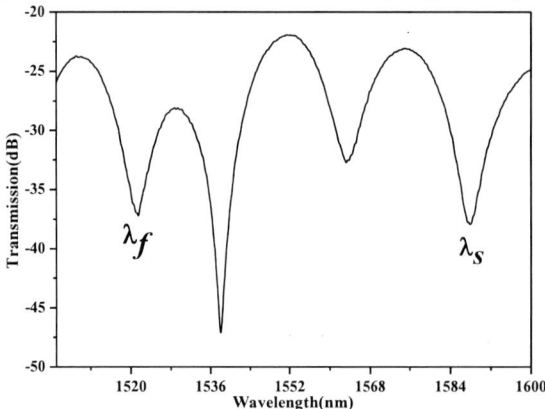

Fig.2. Transmission spectra of the proposed RI sensor.

3. Experiments and results

From Fig.2 we can see that the wavelengths of fast axis and slow axis polarization mode are ~1521.3 nm and ~1587.8 nm, respectively. As expected, values of $n_{eff,f}^{core} - n_{eff,f}^{clad}$ and $n_{eff,s}^{core} - n_{eff,s}^{clad}$ will decreased with the increasing of the surrounding refractive index, so as the corresponding wavelengths to fast axis and slow axis.

The detected solution we selected was sugar solution, and the RI is ranging from 1.333 to 1.381. During the measurement, the sensing head was cleaned by ethanol solution twice each time, and the next test was conducted until the sensing head dried. The transmission spectra variations of the fast axis and slow axis polarization modes are shown in Fig.3.

Fig. 3 illustrates the spectral response of the fast axis and slow axis polarization modes with different RI. It can be seen that with the increasing of the RI, both of the resonant dip wavelengths of the fast axis and the slow axis were move to the direction of the short wavelength. The experimental results are agreed with the theoretical analysis. Since the tapered fiber can enhance the evanescent field and the fast axis and slow axis polarization modes are sensitive to the surrounding environment, the sensitivity of the proposed RI sensor has been greatly improved.

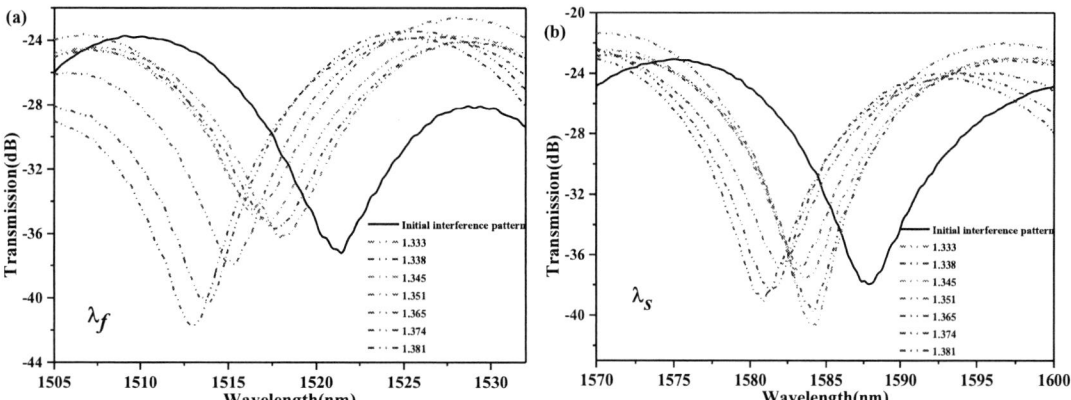

Fig.3. Transmission spectra variations under different RI. (a) fast axis (λ_f), (b) slow axis (λ_s).

Fig.4 shows the fitting line of the resonant dips of the interference patterns corresponding to fast axis and slow axis. It is indicated that there is a good linear relationship between RI and the wavelength. According to the results of the linear fitting, the sensitivity of the fast axis and the slow axis up to ~-107.51 nm/RI and ~-74.54 nm/RI in the RI range between 1.333 and 1.381 is obtained, respectively.

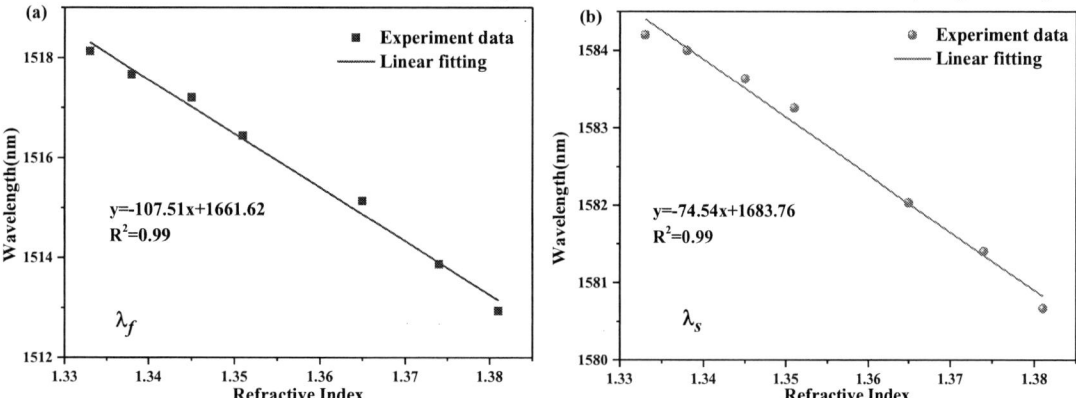

Fig.4. Fitting line of the resonant dips of the interference patterns corresponding to: (a) fast axis (λ_f), (b) slow axis (λ_s).

4. Conclusion

A novel optic fiber sensor constructed by core-offset fusion splicing a tapered polarization maintaining fiber (PMF) with a length of 20 mm between the core-offset and the taper MZI has been proposed for RI measurement. The sensitivity of the fast axis and the slow axis up to ~-107.51 nm/RI and ~-74.54 nm/RI in the RI range between 1.333 and 1.381 is obtained, respectively. The simple fabrication process and the high precision show the proposed sensor has a great potential in chemical and biological sensing fields.

5. References
[1] Allsop T, Reeves R, Webb D J, et al. 2002 *Rev. Sci. Instrum.* **73**, 1702–05
[2] Ding J F, Zhang A P, Shao L Y, Yan J H and He S L 2005 *IEEE Photonics Technol. Lett.* **17** 1247–49
[3] Hatta A M, Farrell G, Wang Q, et al. 2008 *Micro Opt Techn Let* **50(12)** 3036-39
[4] Hatta A M, Semenova Y, Wu Q, et al. 2010 *Appl. Opt.* **49(3)** 536-541
[5] Wu D, Zhu T, Deng M, Duan D W, et al. 2011 *Appl Opt.* **7753(11)** 1548-53
[6] Peng W, Yan F, Li Q, et al. 2013 *Opt Laser Technol.* **45(2)** 348–351
[7] Gao S C, Zhang W G, Zhang H, et al. 2013 *Sens. Actuators, B.* **188** 931-36

[8] Yang J, Jiang L, Wang S, et al. 2011 *Appl Opt.* **50(28)** 5503-07
[9] Wu D, Zhu T, Deng M, et al. 2011 *Appl Opt.* **7753(11)** 1548-53
[10] Shen C Y, Zhong C, You Y, et al. 2012 *Opt. Exp.* **20(14)**, 15406-417

Acknowledgments

This work was supported by the Special Fund for Quality Inspection Research in the Public Interest of China (Grant No. 201510066), and the National Natural Science Foundation of China (Grant No. 61405185).

AOM2015   IOP Publishing
Journal of Physics: Conference Series **680** (2016) 012024   doi:10.1088/1742-6596/680/1/012024

An Incorporate Ultrasonic Coupling Device for Long-Focal-Zone Photoacoustic Imaging System

Dong-qing Peng[1,2], Yuan-yuan Peng[1],Shu-lian,Wu[1] and Hui Li[1]*

1.Key Laboratory of Optoelectronic Science and Technology for Medicine (Ministry of Education of China), College of Photonic and Electronic Engineering, Fujian Normal University, Fuzhou, Fujian, 350007

2.School of Science, Jimei University, Xiamen, Fujian, 361021

Corresponding author's e-mail:hli@fjnu.edu.cn

Abstract. An incorporate ultrasonic coupling device for extended focal zone focused photoacoustic imaging was designed.Phantom experimental imaging was carried out. The results show that, with the incorporate ultrasonic coupling device,accurate imaging position of absorber in the sample can be achieved. And it is worth mentioning that without the large water tank,the photoacoustic imaging experimental system become simple.Our work has a potential application in the noninvasive photoacoustic imaging of bio-tissue

1.Introduction

Photoacoustic imaging(PAI)has attracted great attention in modern medicine since it has already demonstrated its ability to detect breast cancer, skin related diseases, and brain tumors in small-animal studies[1-5]. Taking advantage of rich optical contrast and ultrasonic spatial resolution for deep imaging, a plenty of PA imaging moralities and applications have been proposed, including photoacoustic microscopy (PAM), photoacoustic Doppler flowery (PADF), photoacoustic endoscopy [6-11], etc.Until now, a large water tank were used for acoustic signal couple to reduce sound attenuation between different media, which make the proposed imaging systems inconvenient to clinical test and noninvasive detection.There was a growing need to develop an incorporate ultrasonic coupling device photoacoustic imaging, especially for system with extended focal zone, which will contribute to make photoacoustic imaging for clinical applications such as breast tumor, melanoma imaging and rheumatoid arthritis. In this work we intended to present an incorporate ultrasonic coupling device for extended focal zone focused photoacoustic imaging. The results of phantom experimental imaging show that, with the incorporate ultrasonic coupling device,accurate imaging position of absorber in the sample can be achieved. And it is worth mentioning that without the large water tank,the photoacoustic imaging experimental system become simple. Development of an incorporate ultrasonic coupling device would promote imaging system to transform from laboratory to clinical and has a potential application in the noninvasive photoacoustic imaging of bio-tissue.

2. Design and Experiment

2.1 Device Design

The acoustic wave detection of transducer was need to be immersed in the water for better ultrasonic coupling. But this would make imaging device complicate and bring inconvenience to noninvasive photoacoustic imaging of bio-tissue.Here, an incorporated ultrasonic coupling device was intended to design to used for extended focal zone focused photoacoustic imaging.

Content from this work may be used under the terms of the Creative Commons Attribution 3.0 licence. Any further distribution of this work must maintain attribution to the author(s) and the title of the work, journal citation and DOI.
Published under licence by IOP Publishing Ltd

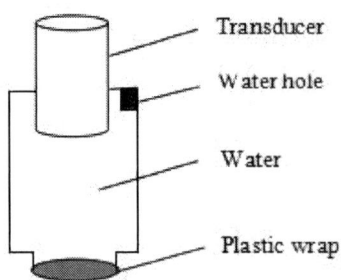

Fig.1 Schematic of incorporated ultrasonic coupling device

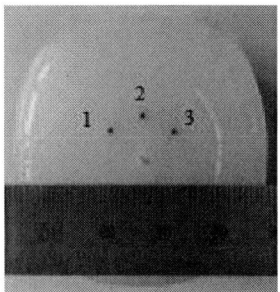

Fig.2 Phantom of experiment

The coupling device was made from a cylindrical plastic bottle with a height 90mm and a diameter 60mm.A circular hole with a diameter 4cm was set in the upper surface for immerse the transducer. and the below surface of the coupling device was covered by a plastic wrap with a diameter 3cm. Water could be added to the middle cavity from the upper right corner. The transducer can be moved up or down by a step adjusting device to ensure the signal interested was located in the focal zone. While imaging, the upper surface of the tissue sample could be coupled by plastic wrap to water, reducing sound attenuation between different media.

2.2 Experimental imaging

To test the feasibility of the designed incorporate ultrasonic coupling device, phantom experimental imaging was carried out. Experimental scheme was illustrated in Fig.2,(a) experimental setup,(b) scanning plan.The tissue-phantom used in the experiment was made with distilled water(100ml),agar (2g), Intralip(10ml), with three carbon rods inserted in the phantom to simulate absorber, as shown in Fig.2.Size of the phantom was 4.5cm $\times$ 4.5cm $\times$ 4cm..

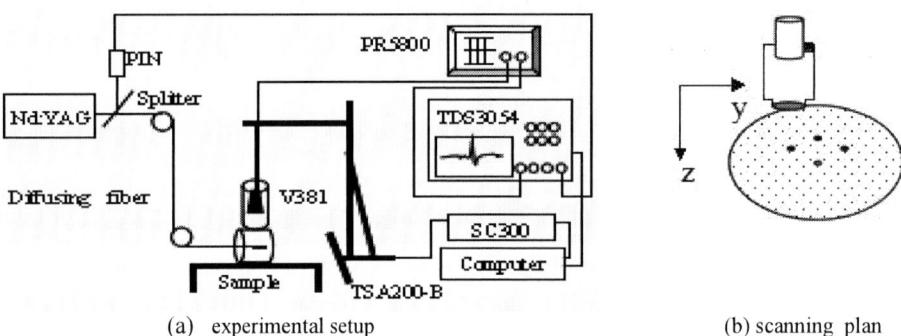

(a) experimental setup (b) scanning plan

Fig.2 schematic of experimental plan

An optical fiber with a 2.0-cm cylindrical diffuse active tip was placed in the center of the tissue,as shown in Fig.2(a).A Nd:YAG laser (Surelite I-10, Continuum, West Newton, MA, USA) with OPO oscillator (Surelite OPO plus, Continuum) was used for photoacoustic wave generation with a wavelength of 532nm, a repetition frequency of 10 Hz, a pulse width of 6 ns, and output energy of ~6mJ. The light beam from OPO was divided into two beams using a splitter mirror. One beam was received by a photo-diode and displayed on an oscilloscope for calibration, while the other was delivered to the phantom by fiber using a core diameter of 600μm to generate the PA signal.The upper surface of the tissue sample was coupled by the designed auxiliary ultrasonic coupling device instead of a large water tank.The depth-resolved photoacoustic signals from the sample were collected by a focused ultrasound transducer (V381, Panametrics, Hamburg, Germany), with a center frequency at 3.5 MHz. The signals were then transferred to an ultrasonic receiver (5800 PR, Parametric-NDT, Hamburg, Germany) for amplitude filtering and amplification. Finally, the signals were displayed on a digital oscilloscope (TDS3054C, Tektronix, Johnston, OH, USA). In order to improve the signal-to-noise ratio, the signals on the digital oscilloscope were averaged 64 times and then saved for follow-up data processing. Ultrasound transducer was fixed at the electronic translation machine (TSA200-B, Zolix), which could be drove by a step motor (SC300-2B, Zolix, Beijing, China) for 2D scanning imaging.Scanning plan of transducer was shown in Fig.2(b).

3. Results and Discussion

The photoacoustic signal intensity profiles through the center of three carbon rods were shown in Fig.3, (a)absorber-1,(b)absorber-2,(c)absorber-3.From the photoacoustic signal intensity profiles, the position of the two sides of carbon rod could be found clearly.Therefore,the position and size of absorber in the sample can be achieved. We use the delay-time(DT) between the Hilbert coefficient of photoacoustic signal at two sides of absorber to estimate the size of absorber. The diameter of absorber was determined by multiplying the DT with the speed c of sound in the medium.DT of the profile of absorber-1,absorber-2,and absorber-3 equal to $0.444us$, $0.438us$ and $0.417us$,respectively.Since the velocity of acoustic in the phantom is $1.5mm/us$, the measured diameter of absorber was $0.67mm$,$0.66mm$ and $0.63mm$,respectively, which was close to the true object size $0.7mm$. In Fig.3(b), the photoacoustic signal of fiber was used to location since it was insert into the middle hole of phantom.The measured diameter of fiber was $1.61mm$,which was close to the true object size $1.5mm$.Similarly,distance between absorber-2 and fiber could be measured,and it was $6.8mm$.The signal-to-noise ratios (SNR) of three curve was 3.25,4.23 and 6.40.The incorporate ultrasonic coupling device was scanned along y-axial direction,as shown in Fig.2(b). The 2D B-scan PAT imaging of cross-section in the y-z plane was gotten,as Fig.3 (d) shown.The three carbon rods could be clearly seen.The position and diameter of absorber in the sample could be obtained easily.The results showed that, with the incorporate ultrasonic coupling device,accurate imaging position of absorber in the sample could be also achieved. And it was worth mentioning that without the large water tank,the photoacoustic imaging experimental system became simple and compact.

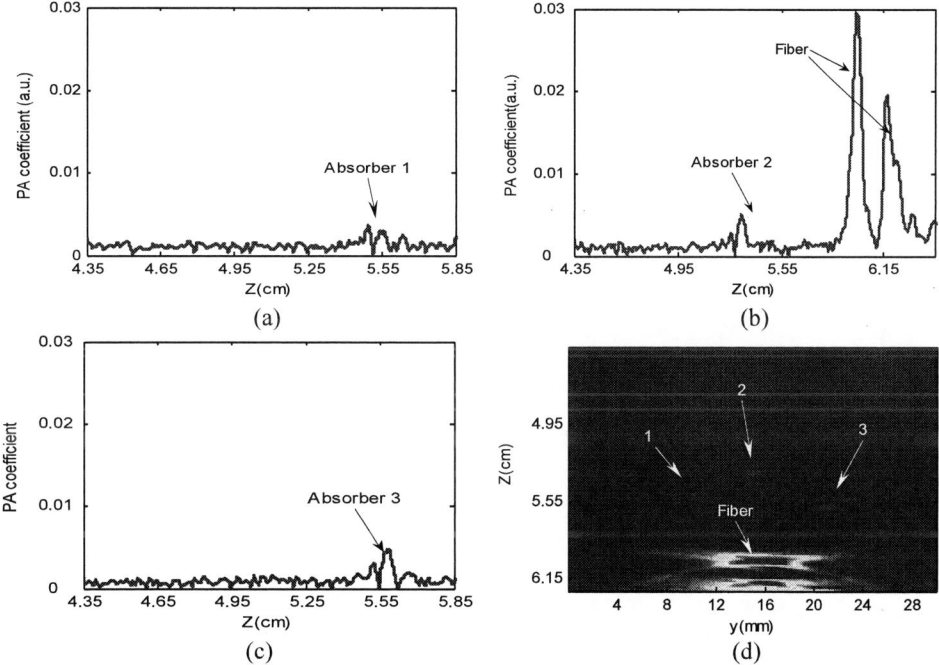

Fig.3 phantom experimental results.(a) profile of absorber-1,(b)profile of absorber-2,(c) profile of absorber-3,(d) 2D scanning imaging

4.Conclusion

In order to reduce sound attenuation between different media,acoustic signal couple was essential especially for imaging system with extended focal zone.However, a large water tank were often adopted, which would make the proposed imaging systems inconvenient to clinical test and noninvasive detection.There was a growing need to develop an incorporate ultrasonic coupling device photoacoustic imaging.In this work we intended to present an incorporate ultrasonic coupling device for extended focal zone focused photoacoustic imaging. The results of phantom experimental imaging show that, with the incorporate ultrasonic coupling device,accurate imaging position of absorber in the sample could be achieved. And it is worth mentioning that without the large water tank,the photoacoustic imaging experimental system became simple. Development of an incorporate ultrasonic coupling device would contribute to the portability of photoacoustic imaging system and has a potential application in the noninvasive photoacoustic imaging of biotissue.

Acknowledgment

This work has been sponsored by Natural Science Foundation of China (No.61178089), Natural Science Foundation of China (No.81201124), Fujian provincial key program of science and technology (No.2011Y0019), Fujian provincial education science research project of young teachers (JA14189). Natural Science Foundation of Fujian Province(No.2014J01226),and supported by the Key Laboratory of OptoElectronic Science and Technology for Medicine of Ministry of Education, Fujian Provincial Key Laboratory for Photonics Technology, Fujian Normal University,China(Grant No.JYG1417)

References

[1] Xu M, Wang L V.2006 *Rev.Sci. Instrum.* **77** 041101.
[2] Xu X,Li H. 2008 *Physics* **37** 111 (In Chinese).
[3] Li C,Wang L V.2009 *Phys. Med. Biol.***54** R59.
[4] Gamelin J,Aguirre A, Maurudis A, Huang F, Castillo D, Wang L V and Zhu Q.2008 *J. Biomed. Opt.* **13** 024007.
[5] Ermilov S A, Khamapirad T, Conjusteau A, Leonard M H, Lacewell R, Mehta K, Miller T, Oraevsky A A.2009 *J. Biomed. Opt.* **14** 024007.
[6] Wang L V.2008 *Med. Phys.* **35** 5758.
[7] Chitnis P V, Brecht H P, Su R, and Oraevsky A A. 2010 *J. Biomed. Opt.* **15** 21313.
[8] Shi W, Kerr S, Utkin I, Ranasinghesagara J,Pan L, Godwal Y, Zemp R J, and Fedosejevs R..2010 *J. Biomed. Opt.* **15** 56017.
[9] Sheaff C,Lau N, Patel H, Huang S W, and Ashkenazi S. 2009 *Annual International Conference of the IEEE Engineering in Medicine and Biology Society. IEEE Engineering in Medicine and Biology Society.*1983
[10] Fang H, Maslov K, and Wang L V. 2007 *Phys. Rev. Lett.* **99** 184501.
[11] Fang H, and Wang L V.2009 *Opt. Lett.* **34** 671.

AOM2015   IOP Publishing
Journal of Physics: Conference Series **680** (2016) 012025   doi:10.1088/1742-6596/680/1/012025

A New Method for Axial Decay Function Calibration of Evanescent Field in Multi-Angle Total Internal Reflection Fluorescence Microscopy

Jian Wu[1,2], Peng Xiu[3], Luhong Jin[1,2], Di Nan[1,2], Cuifang Kuang[3], Xiaoxiang Zheng[1,2,4], Yingke Xu[1,2] and Xu Liu[3]

[1] Key Laboratory for Biomedical Engineering of Ministry of Education, Department of Biomedical Engineering, Zhejiang University, Hangzhou, China 310027

[2] Zhejiang Provincial Key Laboratory of Cardio-Cerebral Vascular Detection Technology and Medicinal Effectiveness Appraisal, Zhejiang University, Hangzhou, China 310027

[3] State Key Laboratory of Modern Optical Instrumentation, Department of Optical Engineering, Zhejiang University, Hangzhou, China 310027

[4] Qiushi Academy for Advanced Studies, Zhejiang University, Hangzhou, China 310027

E-mail: yingkexu@zju.edu.cn

Abstract. Three-dimensional image reconstruction in multi-angle total internal reflection fluorescence microscopy relies on actual penetration depths of evanescent wave at various incidence angles. In this paper, we propose a simple and elegant calibration method to calculate the actual axial decay profile of a given evanescent field, for the analytical solution of theoretical equation is hard to solve in complicated conditions. The results calculated by the proposed method agree with the experimental demands in our research. Our calibration method, together with multi-angle TIRF imaging, permits 3D reconstruction of cell surface in superb axial resolution.

1. Introduction

Total internal reflection fluorescence microscopy (TIRFM) has been widely used in biomedical research to observe a thin layer between the interface of glass cover slip and aqueous solution by exciting an evanescent field that decays exponentially along z-dimension [1]. Its axial resolution can achieve ~100 nm, better than confocal microscopy (300-500 nm). Coherent lasers are routinely used as the light sources for TIRF illumination, and in most of the commercial TIRFM setups, a single laser beam is focused to a tight spot in an eccentric position in the back focal plane (BFP) of the high numerical aperture objective. This arrangement produces imperfection of the TIRF illumination, which leads to the appearance of interference fringes. Time averaging over different evanescent wave propagation directions will produce a more even illumination and facilitate quantitative imaging. In this paper, we customize a ring-illuminated multi-angle TIRFM, which can alter the penetrating depth of evanescent field by varying incidence angle of laser beam and spinning of TIRF illumination using galvanometric mirrors. The incidence angle of TIRF illumination is larger than critical angle θ_c and smaller than θ_{max} (related to NA of objective). Ring of light is used to permit even illumination and

Content from this work may be used under the terms of the Creative Commons Attribution 3.0 licence. Any further distribution of this work must maintain attribution to the author(s) and the title of the work, journal citation and DOI.

Published under licence by IOP Publishing Ltd

remove fringe interference through spinning azimuthal angle of incidence laser beam. A simple and elegant calibration method is proposed to calculate the actual axial decay profile of evanescent field, for the analytical solution of theoretical equation is hard to solve. To obtain quantitative information about finer structure, three-dimensional reconstruction of the biological sample also can be done based on a set of multi-angle TIRFM images and the derived axial decay profile that calculated from the system.

2. Methods

2.1. Multi-angle TIRFM System
In this paper, we establish a multi-angle ring-TIRF microscopy, shown in Figure 1. It is based on a classical objective-type TIRF microcopy, but we apply two-dimension galvanometer (2D-GM) system to control the illumination angle and to provide the ring illumination. The 2D-GM used in our system is for 10 mm beam diameter, the focus distance of scanning lens is 50 mm and the objective lens is 60X (1.45 NA). The full-scale bandwidth of the 2D-GM is connected with its physical size. The highest full-scale sine wave response frequency of 10 mm diameter 2D-GM is 100 Hz and the imaging speed of EMCCD is 10 fps, so in the exposure process the illumination angle can spin at least ten times. The total internal reflection angle is affected by the polarization, here we introduced a quarter wave plate to modulate the laser to circularly polarized light, so the illumination depth of the evanescent wave will not change when the illumination spinning at fixed angle. The angle repeatability of 2D-GM is 15 μrad, and we can subdivide the supercritical angle more than 100 times, which enable superb resolution of the 3D reconstruction image along the Z-axis.

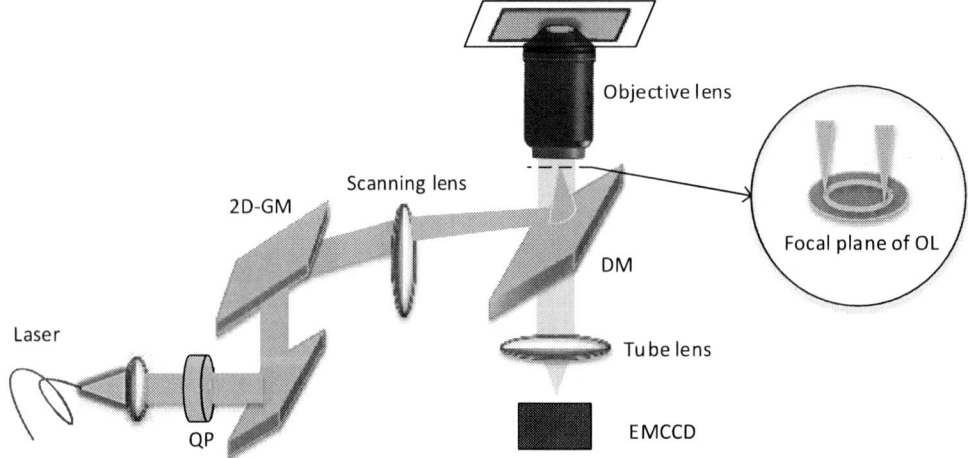

Figure 1. The setup of the multi-angle ring-TIRF microscopy system. QP: quarter wave plate; 2D-GM: two-dimension galvanometer; DM: dichroic mirror.

2.2. TIRFM Imaging Theory
Theoretically, each TIRFM image I with incidence angle θ_i of the actual fluorescence labeled object C is the projection of a 3D volume, can be expressed by an integral along the axial direction depending on the penetrating depth $d_p(\theta_i)$,

$$I(\theta_i) = \phi I(0, \theta_i) \int_0^\infty Q(z) \cdot \mathrm{PSF}(z) \cdot C(z) \cdot e\left[-z/d_p(\theta_i)\right] \mathrm{d}z \tag{1}$$

where $I(0, \theta_i)$ is the intensity at the dielectric surface ($z = 0$), which can be derived according to the Fresnel formula [2],

$$I(0,\theta_i) = \frac{4\cos^2\theta_i\left[2\sin^2\theta_i - \left(\dfrac{n_t}{n_i}\right)^2\right]}{\left(\dfrac{n_t}{n_i}\right)^2\cos^2\theta_i + \sin^2\theta_i - \left(\dfrac{n_t}{n_i}\right)^2} \tag{2}$$

ϕ stands for the quantum efficiency of the EMCCD and fluorophores, $Q(z)$ and $\mathrm{PSF}(z)$ are the photon collection efficiency and point spread function [3].

The analytical solution of penetration depth for evanescent wave is

$$d_p(\theta_i) = \frac{\lambda}{4\pi\sqrt{n_i^2\cdot\sin^2\theta_i - n_t^2}} \tag{3}$$

which deviates from the actual decay function of microscope with other involved complicated factors. After calculating the penetration depth, the actual axial profile of evanescent field behind the dielectric surface can be simply described as

$$I(z,\theta_i) = I(0,\theta_i)\cdot e\left[-z/d_p(\theta_i)\right] \tag{4}$$

A multi-angle TIRFM image sequence taken at different incidence angles (corresponding to different voltages of galvanometric mirrors) is shown in Figure 2, where the epifluorescence image is also given by setting the incidence angle to nearly 0 (far below the critical angle).

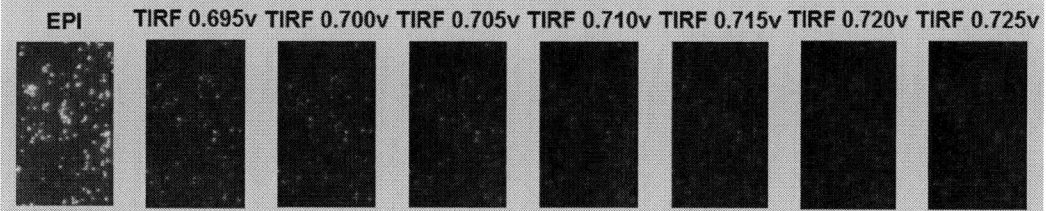

Figure 2. Multi-angle TIRFM images taken at different voltages of galvanometric mirrors

2.3. Axial Decay Function Calibration

Different from previous studies [3-4], we place an inclined plane coated with fluorophores at the glass cover slip (dielectric surface) with inclined angle α, shown in Figure 3. Moving the objective to the field of view, a set of TIRFM images $I_{\mathrm{TIRFM}}(\theta_i)$ can be acquired from different incidence angles. The EPI image I_{EPI} of the same sample at the same field of view can be recorded by setting the incidence angle to nearly 0 (far below the critical angle). The logarithm of intensity ratio at the same site A from both two images is a linear function of the lateral distance x.

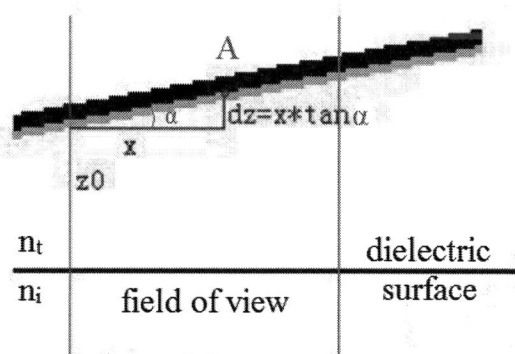

Figure 3. The inclined plane labelled with fluorophores for calibration

$$z_A = z_0 + x\tan\alpha \tag{5}$$

$$\ln\left[\frac{I_{A(TIRFM)}(\theta_i)}{I_{A(EPI)}}\right] = \ln\left[\frac{I_A(0,\theta_i)\cdot e\left[-z_A/d_p(\theta_i)\right]}{I_{A(EPI)}}\right]$$

$$= \ln\left[\frac{I_A(0,\theta_i)}{I_{A(EPI)}}\cdot e\left[-z_0/d_p(\theta_i)\right]\right] - \frac{x\tan\alpha}{d_p(\theta_i)} \tag{6}$$

Where $\ln\left[\dfrac{I_A(0,\theta_i)}{I_{A(EPI)}}\cdot e\left[-z_0/d_p(\theta_i)\right]\right]$ is a constant for the given θ_i. Through curve fitting of

equation (6) with random sampling fluorescence spots from real images, the penetrating depth $d_p(\theta_i)$ can be calculated from the negative reciprocal of slope of the fitting curve.

3. Results

In our experiment, the pixel size of TIRF images is around 133 nm, the inclined angle $\alpha = 0.18^\circ$, the laser wavelength $\lambda = 488nm$, the incidence angle θ_i is set by the voltage of galvanometric scanner from 0.695-0.725 V. Over 800 centroid intensities of detected fluorescence spots are used to fit the equation (6) at 0.725 V, shown in Figure 4(A). The curve fitting result is

$$y = -0.4959x\tan\alpha - 0.5743 \tag{7}$$

with $R^2 = 0.5169$, where the linearity of fitting curve is inevitably blurred by some fluorescence interference exist outside the actual penetrating depth. The penetrating depth $d_p = 268nm$ is the negative reciprocal of slope of the fitting curve (7)

$$d_p = \frac{1}{0.4959}\times 133 = 268nm \tag{8}$$

Other multi-angle penetration depths corresponding to 0.695, 0.700, 0.705, 0.710, 0.715, 0.720, 0.725 V, are also obtained as 900, 715, 559, 372, 369, 323, 268 nm, shown in Figure 4(B). The calculated values agree with the experimental demands in our research.

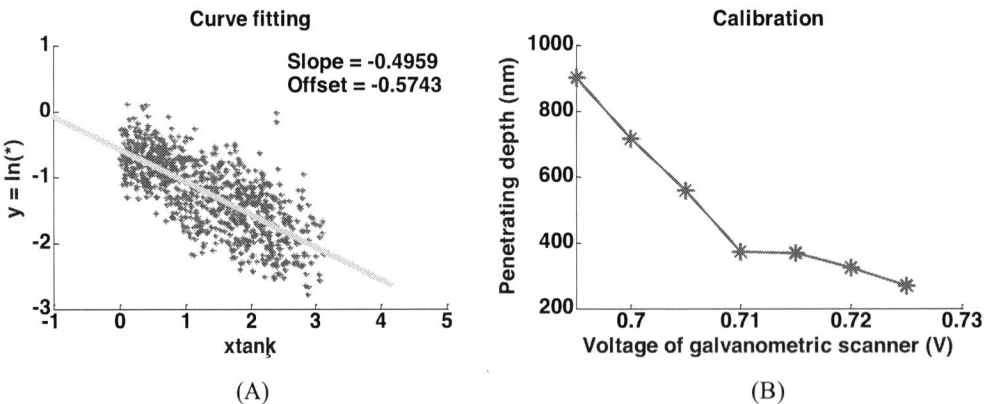

Figure 4. The curve fitting and calibration result of experimental image data.

Acknowledgement

This work is supported by grants from National Basic Research Program of China (2015CB352003), National Natural Science Foundation of China (31301176, 31571480), Zhejiang Provincial Natural Science Foundation of China (LY13C050001), Specialized Research Fund for the Doctoral Program of Higher education (20130101120172) and the Open Foundation of the State Key Laboratory of Modern Optical Instrumentation.

References

[1]   Steyer J A and Almers W 2001 *Nature* **2** 268-276

[2]   Axelrod D, Hellen E H and Fulbright R M 1992 *Topics in Fluorescence Spectroscopy: Principles and Applications* **3** 289-343

[3]   Yang Q, Karpikov A, Toomre D, and Duncan J 2011 *IEEE transaction on Image Processing* **20(8)** 2248-2259

[4]   Boulanger J, Gueudry C, Münch D, Cinquin B, Paul-Gilloteaux P, Bardin S, Guérin C, Senger F, Blanchoin L and Salamero J 2014 *PNAS* **111(48)** 17164-17169

Design of visible/infrared double-band spectral imager

Tang Tian-jin[1], Zhang Zhuo[1], Wang Bao-hua[1]

[1] Beijing Institute of Space Mechanic & Electricity, Beijing 100080

E-mail: tianjin79526@aliyun.com

Abstract. OFFNER hyperspectral imager using convex grating as spectral splitting component has many merits including large relative aperture, no smile, small key stone, compact construction and easier assembly and etc, has widely used in many occasions of various fields. Design of double-band hyperspectral imager using OFFNER structure is presented in this paper. SHAFER fore telescope system and convex grating imaging are adopted, groove etching density is different in different region on the same substrate for different spectral band, and different diffractive order is used respectively to split the two spectral bands. The optical system achieves good image quality in wide spectral range, compact construction and small volume. And all the reflective curved face are sphere, moderate tolerances are good for the processing and alignment.

1. Introduction

Hyperspectral imager, which is used on space sensing platform, combines the imaging technique and spectroscopy to achieve hyper-spectral images of target scene[1]. Since the hyperspectral concept was first put forward by Goetz about twenty years ago, several hyperspectral imagers were developed all over the world, and several imagers were used on orbit successfully. For the eximious information achievement ability, the hyperspectral technique has gained broad attention and fast development, and has extensive prospect in many fields such as military scout, satellite remote sensing, geological survey and resource survey in agriculture, forestry and mining. And in recent years, it has applied many times successfully in deep space exploration[2].

As the continuously deepen of space remote sensing technique and flourish development of small satellite, the requirement for performance and miniaturization of the spectral imager is higher. The hyperspectral imager needs to achieve high resolution in large breadth in order to shorten the access period and achieve more abundant spectral information at the same time. The hyperspectral imager needs also to be miniaturization and light weight to be seasoned with variety of flatform and detection need. The types of hyperspectral imagers consists of prism dispersion spectrometer, grating spectrometer and fourier spectrometer according to the imaging principle. Traditional dispersion method in which prism and grating is insert into the collimated beam has the disadvantages of large smile and heavy weight. OFFNER spectrometer arised in recent years covered the shortages, and has the merits including large relative aperture, no smile, small keystone, compact construction and simplied assembly[4]. OFFNER spectrometer has already used successfully in many explorers for deep space detection[5]-[8].

Content from this work may be used under the terms of the Creative Commons Attribution 3.0 licence. Any further distribution of this work must maintain attribution to the author(s) and the title of the work, journal citation and DOI.

Published under licence by IOP Publishing Ltd

2. Spectrometer based on OFFNER configuration

OFFNER spectrometer was developed form the coaxial three-mirror relay optical system presented by OFFNER in 1973. There are two types of configurations of the OFFNER spectrometer including convex grating and curved face prism, Hyperion and COIS of American adopted convex grating to separate the spectrum, CHRIS of ESA adopted the curved prism to achieve the images of slit for different spectrum bands. The typical configuration of OFFNER spectrometer is shown in fig.1.

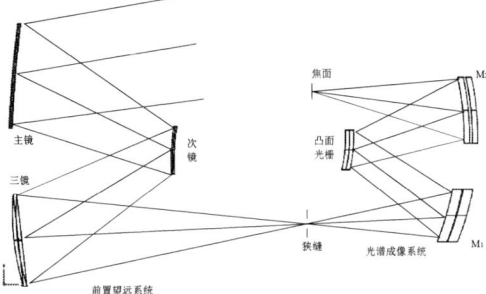

Fig.1 Structural sketch map of Offner imaging spectrometer

Optical system of the hyperspetral imager need to be achieve small smile and keystone, that is to say, the deviation of the curved image of different spectrum from the straight line must be small, the magnification of the image of different spectrum should be consistent. The smile means that the images of slit at different wavelength is not a straight line but a curve when the slit image through the imaging system at different wavelength. The smile is usually expressed by the deviation between the link of two terminal points and the center. The keystone is caused by the different magnification of different wavelength, which will result in that the image points are not on the straight line vertical to the slit but a diagonal. The keystone will result in that the image points of different wavelength is not on the same column of pixels on the detector, which will bring about difficulties to the after data processing[9]. Compared with the traditional methods that grating and prism are insert into the collimated beam, OFFNER system resolves the difficulty of smile, the key stone is also ignored, large field with no field curves and small volume are achieved at the same time[10].

3. Design of double-band spectral imager

3.1. Formatting the titleOptical system and grating design

The wavelength of the spectrometer designed in this paper range from 0.45~5μm，including near infrared spectrum 0.45~1μm and infrared spectrum 1~5μm, the requirements for the image quality, volume and weight is high. And because of the broad wavelength range, the transmission of the lens materials varies as the temperature changes, so reflective optical system is adopted after compared with the optical systems of foreign spectrometers. A small reflective optical system is presented based on considering the domestic processing capability. The optical system designed is shown in fig.2.

The optical system adopted SHAFER fore telescope system and convex grating imaging, in which the SHAFER fore telescope system is composed of the reversal BURCH telescope and OFFNER relay optics, and the BURCH telescope optical system consists of principle mirror, fold mirror and second mirror which are used to eliminate the spherical aberration and coma aberration. We can achieve good image quality at the position of slit after the incident rays pass through the SHAFER fore telescope system, then we can achieve the image of slit on the CCD for the wavelength 0.45~1μm and on the IRFPA for the wavelength 1~5μm. And ,the OFFNER relay optics is laid along the direction Y, the OFFNER spectrometer is laid along the direction X, the two detectors is laid long the direction Y.

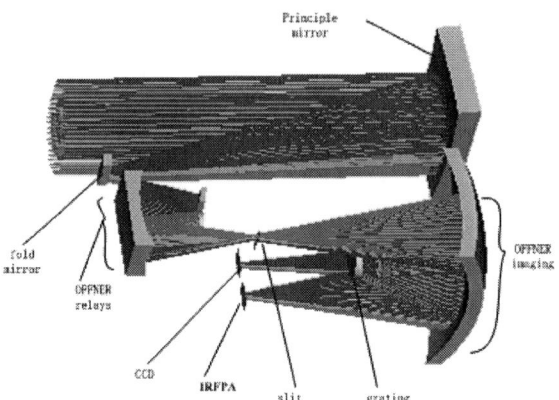

Fig.2 Optical system design result

The field angle of the hyperspectral imager is 3, and the focal length is 150mm, the angle resolution is 0.4mrad, the spectrum resolution can reach 2~10nm. The IRFPA adopted HgCdTe infrared detector with the pixel dimension of 40μm×40μm, and the CCD adopted Thomson-CSF TH7896 detector with the pixel dimension of 20μm×20μm, in which two pixels is combined in order to match the resolution of two spectrum bands.

Table.1 Optical system initial configuration parameters

Surf	Rad/mm	Thi/mm	Glass	Tilt X/°	TiltY/°	Dec X/mm
1	Infinity	150	-	0	-	-
2	-227	-145	MIRROR	-5.7	-	-
3	Infinity	45	MIRROR	20	-	-
4	422	-49	MIRROR	-24	-	-
5	81	42	MIRROR	11	-	-
STO	57	-50	MIRROR	-21	-	-
7	89	107	MIRROR	12	-	-
8	-	0	Break	-	-	10
9	-39	-23	MIRROR	0	-	-
10	-31	23	GRATING	0	-	-
11	-47	-48	MIRROR	0	-	-
12	-	2.2	Break	-	0	0
IMA	Infinity	-	-	0	-	-

The initial configuration parameters are shown in tab.1. the dichroic method in this paper is not sue-fields or prism, but the method in which groove etching density is different in different region on the same substrate for different spectral band, and different diffractive order is used respectively to separate the two spectral band. The simply configuration and compact construction will win more space for supporting and weight. Besides, all the reflective surfaces including the grating face are sphere, the tolerances are moderate, which is good for processing and aligment.

Spectral splitting of two bands is realized through the method that the groove etching density are different in different region, and different diffractive orders are employed for the different bands. The Sketch map of grating etching is shown in fig.4. The etching density for the visible and near visible band id 96lp/mm, the etching density for the visible and near visible band id 60lp/mm, reflective film is coated in order that only infrared spectrum can be reflected in infrared region, and only visible band can be reflected in visible region.

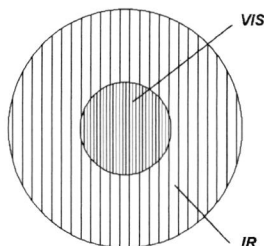

Fig3. Sketch map of grating etching

3.2. Image qulity estimation

The modulation transfer function for the different bands is shown in fig.4 and fig.5, we can see the image quality is very good. The spot diagrams for two different bands in respect fields are shown in fig.6 and fig.5, the diameter of the spot of visible band is within the range of one pixel.Some of the spots' diameters of infrared band have exceeded the range of one pixel, but 95% of the energy is within the dimension of on pixel, which can also satisfied the need of application.

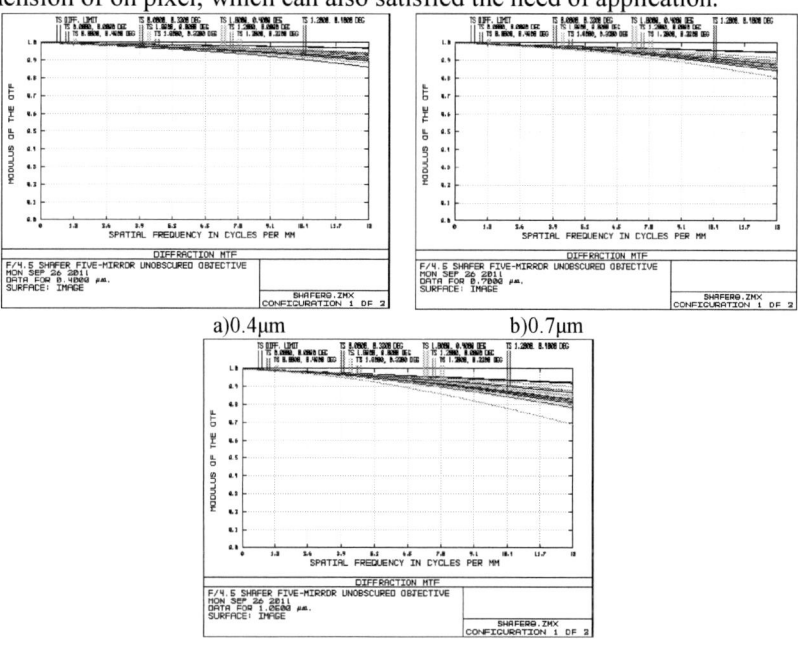

a)0.4μm   b)0.7μm

c)1.05μm

Fig4. MTF curves of VNIR channel

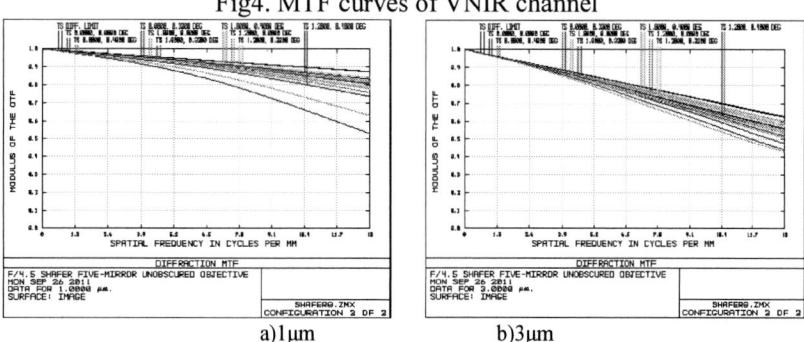

a)1μm   b)3μm

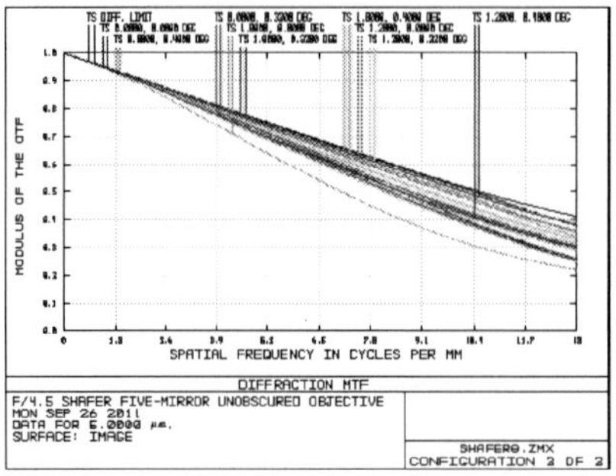

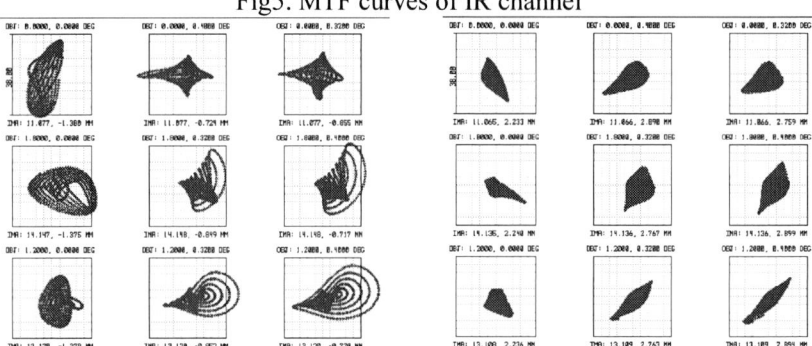

c)5μm

Fig5. MTF curves of IR channel

Fig6. Spot diagram of VNIR channel and IR channel

4. Conclusion

Performance, volume and weight are needed to be taken into consideration for the design of spectrometer optical system. The optical system designed in this paper adopted OFFNER structure which uses convex grating as dispersive element, spectral splitting and diffraction can be realized through on grating. The results shows that the modulation transfer function and spots are better than the requirements. Reflective optical system will be more suitable for the space circumstance on orbit. Small volume and compact structure are all achieved compared with the traditional spectral splitting methods of sue-fields and prism.

5. Reference

【1】 Offner A. Unit power imaging catoptric anastigmat : US, 3748015 [P]. 1973-7-24.

【2】 Mouroulis P. Low-distortion imaging spectrometer designs utilizing convex gratings [J]. SPIE, 1998, 3482: 594-601.

【3】 Lobb D R. Theory of concentric designs for grating spectrometers [J]. Appl. Opt, 1994, 33: 2648-2658.

【4】 Mertz L. Concentric spectrographs [J]. Appl. Opt, 1977, 16: 3122-3124.

【5】 Miller E, Klein G, Juergens D, et al. The visual and infrared mapping spectrometer for Cassini [J]. SPIE, 1996, 2803: 206-220.

【6】 Coradini A, Capaccioni F, Drossart P, et al. VIRTIS: an imaging spectrometer for the ROSETTA mission [J]. Planetary and Space Science, 1998, 46 (9/10): 1291-1304.

【7】 Piccioni G, Drossart P, Suetta E, et al. VIRTIS imaging spectrometer for the ESA/Venus Express mission [J]. SPIE, 2004, 5543: 175-185.

【8】 Silverglate P R, Fort D E. System design of the CRISM (compact reconnaissance imaging spectrometer for Mars) hyperspectral imager [J]. SPIE, 2003, 5159: 283-290.

【9】 Russell C T, Coradini A, Christensen U, et al. Dawn: A journey in space and time [J]. Planetary and Space Science, 2004, 52: 465-489.

【10】 Zheng Yu-quan. Design of compact Offner spectral imaging system. Optics and Precision Engineering. 2005, 13(6): 650-657.

Sacrificial solder based nanowelding of ZnO nanowires

Guoping Liu, Qiang Li, and Min Qiu

State Key Laboratory of Modern Optical Instrumentation, College of Optical Science and Engineering, Zhejiang University, Hangzhou, 310027, China

E-mail: qiangli@zju.edu.cn

Abstract: A new nanowelding approach for joining ZnO nanowires at nanoscale based on a sacrificial solder is demonstrated. This method protects the original junction and gets a strong mechanical property at the welded junction.

1. Introduction:

Nanowires owe excellent optical, electrical and thermal properties at nanoscale. They could be widely used in nanoelectromechanical systems [1-3]. Nanosoldering is a typical method to join nanowires [4-6]. Traditional nanojoining methods always focus on metal nanowires, such as silver, gold and platinum. This is because metal nanowires could be easily applied as transparent electrodes to replace the role of ITO (Indium tin oxide). Many semiconductor nanowires have been demonstrated to form nanodevices (such as splitters, waveguides). The typical plasmonic nanowelding is obviously not feasible for nanowelding of semiconductor nanowires[7]. In this paper, we propose a sacrificial solder based method for nanojoining of ZnO nanowires. This approach of nanojoining could maintain the integrity of nanowire junctions and could be used in fabrication of semiconductor devices.

2. Experiments and discussions:

2.1. Experiments setup:

ZnO nanowires are dropped on a prepared glass substrate. The length of the ZnO nanowires ranges from 20μm to 100μm. The diameter of the nanowires is about 300nm. Figure 1 is the schematic of the experimental set-up. A diode-doped Nd: YAG laser operating at 532-nm wavelength is set as the light source. A beam shutter is put next to the attenuator for controlling the exposure time. Here the exposure time is set as 1 second. The λ/4 plate plays the role of changing the linear polarized beam to circular polarized beam. A CCD is employed to observe and monitor the sample and the nanowires.

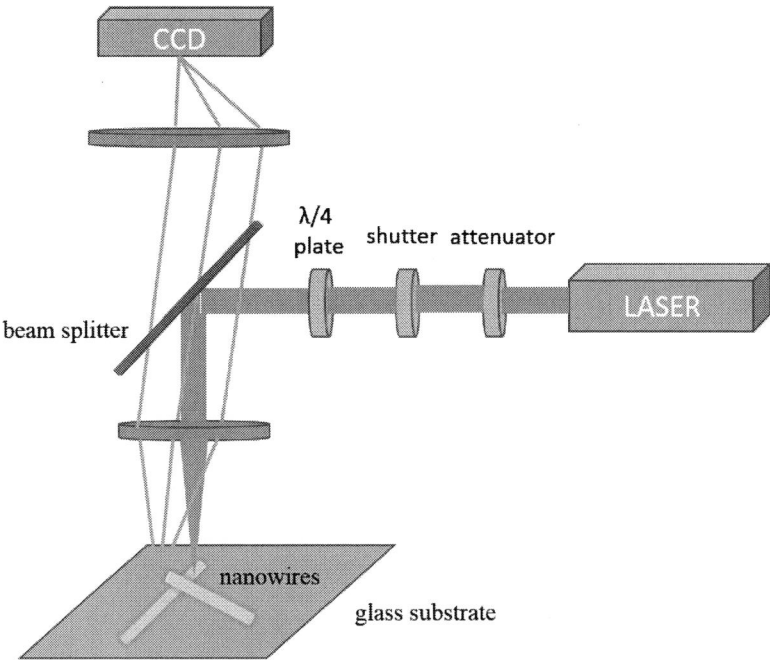

Figure 1. Schematic of experimental set-up

2.2. Results and Discussion:

When the laser is focused on a nanowire, the illumination part of the nanowire gets fused if the raised temperature exceeds the melting point of the nanowire material. Here, the power of the laser is 500mW, which is sufficient for fusing the ZnO nanowires at the focused part. The method we propose in this paper is different from traditional nanojoining methods since a solder is involved in the joining process.

We firstly investigate the phenomenon of illuminating the terminal of one ZnO nanowire. A ZnO nanowire owns a hexagonal cross section, as shown in Fig. 2 (a). Then we focus the center of the laser at the terminal of this nanowire. The terminal shrinks and a sphere is generated. Figure 2 (b) shows the details of the ZnO nanowire after laser fusing process.

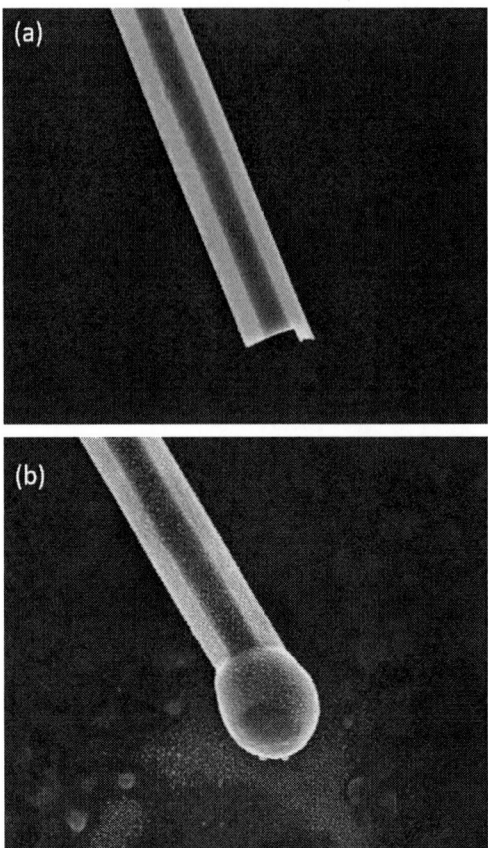

Figure 2. Comparison of two different terminals: (a) before and (b) after laser fusing.

This formed nano-sphere can be exploited as a solder for nanosoldering. Figure 3(a) shows a schematic of nanosoldering of a T-shaped ZnO junction. With the help of a fiber probe, we put two ZnO nanowires together to form a T-shaped junction and then move the focused point to the solder area of the structure. After illumination and fusing process, a sphere generates and shrinks to the remainder nanowire. When the shrinkage part of the nanowire meets the junction, the sphere bonds it, leading to the nanojoining of two nanowires. From the SEM image, we find that the sacrificial ZnO solder perfectly make the two individual ZnO nanowires weld together. The original junction is maintained during the laser illumination.

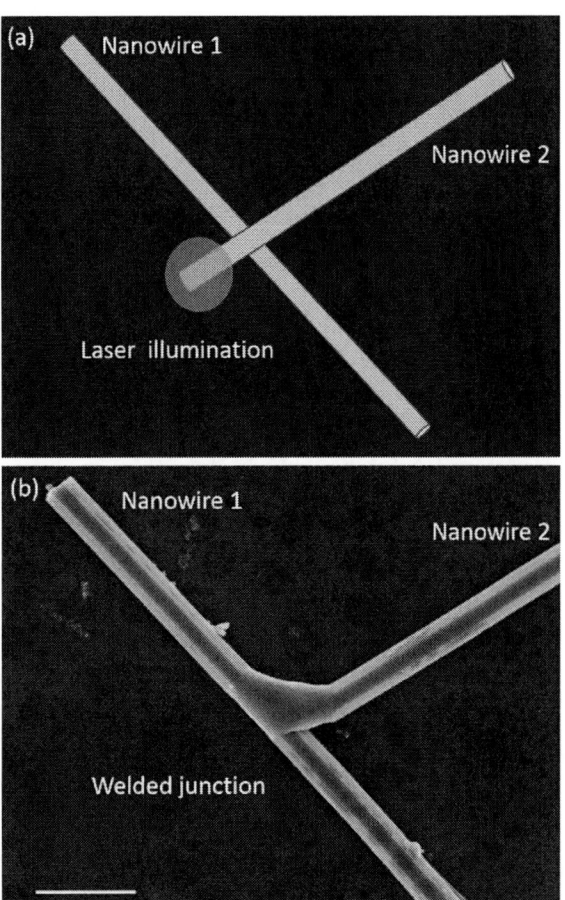

Figure 3. (a) Schematic of nanosoldering of a T-shaped ZnO junction. (b) A SEM image of a welded T-shaped ZnO junction.

Summary

We propose a new sacrificial solder based nanowelding method for ZnO nanowires. This method could perfectly protect the junctions to be linked. As the laser beam is locally focused, we could control it at nanoscale. This method would benefit the development of semiconductor nanowire based nanodevices.

Acknowledgment

This work is supported by the National Natural Science Foundation of China (Grant Nos. 61235007, 61425023, 61575177, 61205030 and 61275030).

Reference

[1]   Tohmyoh H, Imaizumi T, Hayashi H and Saka M 2007 *Scripta Materialia.* **57**(10) 953-956

[2]   Li X, Gao F, and Gu Z 2011 *Open Surf. Sci. J.* **3** 91-104

[3]   Zhou Y and Hu A 2011 *Open Surf. Sci. J.* **3** 32-41

[4]   Huang H C, Walker C R, Nanda A and Rege K 2013 *ACS Nano* **7** 2988-2998

[5]   Jinhwan L, Lee P, Lee H B, Hong S, Lee I, Yeo J, Lee S S, Kim T S, Lee D, and Ko S H 2013 *Adv. Funct. Mater.* **23** 4171-4176

[6]   Peng Y, Cullis T and Inkson B 2008 *Nano Lett.* **9** 91-96

[7]   Garnett E C, Cai W, Cha J J, Mahmood F, Connor S T, Christoforo M G, Cui Y, McGehee M D and Brongersma M L 2012 *Nature Mater.* **11** 241-249

Laser assisted welding of gold nanowires

Lina Zhou, Gongping Liu, Si Luo, Qiang Li, and Min Qiu

State Key Laboratory of Modern Optical Instrumentation, College of Optical Science and Engineering, Zhejiang University, Hangzhou, 310027, China

E-mail: qiangli@zju.edu.cn

Abstract. Miniaturization of optoelectric devices is based on the integration of materials in nanometer scale. Utilizing gold nanowires' good manipulability, electrical and optical conduction performance, we provide an effective laser assisted welding method to implement the joining of materials. This welding method provides a promising and simple technology for future integration techniques.

1. Introduction

Integrating of nanoscale devices is an inevitable trend in photonics community. The integration of nanostructures (such as nano-rods, nano-particles, nano-disks, and nano-wires[1, 2, 3, 4]) can be reinforced by nanojoing[5, 6, 7]. So far, several methods of nanojoining have been proposed to weld nano-objects together. These nanojoinging methods include ion irradiation[8], thermal annealing[9], electrics current welding[10], and laser beam[11]. In this paper, we use a sacrificial gold nanowire as a solder to nanojoin the gold nanowires.

2. Materials preparation and devices setup

Gold nanowires used in the experiment are synthesized by wet-chemical approach[12]. Initially 50ml of ethanol 0.72mmol/L is added to the conical flask with water bath of 20mins under 95℃. Then 710ul of 0.1mol/L ethanol aniline is added with vigorous stirring under constant temperature of 95℃. As the color of solution changes from brown to dark violet, we obtain required solution and then wash the solution with ethanol by paper-filter. Finally, we get the mixture of gold nanoplates and nanowires. The nanowires are about 300nm in diameter. Then 1.5ul of solution is dropped onto a cleaned glass wafer for further experiment.

In the welding, we focus the 532nm laser beam (continuous laser) with a power of 1.5mw into the microscope. CCD is used to record the process of welding. The spot size of the laser beam is about 1500nm in diameter.

3. Experiments and discussion

For a single nanowire, the laser spot is focused on the middle part, as shown in figure 1. We can see that the original nanowire (296.7nm in diameter) breaks into two short nanowires with ball-like ends. The diameters of the two balls are about 477 nm and 348 nm.

Content from this work may be used under the terms of the Creative Commons Attribution 3.0 licence. Any further distribution of this work must maintain attribution to the author(s) and the title of the work, journal citation and DOI.

Published under licence by IOP Publishing Ltd

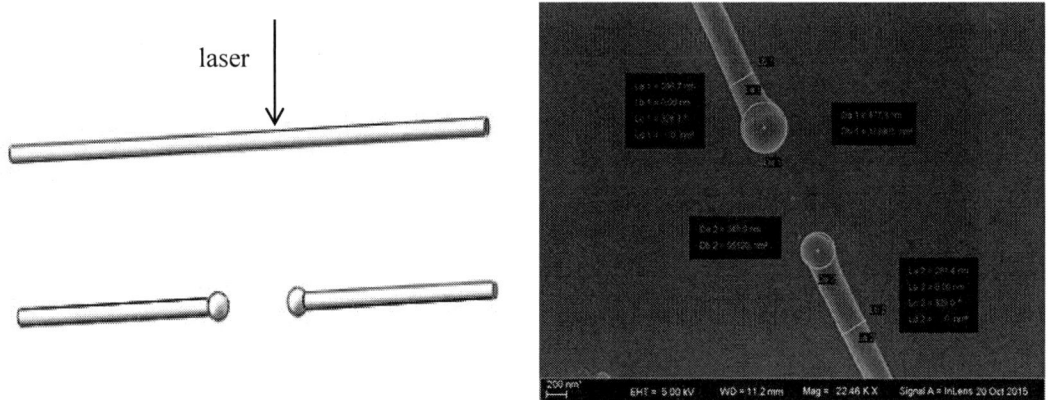

Figure 1.SEM image of a single nanowire after laser illumination.

As shown in Figure 2 two nanowires are put together to form a junction with a fiber probe. When the laser is focused at the extended part of one nanowire, it melts into a ball to bond the junction.

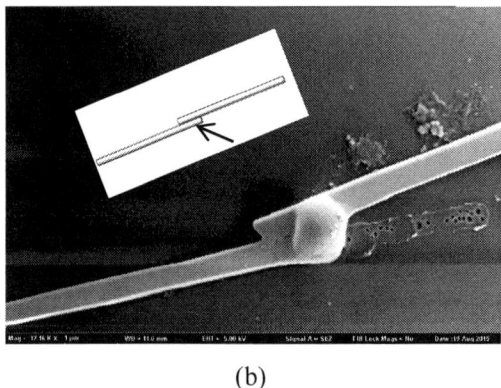

(b)

Figure 2. Two nanowires forming a coupler after welding.

Apart from nanowire-nanowire welding, the nanowires can also be welded with the nanoplates. As shown in Fig. 3(a), one nanowire is placed around the corner of one pentagon nanoplate by fiber probe. After the laser welding, a ball with a diameter of around 710 nm is formed, bonding the nanowire and the nanoplate.

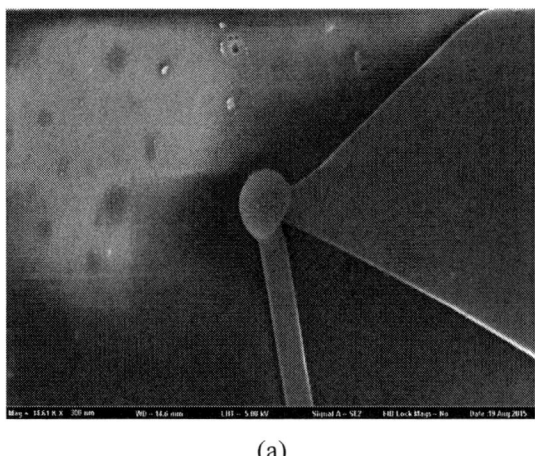

(a)

Figure 3. Nanowelding between an Au nanowire and an Au nanoplate

4. Conclusion

In summary, the light controlled plasmonic nanowelding of gold nanostructures is demonstrated. Utilizing this technology, we effectively weld the nanowire-nanowire structure and nanowire-nanoplate structure. This method provides a promising and simple technology for future integration techniques.

References

[1] Guo S W 2010 *Nanoscale.* **2** 2521-29
[2] Zhang Y and Dai H 2000 *Appl. Phys. Lett.* **77** 3015−7
[3] He X, Zhao X, Li Y and Sui X 2009 *J. Mater. Res.* **24** 2201
[4] Wei G, Zhou H, Liu Z, Song Y, Wang L, Sun L and Li Z 2005 *J. Phys. Chem.* B **109** 8738−43
[5] Hong B H, Bae S C, Lee W, Jeong S and Kim K S 2001 *Science.* **294** 348−351
[6] Zhou Y and Hu A 2011. *Open Surf. Sci. J.* **3** 32−41
[7] Fang Y R, Wei H, Hao F, Nordlander P and Xu H X 2009 *Nano Lett.* **9** 2049–53
[8] Johannes A, Noack S, Wesch W, Glaser M, Lugstein A and Ronning C 2015 *Nano Lett.* **15** 3800-7
[9] Langley D P, Lagrange M, Giusti G, Jiménez C, Bréchet Y, Nguyen N D and Bellet D 2014 *Nanoscale.* **6** 13535-43
[10] Ni C J and Hong F C N 2014 *RSC Adv.* **4** 40330-8
[11] Yang L J, Cui J, Wang Y, Hou C J, Xie H, Mei X S, Wang W J and Wang K D 2015 *RSC Adv.* **5** 56677
[12] Kang X, Rui G Z and Ning G 2009 *Chinese Chemical Letters.* **20** 241-4

AOM2015   IOP Publishing

Journal of Physics: Conference Series **680** (2016) 012029   doi:10.1088/1742-6596/680/1/012029

Optical microfiber-based photonic crystal cavity

Yi-zhi Sun[1,3], Yang Yu[2], Hui-lan Liu[1], Zhi-yuan Li[2] and Wei Ding[2,3]

[1]School of Instrumentation Science and Opto-electronics Engineering, Beihang University, Beijing 100191, China

[2]Laboratory of Optical Physics, Institute of Physics, Chinese Academy of Sciences, Beijing 100190, China

E-mail: 375816019@qq.com, wding@iphy.ac.cn

Abstract. Using a focused ion beam milling technique, we fabricate broad stop band (~10% wide) photonic crystal (PhC) cavities in adiabatically-tapered silica fibers. Abrupt structural design of PhC mirrors efficiently reduces radiation loss, increasing the cavity finesse to ~7.5. Further experiments and simulations verify that the remaining loss is mainly due to Ga ion implantation. Such a microfiber PhC cavity probably has potentials in many light-matter interaction applications.

1. Introduction

Efficient coupling of light to subwavelength structures (either quantum emitters or dielectric particles) recently attracts great interests in fields ranging from quantum information science [1], to nonlinear optics [2], and to biochemical sensing [3]. Tightly confining light to small volume, as various optical microcavities pursue [4], is the key issue. In such a context, a tapered silica fiber enclosed by two mirrors exhibits great virtues because of moderate refractive index of silica and fabrication superiority of fiber taper. Firstly, the refractive index of silica is low enough for allowing intense evanescent field to spread out across silica boundary. Such a tapered fiber is usually referred to as micro/nano-fiber [5]. Secondly, refractive index of silica is also large enough not only for providing tight transverse field confinement but also for shortening cavity length in the longitudinal direction. For the former, when the guided mode of a tapered fiber is shrunk to the order of wavelength square, efficient light-particle interaction inside a cavity does not rely on ultra-high-finesse mirrors any more [6]. For the latter, if the index contrast between silica and air is efficiently exploited, rather than being shallowly utilized [7], penetration depth in the photonic crystal (PhC) mirror in a microfiber can be substantially shortened with the increase of its stop band [8]. Thirdly, thanks to the maturity of fiber tapering technique, the linkage between microfiber cavity and external optical network can be created conveniently and with low loss. Surface of silica fiber keeps atomically smooth during fiber tapering, and adiabatic taper transition can efficiently eliminate in-coupling and out-coupling losses. Lastly but not least, different from other on-chip photonic wires [9], a tapered fiber can maintain free-suspension configuration, which is flexible for many practical applications.

Achieving all these merits in a single device is a big challenge because of both optical and mechanical reasons. Compared with higher-index semiconductor waveguides [10], PhC structures in a silica fiber taper will produce much higher radiation loss, undermining cavity finesse. To elucidate this

---

[3] To whom any correspondence should be addressed.

Content from this work may be used under the terms of the Creative Commons Attribution 3.0 licence. Any further distribution of this work must maintain attribution to the author(s) and the title of the work, journal citation and DOI.

Published under licence by IOP Publishing Ltd

radiation generation process, we had developed an analytical model and found that significant part of radiation light could be suppressed and recycled via an interferometric means [11]. This radiation suppression effect was later verified in experiment [8] and inspired us to envision deploying high-reflection, short-penetration-depth microfiber PhC mirrors on both ends of a central nanofiber cavity. Our experiment also verifies that a micron-meter-sized silica fiber has sufficient mechanical strength to survive after being perforated, which is not feasible in a nanometer-sized fiber. Therefore, in an ideal situation, the envisaged fiber taper cavity has no mechanical problem and will simultaneously possess the following 4 advantages: substantial evanescent field spread outside the central nanofiber; two short but adiabatic taper transition sections connecting the central nanofiber and the extremity microfibers; tightly compressed cavity mode volume in both lateral and longitudinal directions; sufficient mechanical strength for free-suspension configuration. In contrast, two recently reported nanofiber microcavities cannot eliminate long penetration tails in their mirror sections (either nanofiber Bragg grating [12] or conventional fiber Bragg grating [13]). Intrinsically limited by mechanical strength and UV-photosensitivity technique, broad band PhC mirror in a nanofiber or in a conventional fiber is not feasible, disabling small mode volumes in such fiber taper cavities. In another work, although authors exhibit the minimal mode volume in a microfiber cavity, the fiber was buried in a layer of polymer for mechanical support [14]. As a result, the free-suspension configuration is lost.

In this work, ignoring the central nanofiber cavity temporarily, we focus our study on the design, fabrication, and characterization of PhC cavity in a freely-suspended microfiber. Investigations are expected to answer the possibility of the scheme proposed above, to test the fabrication technique, and to elucidate physical origins of the problems appearing in practical implementations.

2. Design of microfiber PhC cavity

A microfiber PhC cavity is schematically illustrated in Figure 1(a). We design a resonance at $\lambda_0 = 1.53$ μm, and the fiber diameter is set to be 1.7 μm. Without specially mentioned, in the following, we fix the refractive index of silica to be 1.444 with no dispersion. Firstly, the PhC mirror sections can be split to individual microfiber segment as shown in the inset of Figure 1(a). In each segment, an etched rectangular through-hole with the dimension of $t \times 1.2$ μm lies in the middle, longitudinal pitch of the microfiber segment is Λ, and Bloch boundaries [the green planes in Figure 1(a)] are applied on the two facets. We allow the two innermost microfiber segments to vary with their parameters t and Λ. The whole PhC cavity is symmetric with respect to its center. Secondly, using Bloch boundary conditions, we quickly calculate wavevector diagram of each microfiber segment as Figure 1(b) shows, in which a stop band appears under the air light cone (the pink line), and the middle of the stop band is kept at λ_0 by tuning the pitch Λ. Figure 1(c) plots the variations of the stop band and the pitch length as a function of the etch hole thickness t. Thirdly, choosing parameters of Figure 1(c), we can compose a whole cavity structure with $N = 9$, $t3 = 300$ nm, and $L_{cav} = 50$ μm, leaving $t1$ and $t2$ varied. All the geometrical parameters and the polarization direction have been defined in Figure 1(a), and our previous paper [11] has described this design procedure in detail.

Using finite-difference time-domain (FDTD) simulation with excitation of an internal dipole source, we obtain quality factor of the cavity (defined as $Q = \omega_0 \dfrac{\text{Energy stored}}{\text{Power loss}}$) as a function of $t1$ and $t2$. The maximum Q-factor appears at $t1/t2 \approx 200/100$ nm, which represents a reversely varied structure, rather than a gradually varied structure ($t1 < t2$) which is predicted by conventional impedance-matching picture [15]. It is therefore manifest that an interferometric effect plays an important role in the radiation generation and suppression process [11]. More importantly, simulation shows a finesse value of ~80 (defined as $\Im = Q \dfrac{\text{Free Spectral Range}}{\text{Cavity Frequency}}$), which is believed enough for many cavity quantum electrodynamics applications [6].

AOM2015   IOP Publishing

Journal of Physics: Conference Series **680** (2016) 012029   doi:10.1088/1742-6596/680/1/012029

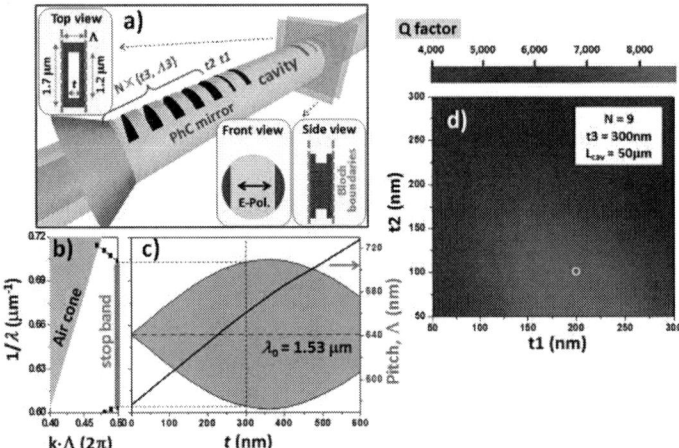

Figure 1. (a) Illustration of a microfiber PhC cavity. Individual microfiber segment is defined in the inset. (b) Wavevector diagram of one microfiber segment under Bloch boundary condition. The electric polarization is denoted in (a). (c) Stop band and pitch length as a function of the hole thickness. (d) Simulated Q-factor with variations of *t1* and *t2*.

To verify above results, we carry out another FDTD simulation by sending an optical pulse into the cavity [the red arrow in Figure 1(a)]. Normalized transmission and reflection spectra are presented in Figure 2(a). Q-factor can be read out from the spectra, $Q = \lambda_0 / \Delta\lambda$, agreeing well with the result in Figure 1(d), where $\Delta\lambda$ is the full width at half-maximum. Figure 2(a) also shows the reflectance spectra of the single PhC mirror at the two incidence directions. At the resonant wavelength 1.53 μm, a higher reflectance is observed when radiation suppression effect of PhC mirror acts. Additionally, Figure 2(b) compares the simulated spectra at the two incidence polarizations, indicating that our microfiber PhC cavity has small polarization dependence in terms of resonance wavelength. In contrast, the works based on nanofiber cavity [12, 13] exhibit much greater polarization dependence.

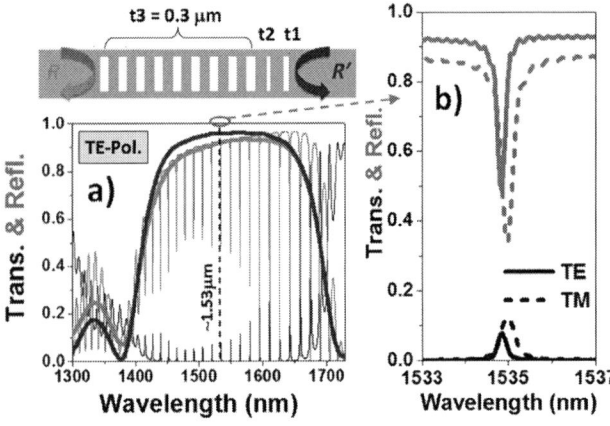

Figure 2. (a) Simulated transmittance and reflectance spectra of a microfiber PhC cavity with *t1*= 200 nm and *t2*=100 nm. Reflectance's of the single PhC mirror at different incidence directions are also shown. (b) Comparison of the reflectance spectra of the cavity at the two different polarizations.

3. Fabrication and characterization

3.1. Long cavity

In order to fabricate above devices, we adopt adiabatic fiber tapering and focused Ga ion beam (FIB) milling techniques. A scanning electron microscope (SEM) image of a microfiber PhC mirror is shown in the inset of Figure 3(a). Microfiber was firstly gently adsorbed on a flat silicon wafer via Van der Waals bonding, which allowed fixture of fiber during focused ion beam milling and detachment of fiber after fabrication. As a conducting substrate, the n-type silicon wafer also undertook the role of charge leakage. Our microfiber-based fabrication can be implemented in normal

152

optical room environment. In contrast, maintaining extremely clean environment is essential in nanofiber fabrication. Thanks to the dust-contamination resistance capability, our subsequent spectrum measurement to a microfiber device could be more reliable and repeatable.

To measure the transmission and reflection, a broadband fiber-based superluminescence light emitting diodes (SLED), an in-line optical circulator, and an optical spectrum analyzer were used. As shown in Fig. 3(b), we splice a passive fiber depolarizer [16, 17] after SLED source to eliminate polarization fluctuation. We also tested using in-line polarizer and polarization controller in our setup, but it is hard to resolve different polarization states in experiment. Figure 3(c) shows the measured transmittance and reflectance with a spectral resolution of 0.2 nm under unpolarised condition. The spectra can be interpreted by a Fabry-Perot picture [15], which relates Q-factor with the mirror reflectance in the inner side, r, $Q = r^{1/2}(1-r)^{-1}k_0 n_g (L_{cav} + 2L_p)$, where n_g is the group index of the guided mode, L_p is the penetration depth into the mirrors, and Q can be read out from the dips in the reflection spectrum. All of n_g, L_{cav}, and L_p can be calculated from the geometrical parameters, the Q-factor is therefore only determined by r. On the other hand, the reflectance of the same PhC mirror in its outer side (labeled R) can be estimated from the envelope of the spectrum.

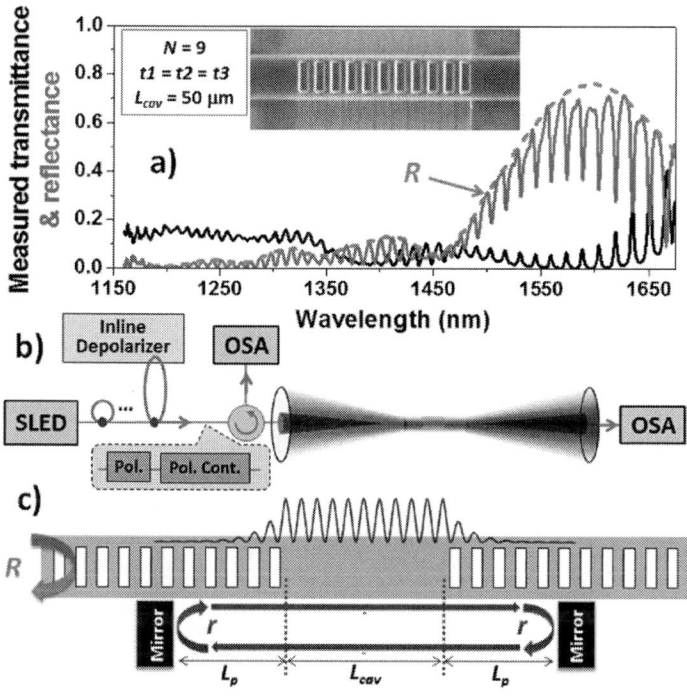

Figure 3. (a) Measured transmittance and reflectance spectra of a microfiber PhC cavity with L_{cav} = 50 μm and two PhC mirrors each having 11 etched holes (t = 300 nm). The pink dashed line is a guide to the eye and represents the reflectance envelope. Inset: SEM image of a PhC mirror. (b) Schematic of our measurement setup. (c) Fabry-Perot picture of the microfiber cavity.

Using the analysis method described above, we compare the inner and outer reflectances of a single PhC mirror without influence from fabrication variations between different devices. To test this idea, we fabricated three microfiber cavities with their geometries respectively illustrated in Figure 4. The Q-factors read out from the measured spectra are compared with the deduced Q-factors, which are derived from envelopes of reflectance. It is manifest that a gradually varied PhC mirror ($t1$ = 100 nm, $t2$ = 200 nm, $t3$ = 300 nm) can suppress radiation loss [Figure 4(b)], but, a reversely varied PhC mirror ($t1$ = 200 nm, $t2$ = 100 nm, $t3$ = 300 nm) suppress radiation more substantially [Figure 4(c)]. As a result, the inner reflectivity in Figure 4(c) is the maximum one because of the most efficient radiation suppression. All the comparisons agree well with the simulation of Figure 1(d) and can be interpreted by our theory [11]. Note that, in obtaining results of Figure 4, we insert the in-line polarizer and polarization controller in our measurement setup.

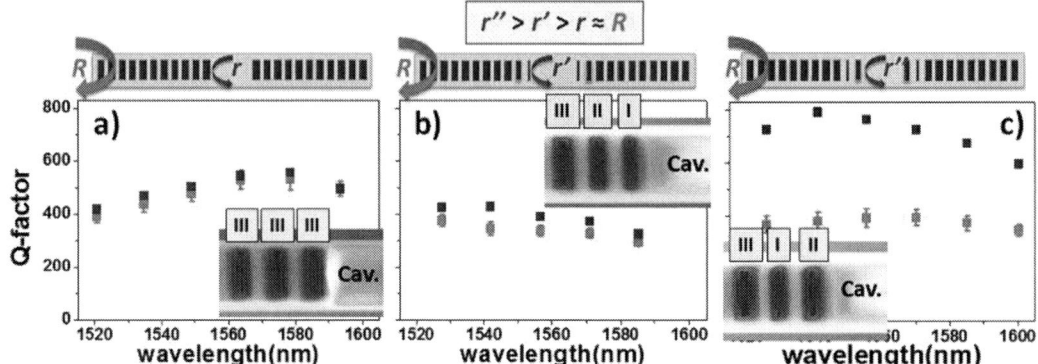

Figure 4. Measured Q-factors from reflection spectra (blue squares) and deduced Q-factors from envelopes of reflectance (pink squares) of three microfiber PhC cavities. Symbol I, II, III represent etched microfiber segments with t = 100/200/300 nm, respectively. SEM images and schematics of these three devices are depicted in the insets of (a), (b), (c).

However, our experimentally measured Q-factor (~800), hence cavity finesse (~7.5), is only one tenth of the theoretically predicted value. To find out physical origins of reminding cavity loss is our next task (see below).

3.2. Short cavity
In above analysis, we regard microfiber structure to be ideal, i.e. the geometry is symmetric and the material is pure silica. As we have pointed out [8], after FIB milling, the etched holes in microfiber cannot be vertical, and the inner walls have a tilt angle of ~9°. However, taking into account this nonideality in FDTD simulation cannot fully explain the newly added cavity loss. To elucidate this problem, we study a micron-meter long cavity whose spectral details should be manifested more clearly than a macroscale cavity because of much sparser spectral distribution. Figure 5(a) shows the SEM image at tilt and normal view angles, indicating a good fabrication with center-to-center distance of the center two etched holes of 1.02 μm. Transmittance and reflectance spectra as shown in Figure 5(b) are measured under nonpolarized condition. In simulation, we assume a non-zero imaginary part (κ) of refractive index for silica material. It is seen that, when κ is set to be 0.004 [Figure 5(d)], the simulated spectra fit well with the measured spectra with respect to both magnitude and shape. The Q-factor agrees with the experimental value as well. This result implies that silica material of microfiber has been modified during FIB milling.

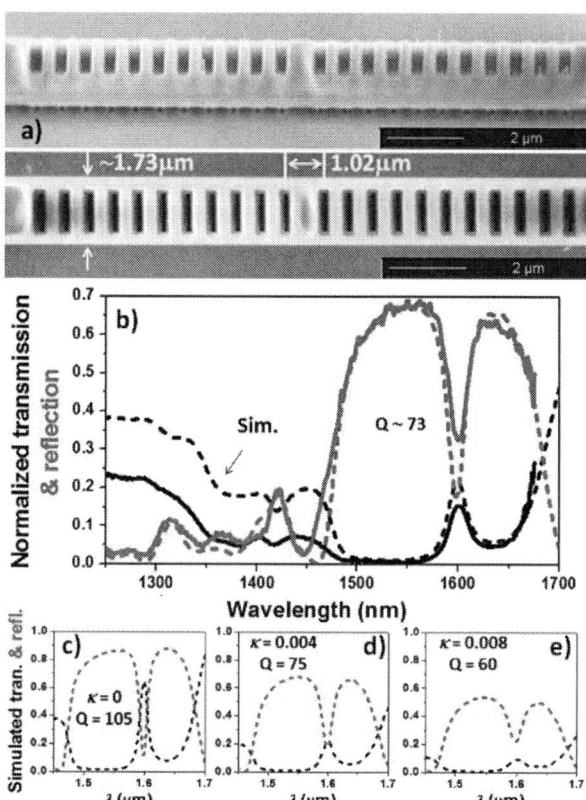

Figure 5. (a) SEM images of a microfiber PhC cavity with micron-meter cavity length. (b) Measured and simulated transmittance and reflectance spectra. (c), (d), (e) Simulations under assumptions of different imaginary parts of silica refractive index (κ), respectively. For both experiment and simulation, input sources are assumed to be unpolarised.

3.3. Ga ion contamination

It is known that Ga ion implantation in silica, which cannot be avoided in our fabrication, can cause significant loss. This may be the main origin of the above material loss. To test it, we expose different doses of Ga ions into bare microfibers, as illustrated in the inset of Figure 6, and then measure their insertion losses. Figure 6 shows an increased imaginary part of the effective modal index, $\kappa_{eff} = -\ln T \cdot \left(2k_0 L\right)^{-1}$, where T is the transmittance, L is the microfiber length having Ga ion implantation, and k_0 is the vacuum wavevector. We also present in Figure 6 the calculated k_{eff} based on the data in Reference [18]. Ga ions are assumed to concentrate in the surface layer of about 50 nm depth and overlap with the fundamental mode profile. All the curves in Figure 6 indicate that the inclusion of 0.004 into the imaginary part of silica refractive index in Figure 5 is reasonable and may be due to Ga ion contamination. We therefore conjecture that some newly appearing ion beam milling technology, such as Helium ion beam milling [19], may improve performance of microfiber PhC cavity in the future.

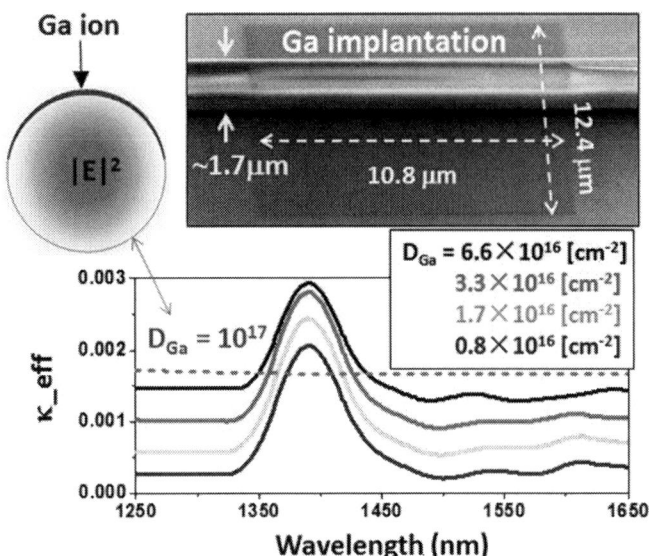

Figure 6. Measured imaginary part of the effective modal index of Ga ion contaminated microfiber. Different colours represent different Ga ion exposure doses. The peaks around 1.4 µm are due to water absorption. The grey curve is calculated based on the data in Reference [18] and fundamental modal profile.

4. Discussion and Conclusion

In summary, in an adiabatically-tapered optical fiber with diameter of a few microns, we achieve tight light confinement in PhC cavities. Our devices realize broad PhC stop bands ($\Delta\lambda/\lambda \sim 10\%$) and modest cavity finesse ($\Im \sim 7.5$) in the most popularly used optical waveguide, i.e. silica fiber. The inherent merits of silica tapered optical fibers for applications, e.g. seamless connection with conventional fiber network, excellent transparency in visible and near-infrared region, substantial evanescent field spread into environment, etc, make light confinement in microfiber of critical importance.

We present a PhC mirror design to efficiently suppress radiation loss. Also, we identify that the remaining loss in PhC cavity is the material loss caused by Ga ion implantation. Eliminating this loss may increase cavity finesse greatly. Our explorations to microfiber based PhC devices may inspire great number of interests in light-matter interaction research.

Acknowledgments

This work was supported by the National Natural Science Foundation of China (No. 61275044, 61575218, and 11204366), the Instrument Developing Project of the Chinese Academy of Sciences (No. YZ201346).

5. References

[1] Claudon J, Bleuse J, Malik N S, Bazin M, Jaffrennou P, Gregersen N, Sauvan C, Lalanne P and Gérard J- M 2010 *Nature Photonics* **4** 174
[2] Belotti M, Galli M, Gerace D, Andreani L C, Guizzetti G, Md Zain A R, Johnson N P, Sorel M and De La Rue R M 2009 *Opt. Express* **18** 1450
[3] Foreman M R, Swaim J D, and Vollmer F 2015 *Adv. Opt. Photon.* **7** 632
[4] Vahala K J 2003 *Nature* **424** 839
[5] Wu X-Q and Tong L-M 2013 *Nanophotonics* **2** 407
[6] Le Kien F and Hakuta K 2009 *Phys. Rev. A* **80** 053826
[7] Ding W, Andrews S R and Maier S A 2007 *Opt. Lett.* **32** 2499
[8] Yu Y, Ding W, Gan L, Li Z-Y, Luo Q and Andrews S 2014 *Opt. Express* **22** 2528
[9] Gong Y and Vukovi J 2010 *Appl. Phys. Lett.* **96** 031107
[10] Foresi J S, Villeneuve P R, Ferrera J, Thoen E R, Steinmeyer G, Fan S, Joannopoulos J D, Kimerling L C, Smith H I and Ippen E P 1997 *Nature* **390** 143
[11] Ding W, Liu R-J and Li Z-Y 2012 *Opt. Express* **20** 28641

[12] Nayak K P, Le Kien F, Kawai Y, Hakuta K, Nakajima K, Miyazaki H T and Sugimoto Y 2011 *Opt. Express* **19** 14040
[13] Wuttke C, Becker M, Brückner S, Rothhardt M and Rauschenbeutel A 2012 *Opt. Lett.* **37** 1949
[14] Ding M, Wang P-F, Lee T and Brambilla G 2011 *Appl. Phys. Lett.* **99** 051105
[15] Lalanne P, Sauvan C and Hugonin J P 2008 *Laser & Photon. Rev.* **2** 514
[16] Shen P and Palais J C 1999 *Appl. Opt.* **38** 1686
[17] Wan F, Ding W and Wang Z-Y 2006 *Chinese J. Electron.* **15** 619
[18] Karge H and Mühle R 1992 *Nuclear Instruments & Methods in Physics Research B* **65** 380
[19] Morgan J, Notte J, Hill R and Ward B 2006 *Microscopy Today* **14** 24

Identification of high explosive RDX using terahertz imaging and spectral fingerprints

Jia Liu[1,2], Wen-Hui Fan*[,1], Xu Chen[1,2] and Jun Xie[1,2]

[1] State Key Laboratory of Transient Optics and Photonics, Xi'an Institute of Optics and Precision Mechanics, Chinese Academy of Sciences, Xi'an 710119, China

[2] Graduate University of Chinese Academy of Sciences, Beijing 100049, China

E-mail: fanwh@opt.ac.cn

Abstract. We experimentally investigated the spectral fingerprints of high explosive cyclo-1,3,5-trimethylene-2,4,6-trinitramine (RDX) in terahertz frequency region. A home-made terahertz time-domain spectroscopy ranging from 0.2 THz~3.4 THz was deployed. Furthermore, two sample pellets (RDX pellet and polyethylene pellet), which were concealed in an opaque envelop, could be identified by using terahertz pulse imaging system. For the purpose of distinguishing the RDX between two pellets, we further calculated the THz frequency-domain map using its spectral fingerprints. It is demonstrated that the high explosive RDX could similarly be identified using terahertz frequency-domain imaging.

1. Introduction

Over the past several years, terrorist attacks have been substantially increasing worldwide. For example, mail-bombs, anthrax hidden in envelops and concealed metallic or non-metallic (ceramic, fiberglass, plastic) weapon threaten people's daily life. Counter-terrorist has become a significant mission of local government in every country. Hence, a rapid detection and effective identification are needed. Terahertz spectroscopy and imaging have been identified as very promising techniques in a wide area of security applications, such as chemical identification [1-2], imaging of concealed weapon [3-4], biological agent [5-6], and the detection of illicit drugs [7-8] and explosives [9-12].

Terahertz frequency radiation allows for non-destructive and non-invasive inspection of most opaque barrier materials, such as cardboard, wood, plastic, leathern, and textile [13-15]. Many solid-state explosives exhibit distinctive absorption characteristics in THz frequency [16-20]. Each explosive has its own spectral fingerprint, which is essential in the process of identifying the unknown target. These spectral fingerprints arise from the intramolecular and intermolecular vibrational modes or photon modes of the materials.

This paper was experimentally investigated that the high explosive cyclo-1,3,5-trimethylene-2,4,6-trinitramine (RDX) could be identified using terahertz imaging and spectral fingerprints. A home-made terahertz time-domain spectroscopy (THz-TDS) was deployed to measure the spectral fingerprints of RDX in the range from 0.2THz to 3.4THz. It could be clearly observed that the RDX

Content from this work may be used under the terms of the Creative Commons Attribution 3.0 licence. Any further distribution of this work must maintain attribution to the author(s) and the title of the work, journal citation and DOI.

Published under licence by IOP Publishing Ltd

has the absorption features at 0.80THz, 1.05THz, 1.30 and 1.91THz respectively. Furthermore, a sample consisting of the RDX and high density polyethylene (PE) pellets, which are mounted side by side in an opaque envelop, was imaged using terahertz pulse imaging system. For the purpose of identifying the high explosive RDX pellet, we further calculated the THz frequency-domain map using its spectral fingerprints. It was well demonstrated that the high explosive RDX could be distinguished between the two pellets by using terahertz frequency-domain imaging.

2. Experimental setup

A home-made terahertz time-domain spectroscopy (THz-TDS) we deployed in experiments is shown in Figure 1. THz pulses are generated by the Ti: Sapphire oscillator (Spectra Physics) with pulse duration of 20fs, repetition rate of 76MHz, and average power of 600 mW. The laser beam is splited into two beams by a splitter, one for exciting the 0.5mm thick low temperature-grown GaAs photoconductive antenna (emitter), and the other for measuring the THz signal at 1mm thick <110> ZnTe crystal (detector). The emitted THz pulses are collimated and focused by a pair of off-axis parabolic mirrors. The sample is placed right at the THz focus point, perpendicular to the incident THz beam. The transmitted THz beam is collected and focused by using the other pair of off-axis parabolic mirrors onto ZnTe crystal, in which the probe beam detected the THz field by electro-optic sampling (EOS) [21]. To avoid the absorption of water vapor, the THz radiation region is purged by dry air. The THz spectrum without passing through samples is obtained by applying the fast Fourier transform to THz waveform in the ranging of 0.2 THz~3.4 THz, which is shown in Figure 2.

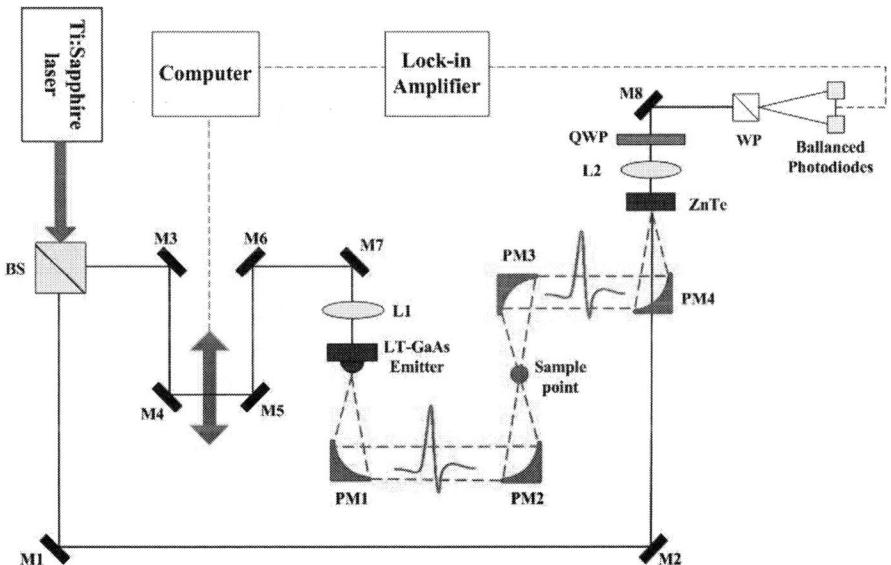

Figure 1. The schematic setup of home-made THz-TDS; BS is beam splitter, M1-M8 is optical mirror, PM1-PM4 is off-axis parabolic mirror, L1-L2 is optical lens, QWP is quarter wave plate, and WP is Wollaston prism.

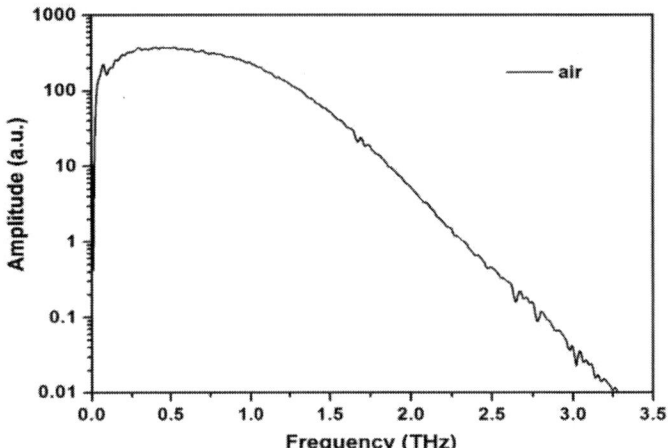

Figure 2. The THz spectrum without passing through samples in dry air.

3. Terahertz Spectra

3.1 Theoretical background

Unlike others spectroscopy, THz-TDS can use to measure the spectra of the explosive related compounds because a single measurement of THz electric field can provide both the amplitude and phase information. Hence, the THz-TDS is capable of obtaining the absorption coefficient and the refractive index without using Kramers-Kronig (K-K) relation [22]. The field of transmitted THz pulse is changed by the dispersion and absorption of sample. The amplitude transmittance of THz field can be described as Eq. (1) [23]:

$$T = \frac{\tilde{E}_{sam}(\upsilon)}{\tilde{E}_{ref}(\upsilon)} = Ae^{i\varphi} = \frac{4\tilde{n}(\upsilon)}{[1+\tilde{n}(\upsilon)]^2} \cdot \left\{ 1 + \frac{[\tilde{n}(\upsilon)-1]^2}{[\tilde{n}(\upsilon)+1]^2} e^{i2\tilde{n}(\upsilon)\upsilon d/c} \right\}^{-1} e^{i2\pi[\tilde{n}(\upsilon)-1]\upsilon d/c} \tag{1}$$

Where $\tilde{E}_{sam}(\upsilon)$ and $\tilde{E}_{ref}(\upsilon)$ are the THz transmitted field and the incident field respectively; A is the amplitude of THz transmitted field; φ is the phase difference between sample and reference waveform; $\tilde{n}(\upsilon)=n(\upsilon)+i\kappa(\upsilon)$ is the complex refractive index; d is the thickness of sample; υ is the frequency of THz field radiation; $\kappa(\upsilon)$ is the extinction coefficient; c is the speed of light in vacuum. In the condition of $\kappa(\upsilon) << n(\upsilon)$, the multiple internal refraction and the Fabry-Perot effect can be neglected. Hence the refractive index $n(\upsilon)$ and extinction coefficient $\kappa(\upsilon)$ can be obtained after a single measurement [24]:

$$n(\upsilon) = \frac{c\varphi}{2\pi d\upsilon} + 1 \tag{2}$$

$$\kappa(\upsilon) = \frac{c}{2\pi d\upsilon} \ln(\frac{4n}{A(1+n)^2}) \tag{3}$$

Refer to Eq. (2) and Eq. (3), the absorption coefficient can be calculated as Eq. (4):

$$\alpha(\upsilon) = \frac{4\pi\upsilon\kappa(\upsilon)}{c} = \frac{2}{d}\ln\left\{\frac{4n(\upsilon)}{A[1+n(\upsilon)]^2}\right\} \tag{4}$$

Therefore, after measuring the THz transmitted field $\tilde{E}_{sam}(\upsilon)$, the incident field $\tilde{E}_{ref}(\upsilon)$ and the thickness of sample d, the refractive index and absorption coefficient can be obtained.

3.2 Sample preparation

The samples of the high explosive RDX (purity > 99%) used in experiments are powder samples. In order to eliminated the influence of scattering, the sample are crash into fine powders using mortar and pestle to reduce the particle size less than 80μm. Considering that the RDX is sensitive and unsafe to high pressure, we mix the RDX with high density polyethylene (PE) which is transparent in THz frequency region. The mixing ratio is about 1:5 (40mg RDX vs. 200mg PE). Then, the RDX/PE samples are compressed into pellets with the thickness of 1.5mm and diameter of 13mm under the pressure of 1 ton.

3.3 Results and discussions

As shown in Figure 3, the fast Fourier Transform (FFT) amplitudes of THz time-domain waveform transmitting through the RDX/PE (sample) and pure PE pellets (reference) were obtained. In order to eliminate the influence of the sample surface effects, we measured at least five different spot on the surface of pellet. The data of the spectra we obtained were the average of five spot with 1800 scans at each spot. It is clearly observed that the FFT amplitude of THz filed is dropped down dramatically due to the strong absorption of RDX, especially at about 0.8 THz.

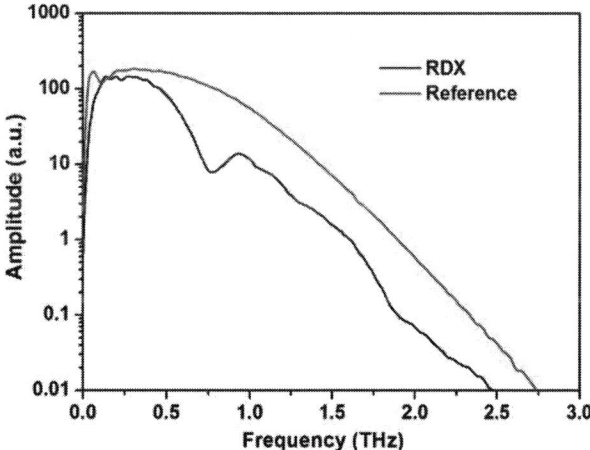

Figure 3. The spectra of THz field transmitted through the RDX/PE pellet and pure PE pellet (reference), respectively.

In order to calculate the absorption coefficient of RDX, we compressed the pure PE powder into pellets, which are taken as the reference. These PE pellets have the same size like the RDX/PE pellets. The absorption spectrum of RDX is calculated using Eq. (2), Eq. (3) and Eq. (4), which is plotted in

Figure 4. It is shown that the RDX has a strong absorption at 0.8 THz and three other relative weak absorptions at 1.05THz, 1.30THz, 1.91THz respectively. This result agrees well with the previous researches [25-26]. Hence, we can easily identify the high explosive RDX using THz spectra fingerprints.

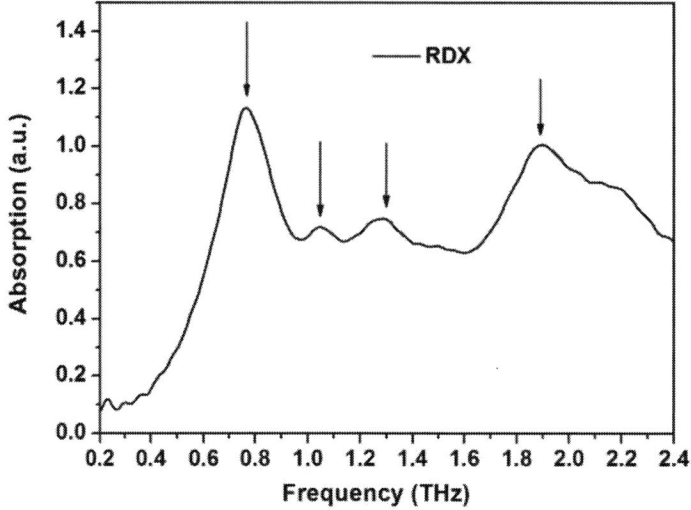

Figure 4. The absorption spectrum of RDX in THz frequency-domain.

4. Terahertz Imaging

Terahertz radiation can penetrate most of non-polarization materials, including cardboard, wood, plastic, leathern, and textile. Hence, terahertz imaging has been identified as a very promising technique in a wide area of security applications, especially for detecting explosive hidden in opaque package. Recently, terahertz imaging using peak-to-peak value of terahertz time-domain waveform is commonly used in nondestructive detection.

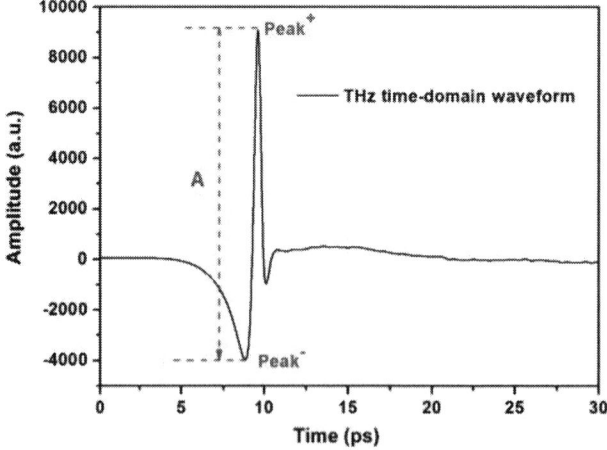

Figure 5. The terahertz time-domain waveform, where Peak[+] and Peak[-] are the maximum and minimum value of waveform.

Figure 5 shows the terahertz time-domain waveform obtained in experiments. The amplitude A of peak-to-peak value can be calculated by Eq. (5):

$$A = \left| Peak^+ - Peak^- \right| \tag{5}$$

As shown in Figure 6(a), the RDX/PE and pure PE pellets mounted side by side were concealed in an opaque envelop. Both the two pellets have the same size in geometry: 13mm diameter and 1.5mm thickness. Figure 6(b) show the measured terahertz map using peak-to-peak value of terahertz time-domain waveform. The sample was fixed at the THz radiation focus and the map was obtained by raster scanning the THz transmission. The scanned area was 30mm×18mm, which corresponds to 120 pixels×72 pixels with 0.25mm spacing. Each pixel of terahertz map used the calculated value A of terahertz time-domain waveform passing through the imaging object. It is clearly observed that the two "invisible" pellets concealed in envelop can be detected by terahertz time-domain imaging. The bright two circles in map are much caused by edge scattering effects [27].

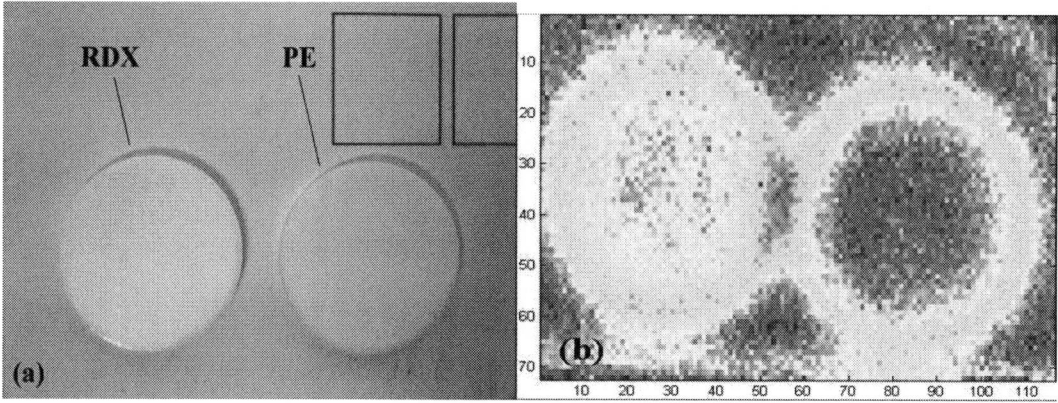

Figure 6. The image of samples concealed in envelop: (a) the photograph of RDX and PE pellets; (b) the false-color THz time-domain image with peak-to-peak amplitude.

From the false-color THz time-domain image alone, it was well demonstrated that the RDX and PE pellets concealed in opaque envelop could be detected by using THz time-domain imaging. However, it couldn't tell which one is the high explosive RDX or the safe PE pellet. For the purpose of identifying the RDX, we further imaged the sample in frequency-domain using its spectral fingerprints. We firstly Fourier transformed the time-domain waveform of each pixel in THz time-domain map. Here we took the THz waveform passing through the hollow envelop as the reference. Refer to Eq. (1)~Eq. (3), the refractive index $n(\upsilon)$ and extinction coefficient $\kappa(\upsilon)$ were calculated. Finally, the THz frequency-domain imaging using the absorption spectra could be obtained. Figure 7(a) shows the process chart of calculated absorption coefficient $\alpha_{i,j}$ responding to the grey scale of each pixel in THz map, and Figure 7(b) shows the calculated THz frequency-domain map using the absorption coefficient at $\upsilon=0.80$ THz. It is clear that only one pellet (left) in Figure 7 (b) can be observed and this circular area has a great absorption. Because high explosive RDX has a strong absorption at 0.8THz, while PE not. It was well demonstrated that the high explosive RDX could be easily distinguished between the two pellets by using the terahertz frequency-domain imaging.

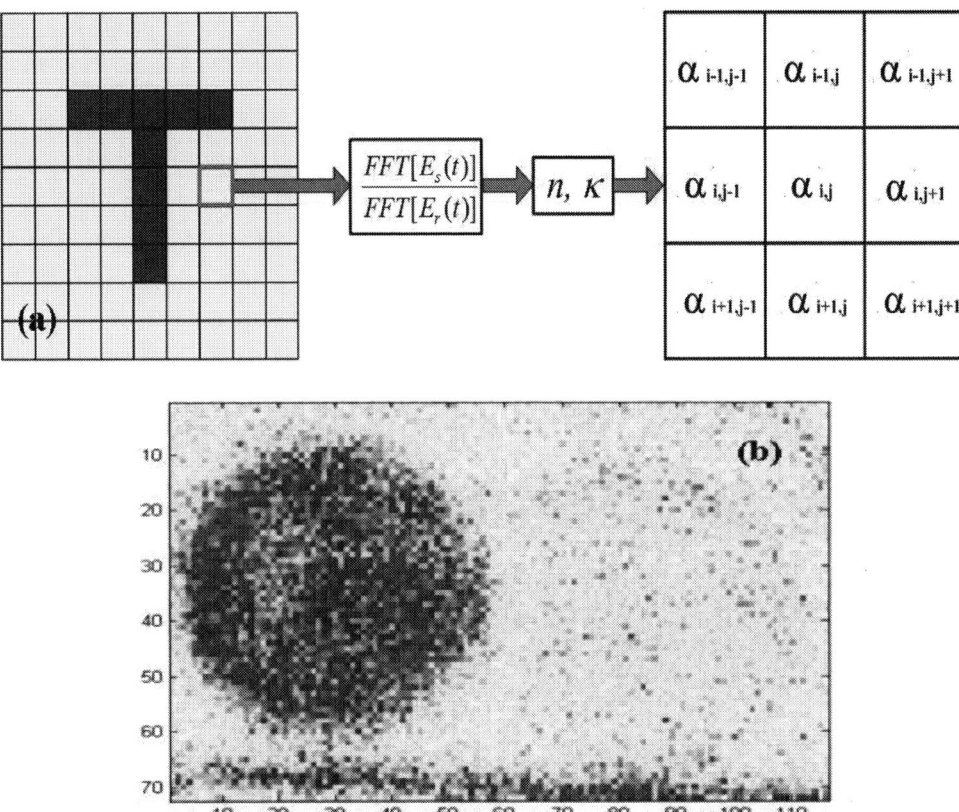

Figure 7. Terahertz frequency-domain imaging: (a) the process chart of calculated absorption coefficient $\alpha_{i,j}$ responding to the grey scale of each pixel in THz map; (b) THz frequency-domain map using the absorption coefficient at $\upsilon=0.80$ THz.

5. Conclusion

We experimentally investigated the spectra fingerprints of the high explosive RDX in terahertz frequency region. A home-made terahertz time-domain spectroscopy in the range of 0.2THz~3.4THz was deployed. It is clearly observed that the RDX has a strong absorption at 0.8 THz and three other relative weak absorptions at 1.05THz, 1.30, 1.91THz respectively. Furthermore, a sample consisting of the RDX/PE and pure PE pellets concealed in the opaque envelop, was imaged by using THz pulse system. For the purpose of distinguishing the RDX between the two pellets, we further calculated the THz frequency-domain map using its spectral fingerprints. It is clearly observed that only the RDX pellet can be observed at $\upsilon=0.80$ THz. This result well demonstrates that the high explosive RDX could similarly be identified by using the terahertz frequency-domain imaging.

6. References
[1]   Jackson J B, Mourou M, Whitaker J F, Duling III I N, Williamson S L, Menu M and Mourou G

A 2008 Terahertz imaging for non-destructive evaluation of mural paintings *Optics Communications* **281** 527

[2] Adam A J L, Planken P C M, Meloni S and Dik J 2009 TeraHertz imaging of hidden paint layers on canvas *Opt. Exp.* **17** 3407

[3] Zimdars D and White J S 2004 Terahertz reflection imaging for package and personnel inspection *Proc. of SPIE* **5411** 78

[4] Creeden D, McCarthy J C, Ketteridge P A, Schunemann P G, Southward T, Komiak J J and Chicklis E P 2007 Compact, high average power, fiber-pumped terahertz source for active real-time imaging of concealed objects *Opt. Exp.* **15** 6478

[5] Hoshina H, Hayashi A, Miyoshi N, Miyamaru F and Otani C 2009 Terahertz pulsed imaging of frozen biological tissues *Applied Physics Letters* **94** 123901

[6] Fischer B M, Hoffmann M, Helm H, Wilk R, Rutz F, Kleine-Ostmann T, Koch M and Jepsen P U 2005 Terahertz time-domain spectroscopy and imaging of artificial RNA *Opt. Exp.* **13** 5205

[7] Kawase K, Ogawa Y and Watanabe Y 2003 Non-destructive terahertz imaging of illicit drugs using spectral fingerprints *Opt. Exp.* **11** 2549

[8] Lu M H, Shen J L, Li N, Zhang Y, Zhang C L, Liang L S and Xu X Y 2006 Detection and identification of illicit drugs using terahertz imaging *Journal of Applied Physics* **100** 103104

[9] Davies A G, Burnett A D, Fan W H, Linfield E H and Cunningham J E 2008 Terahertz spectroscopy of explosives and drugs *Materials Today* **11** 18

[10] Fan W H, Burnett A, Upadhya P C, Cunningham J E, Linfield E H and Davies A G 2007 Far-Infrared Spectroscopic Characterization of Explosives for Security Applications Using Broadband Terahertz Time-Domain Spectroscopy *Applied Spectroscopy* **61** 638

[11] Yamamoto K, Yamaguchi M, Miyamaru F, Tani M, Hangyo M, Ikeda T, Matsushita A, Koide K, Tatsuno M and Minami Y 2004 Noninvasive inspection of C-4 explosive in mails by terahertz time-domain spectroscopy *Jpn. J. Appl. Phys.* **43** 414

[12] Shen Y C, Lo T, Taday P F, Cole B E, Tribe W R and Kemp M C 2005 Detection and identification of explosives using terahertz pulsed spectroscopic imaging *Appl. Phys. Lett.* **86** 241116

[13] Bjarnason J E, Chan T L J, Lee A W M, Celis M A and Brown E R 2004 Millimeter-wave, terahertz, and mid-infrared transmission through common clothing 2004 *Applied Physics Letters* **58** 519

[14] Morita Y, Dobroiu A, Kawase K and Otani C 2005 Terahertz technique for detection of microleaks in the seal of flexible plastic packages *Optical Engineering* **44** 019001

[15] Dunayevskiy I, Bortnik B, Geary K, Lombardo R, Jack M and Fetterman H 2007 Millimeter- and submillimeter-wave characterization of various fabrics *Applied Optics* **46** 6161

[16] Liu H B, Chen Y Q, Bastiaans G J and Zhang X C 2006 Detection and identification of explosive RDX by THz diffuse reflection spectroscopy *Opt. Exp.* **14** 415

[17] Hooper J, Mitchell E, Konek C and Wilkinson J 2009 Terahertz optical properties of the high explosive b-HMX *Chemical Physics Letters* **467** 309

[18] Konek C, Wilkinson J, Esenturk O, Heilweil E and Kemp M 2009 Terahertz Spectroscopy of Explosives and Simulants – RDX, PETN, Sugar and L-Tartaric Acid *Proc. of SPIE* **7311**

73110K

[19] Allis D G, Hakey P M and Korter T M 2008 The solid-state terahertz spectrum of MDMA (Ecstasy)–A unique test for molecular modeling assignments *Chemical Physics Letters* **463** 353

[20] Chen J, Chen Y Q, Zhao H W, Bastiaans G J and Zhang X C 2007 Absorption coefficients of selected explosives and related compounds in the range of 0.1-2.8 THz *Opt. Exp.* **15** 12060

[21] Wu Q, Litz M and Zhang X C 1996 Broadband detection capability of ZnTe electro-optic field detectors *Appl. Phys. Lett.* **68** 2924

[22] Chen Y, Liu H, Deng Y, Schauki D, Fitch M J, Osiander R, Dodson C, Spicer J B, Shur M and Zhang X C 2004 THz spectroscopic investigation of 2,4-dinitrotoluene *Chem. Phys. Lett.* **400** 357

[23] Born M and Wolf E 1964 *Principles of Optics, 2ed* (Oxford: Pergamon Press)

[24] Hu Y, Huang P, Guo L T, Wang X H and Zhang C L 2006 Terahertz spectroscopic investigations of explosives *Physics Letters A* **359** 728

[25] Fitch M J, Leahy-Hoppa M R, Ott E W and Osiander R 2007 Molecular absorption cross-section and absolute absorptivity in the THz frequency range for the explosives TNT, RDX, HMX, and PETN *Chem. Phys. Lett.* **443** 284

[26] Huang F, Schulkin B, Altan H, Federici J F, Gary D, Barat R, Zimdars D, Chen M H and Tanner D B 2004 Terahertz study of 1,3,5-trinitro-s-triazine by time-domain and Fourier transform infrared spectroscopy *Applied Physics Letters* **85** 55535

[27] Zhang Z W, Zhang Y, Zhao G Z and Zhang C L 2007 Terahertz time-domain spectroscopy for explosive imaging *Optik* **118** 325

Acknowledgements

This work was supported by the Hundred Talent Program of the Chinese Academy of Sciences (Grant No. J08-029), the Main Direction Program of Knowledge Innovation of the Chinese Academy of Sciences (Grant No. YYYJ-1123-4), the National High Technology Research and Development Program of China (Grant No. 2011AAxxx2008A), the National Basic Research Program of China (Grant No. 2007CB310405) and the CAS/SAFEA International Partnership Program for Creative Research Teams.

Micro-nano Structurized Gold Chip for SPR Imaging Sensor

Bing Zhang[1], Kai Pang[1], Chunfei Shi[1], Yi Sun[1], Wei Dong[2,*] and Xiaoping Wang[1,*]

[1]State Key Laboratory of Modern Optical Instrumentation, College of Optical Science and Engineering, Zhejiang University, Hangzhou, China
[2]Center for Bioelectronics and Biosensors, Biodesign Institute, Arizona State University, Tempe, USA

xpwang@zju.edu.cn and starkiller2006@163.com

Abstract A micro-nano structurized gold chip was developed and applied to a surface plasmon resonance imaging (SPRi) sensor with polarization contrast method. Compared with the planar gold film, a total sensitivity enhancement (SEF=287%) was obtained.

1. Introduction

The demand for rapid high-throughput chemical sensor and biosensor technologies is increasing in many important areas such as life science, environment monitoring, and food safety. Due to the ability of multi-analyte interrogation and the high-throughput applications, surface plasmon resonance imaging (SPRi) has gained considerable interest in these areas. The general implementation for SPRi is based on analysis of the distribution of light intensity reflected from an SPR surface containing multiple sensing arrays. However, the main limitation of that is its poor refractive index resolution, which is lower than conventional angular or phase modulation SPR systems. So improving the performance of these sensors became particularly important. In recent years, a polarization contrast method for high-throughput SPR sensing has been developed. It converts changes of light polarization (phase and amplitude) into changes of light intensity, which was measured directly and thus it gains the same level of sensitivity as the phase modulation method.

In order to further improve the performance of the SPRi sensor, we proposed a micro-nano structurized gold chip for the SPRi sensor operated with polarization contrast method. The key innovations have two aspects. One is that a microwell structure is used to confine surface plasmon wave (SPW) inside the well, leading to a preliminary enhancement of sensitivity. The other is that a nano-gratings structure is designed inside the microwell to generate enhanced electric field due to localized surface plasmon resonance (LSPR) and SPR coupling, realizing a second enhancement of sensitivity. Bulk refractometric experiments are implemented to establish the real function of the micro-nano structure, and performance of the SPRi sensor is evaluated.

2. Micro-nano Structurized Gold Chip

2.1. Numerical Simulations

Numerical simulations were performed using finite-difference time-domain (FDTD) method. Non-uniform meshings were implemented during simulations. Models of microwell structure and nano-gratings structure are presented in Fig. 1. Both of the structures were illuminated with an incident plane wave ($\lambda = 633$ nm, the incident angle $\theta 1 = 75°$). The optical parameters of gold are derived from literature, and prism material is set as BK7 (n = 1.515).

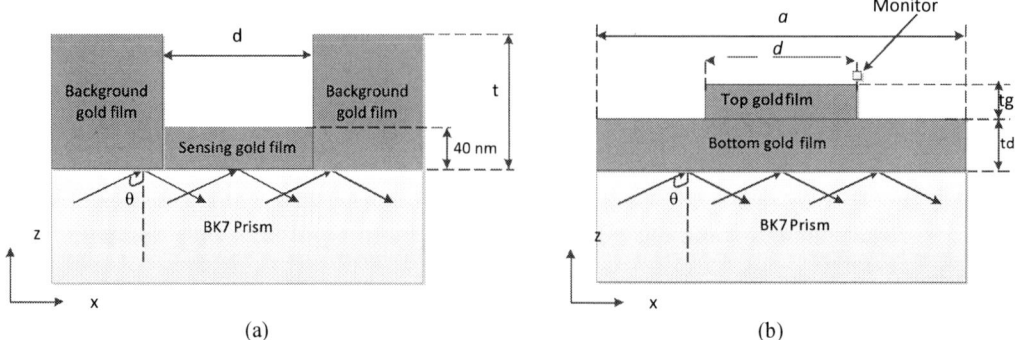

Figure 1. FDTD models of (a) microwell structure and (b) nano gratings structure

Figure 1(a) shows the xz cross-section of one circular sensing spot of the microwell structure. According to the results, microwell with a diameter of d = 10 μm was simulated in Perfectly Matched Layer (PML) absorbing boundary conditions. The thickness of the sensing gold film is 40 nm, and the optimal thickness for the gold film in background area t = 150 nm, considering difficulties in the fabrication of thicker patterns.

Figure 1(b) shows the xz cross-section of one period of the nano-gratings structure. This is a 2D model which is infinite in y axis since the length of the gratings is much larger than the period. It has a thin bottom gold film covered by a thin top gold film. The structure has parameters of period a, width d, thickness of the bottom film td and thickness of the top film tg. Due to its periodic property, the model was simulated in "Periodic" absorbing boundary conditions in x direction. When the structure parameters are a = 200 nm, d = 100 nm, td = 30 nm and tg = 10 nm, there is a resonant peak at 635 nm, which coincides with the working wavelength of the SPRi sensor.

2.2. Fabrication of micro-nano structurized chip

Fabrication of microwell structure was realized by photolithography and two-layer deposition of gold film. The thick background film was firstly deposited on a thin, polished BK7 glass substrate by magnetron sputtering of gold with a 2-nm-thick chromium layer as an adhesion layer followed by a photolithography process. After etching the unprotected thick gold film, and then lifting off the photoresist using acetone, the microwell patterns were obtained. The exposed glass surface was cleaned by plasma, and then a thin gold film as the SPR-active layer and another 2-nm-thick chromium layer as an adhesion layer were sputtered onto the thick patterned gold film to finally form the microwell structure.

Nano-gratings structure was fabricated just inside the microwell structure by etching the thin sensing gold film using a focused-ion-beam (FIB) system (30 K eV Ga ions, resolution 5 nm). Periodic hollows were obtained by etching and the top gold films were formed, leaving the rest as the bottom gold film.

3. Experimental and setup

A home-built SPRi sensor comprising of a 635 nm LED light source and a 12-bit CCD with a resolution of 1392 × 1040 pixels as the detector, was established based on the polarization contrast method. The LED light was first collimated, and then passed through a polarizer. The transmitted light contains both p (vector of electric intensity parallel to the plane of incidence) and s (perpendicular to the plane of incidence) polarizations. It illuminated a right-angle BK7 prism with an incident angle of 75 degree. After reflected by the sensing surface, it passed through a λ/4 waveplate and an analyzer, and was finally captured by the CCD camera. The micro-nano structurized gold chip was mounted on the surface of prism by matching liquid (n = 1.515). A 350 uL flow cell was attached on the chip.

To study the refractive index sensitivity of the micro-nano structurized chip, a series of NaCl (0, 20, 50, 100, 150, and 200 g/L) solutions were prepared and measured. The refractive indices of these solutions cover the range of 1.3315–1.3632 RIU, which had been previously measured by an Abbe refractometer. Before measurement, deionized water was pumped through the flow-cell to the sensor surface until a stable baseline was established. Liquid samples were then injected sequentially with a loop of water-NaCl-water at a controllable flow rate. Each concentration was injected thrice to reduce measurement error.

4. Results and discussion

The sensitivity enhancement factor (SEF) to evaluate the improvement of sensitivity, which is defined as the ratio of sensitivity of the micro-nano structure to that of the reference chip, i.e. SEF = $\Delta I_{micro-nano} / \Delta I_{reference}$.

4.1. Microwell chip

The gold microwell chip was characterized by high sensitivity to bulk refractive index changes in NaCl solution. Figure 2 depicts sensor response to sequential concentration changes of NaCl (0–200 g/L) from the sensing spots.

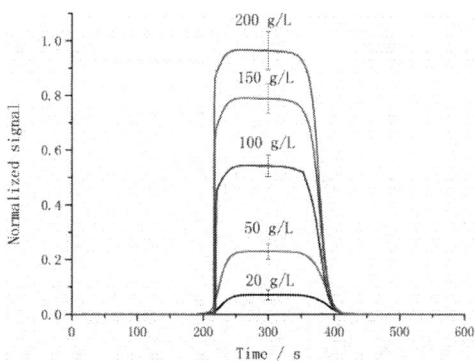

Figure 2. Sensor response to NaCl (0 – 200 g/L)

The signal value was normalized by the intensity of air to eliminate the influence of light intensity fluctuations. Using the refractive indices obtained from the Abbe refractometer as horizontal coordinates, a refractive index calibration curve was established, and linear fit of the curve can be achieved. Sensitivity can be obtained as the slope of the linear curve, which was 3521 %/RIU. The responses of NaCl solutions from the reference chip were also recorded for comparing, and the calculated sensitivity of the reference chip is 2237 %/RIU. So the SEF here is 157%.

4.2. Nano gratings chip

Sensitivities of gold microwell spots with nano gratings and without nano gratings in detection of NaCl are compared. Linear fits of the curves are implemented within the low refractive index range (n=1.3315–1.3394 RIU, corresponding to 0–50 g/L of NaCl), which is shown in Figure 3.

Sensitivity of microwell spot without nano gratings is 2526 %/RIU, while with nano gratings is 4619 %/RIU (SEF=183%). Combined with the afore obtained sensitivity enhancement of microwell to planar gold film (SEF = 157%), a total enhancement of the micro-nano chip can be calculated by SEFtotal = 157%×183% = 287%.

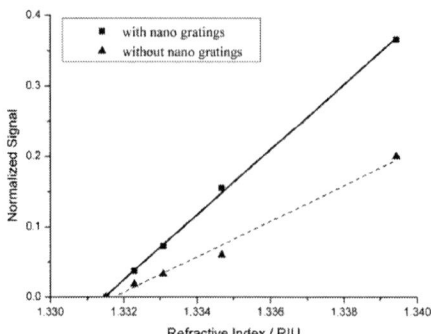

Figure 3. Linear fit of the microwell spots with nano gratings

5. Conclusion

A micro-nano structurized gold chip was developed and applied to a SPRi sensor with polarization contrast method. A total sensitivity enhancement (SEF = 287%) was obtained. These improvements are attributed to: (1) sensitivity enhancement (SEF = 157%) due to the electric field enhancement in the microwell caused by SPW interference; and (2) a second sensitivity enhancement (SEF = 183%) due to LSPR-SPR coupling in the nano gratings structure.

Acknowledgment

This research was financially supported by the National Natural Science Foundation of China (NSFC) (No.61036012 and No.21277118).

References

[1]   Piliarik M et al 2007 Novel polarization control for high-throughput surface plasmon resonance sensors *Optical Sensing Technology and Applications* 6585

[2]   Huang Z et al 2012 Contrast-enhancing polarization control method for surface plasmon imaging sensor Opt. Eng. 51

[3]   Law W C et al 2007 Wide dynamic range phase-sensitive surface plasmon resonance biosensor based on measuring themodulation harmonics Biosens. Bioelectron 23 627-632

[4]   Zeng S W et al 2013 Size dependence of Au NP-enhanced surface plasmon resonance based on differential phase measurement Sensor. Actuat. B-Chem 176 1128-1133

[5]   Zeng S W et al 2014 Nanomaterials enhanced surface plasmon resonance for biological and chemical sensing applications Chem. Soc. Rev. 43 3426-3452

[6]   Im H. et al 2014 Label-free detection and molecular profiling of exosomes with a nano-plasmonic sensor Nat. Biotechnol. 32 490-495

[7]   Dong W et al 2015 Improved polarization contrast method for surface plasmon resonance imaging sensors by inert background gold film extinction Opt. Commun. 346 1-9

[8]   Kim K et al 2009 Localized surface plasmon resonance detection of layered biointeractions on metallic subwavelength nanogratings Nanotechnology 20 315501

[9]   Palik E D et al 1998 Handbook of optical constants of solids, Academic Press, San Diego; London

Optical microfiber knot resonator (MKR) and its slow-light performance

Liyong Ren[1], Yiping Xu[1,2], Chengju Ma[1,2], Yingli Wang[1], Xudong Kong[1,2], Jian Liang[1], Haijuan Ju[1], Kaili Ren[1,2], Xiao Lin[1,2]

[1]State Key Laboratory of Transient Optics and Photonics, Xi'an Institute of Optics and Precision Mechanics, Chinese Academy of Sciences, Xi'an 710119, China
[2]University of Chinese Academy of Sciences, Beijing 100049, China

renliy@opt.ac.cn

1. Introduction

There has been increasing interest in the development of optical microring resonators owing to their simple structure, compact size, and extensive applications in optical signal processing, communication, and active devices, such as optical filters, wavelength multiplexers, and lasers [1-3]. Planar waveguide microring resonators fabricated lithographically have been well developed, but suffer from larger internal and connection losses, higher cost and more complicated fabrication methods. Recently, research on low-loss micro- and nanofibers has opened up new opportunities for developing microphotonic devices such as resonators [4], couplers [5], and sensors [6]. As one of the basic functional elements, various structures of microfiber resonators, including loop, knot and coil, have been investigated [4, 7, 8] and extensively applied to optical filters [9], optical sensors [10], microfiber lasers [11], nonlinear resonators [12] and slow or fast light systems [13, 14]. Among these microfiber resonators, the microfiber knot resonator (MKR) has been regarded as one of the most attractive resonators [15-17], due to its many advantages, including easy fabrication, high stability, good compatibility with the available communication system, compactness, low loss, high Q value and high finesse. Since, Jiang et al. [4] firstly proposed the MKR, the resonator has been extensively applied to add-drop filters [9], miniature lasers [18], fast-light system [14], and so on. Recently, Xiao et al. proposed a new method to directly fabricate a MKR from a double-ended tapered fiber [15], which benefits the high finesse. We also proposed another approach to fabricate MKR with different structures [19], which might prompt the resonator to have a more extensive application.

In this paper, we investigate the MKRs with different structures and the slow-light performance of them from 6 sections. In Section 1, we make a simple introduction about microfiber resonators and their applications, and the structure of the whole paper. In Section 2, we give the mathematical expressions of the output light field with respect to the input one in the MKR, in the microfiber multi-knot resonator with a parallel structure, and in the microfiber multi-knot resonator with a serial structure. The slow-light performances of them are also investigated. It is found that a large slow-light time delay with a narrow bandwidth can be obtained in the microfiber multi-knot resonator with a parallel structure, while a slow-light time delay with a wide bandwidth can be obtained in the microfiber multi-knot resonator with a serial structure. In Section 3, we theoretically analyze and experimentally investigate how to design and fabricate a tapered microfiber with good optical and mechanical performance. In Section 4, we introduce a simple, polymer-microfiber-assisted approach to fabricating the silica MKRs with different structures. Comparing with other fabrication methods, this technique is quite simple and is easy to fabricate much more complicated multi-ring MKRs. In Section 5, we demonstrate a wide-bandwidth and zero-dispersion slow light in the microfiber double-knot resonator with a parallel structure based on an analogue of EIT through changing the correlated parameters, such as the coupling coefficients, and the diameters of the two knot rings. In Section 6, we make a conclusion about the whole paper.

Content from this work may be used under the terms of the Creative Commons Attribution 3.0 licence. Any further distribution of this work must maintain attribution to the author(s) and the title of the work, journal citation and DOI.
Published under licence by IOP Publishing Ltd

2. Theoretical study on slow light in different structures of optical microfiber knot resonators (MKRs)

2.1. A single-ring MKR

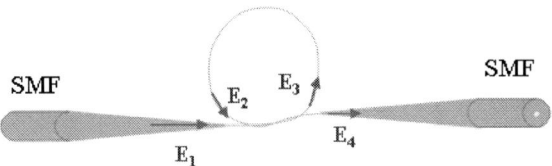

Figure 1. A single-ring MKR.

Figure 1 shows the schematic of a single-ring MKR. The theory of the MKR can be developed by combining the theory of ring resonators [20-23] with that of directional couplers [24-26]. The dependence of the output light fields, E_3 and E_4, on the input one E_1 (the detailed deduced process is given in Ref. [27].) can be expressed respectively as

$$E_3 = \frac{\sqrt{(1-r)(1-k)}}{1-j\sqrt{k(1-r)}\exp(j\beta L)}E_1, \tag{1}$$

$$E_4 = \left(\frac{j\sqrt{k(1-r)}+(1-r)\exp(j\beta L)}{1-j\sqrt{k(1-r)}\exp(j\beta L)}\right)E_1 = TE_1, \tag{2}$$

where T is the amplitude transmission coefficient of the single-ring MKR, k is the coupling coefficient, r is the coupling loss coefficient, L is the circumference of the ring resonator, and β is the propagation constant given in Ref. [28]. The phase of the transmission amplitude can be expressed by

$$\phi_T = \text{Im}\left(\ln(T)\right). \tag{3}$$

And the group time delay is obtained by [29]

$$\tau_d = \frac{d\phi_T}{d\omega} = \frac{n_{eff}}{c}\left(\frac{d\phi_T}{d\beta}\right), \tag{4}$$

where c is the speed of light in vacuum and n_{eff} is the effective refractive index of the microfiber. And the group delay dispersion (GDD) is obtained by differentiating Eq. (4) with respect to the angular frequency ω [30] and can be expressed by

$$GDD = \frac{d^2\phi_T}{d\omega^2} = \frac{d\tau_g}{d\omega} = -\frac{\lambda^2}{2\pi c}\frac{d\tau_g}{d\lambda}. \tag{5}$$

Note that GDD is the counterpart of the group velocity dispersion (GVD), which also determines the spreading degree of the optical pulse.

2.2. Microfiber multi-knot resonator with a parallel structure

Let us first analyze the case of microfiber double-knot resonator with a parallel structure. As shown in Fig. 2, it consists of two rings labeled 1 and 2 that are realized by a microfiber. The smaller ring, MKR 2, is embedded into the larger one, MKR 1. The amplitude transmission coefficients of ring 2 and ring 1 can be obtained by

AOM2015   IOP Publishing
Journal of Physics: Conference Series **680** (2016) 012032   doi:10.1088/1742-6596/680/1/012032

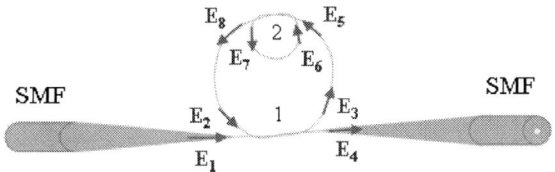

Figure 2. Microfiber double-knot resonator with a parallel structure.

$$T_2 = \frac{E_8}{E_5} = \frac{j\sqrt{k_2(1-r_2)} + (1-r_2)\exp(j\beta L_2)}{1 - j\sqrt{k_2(1-r_2)}\exp(j\beta L_2)}, \tag{6}$$

$$T_1 = \frac{E_4}{E_1} = \frac{j\sqrt{k_1(1-r_1)} + (1-r_1)T_2\exp(j\beta L_1)}{1 - j\sqrt{k_1(1-r_1)}T_2\exp(j\beta L_1)}. \tag{7}$$

Where k_1, k_2 are the coupling coefficients, r_1, r_2 are the coupling loss coefficients and L_1, L_2 are the circumferences of ring 1 and ring 2, respectively. Note that T_1, in fact, is the overall amplitude transmission coefficient T, i.e., $T=T_1$.

Similarly, the amplitude transmission coefficients of microfiber multi-knot resonator with a parallel structure can be calculated as

$$\left.\begin{aligned}
T_n &= \frac{j\sqrt{k_n(1-r_n)} + (1-r_n)\exp(j\beta L_n)}{1 - j\sqrt{k_n(1-r_n)}\exp(j\beta L_n)}, \\[2mm]
T_{n-1} &= \frac{j\sqrt{k_{n-1}(1-r_{n-1})} + (1-r_{n-1})T_n\exp(j\beta L_{n-1})}{1 - j\sqrt{k_{n-1}(1-r_{n-1})}T_n\exp(j\beta L_{n-1})}, \\
&\qquad\qquad\vdots \\
T_2 &= \frac{j\sqrt{k_2(1-r_2)} + (1-r_2)T_3\exp(j\beta L_2)}{1 - j\sqrt{k_2(1-r_2)}T_3\exp(j\beta L_2)}, \\[2mm]
T_1 &= \frac{j\sqrt{k_1(1-r_1)} + (1-r_1)T_2\exp(j\beta L_1)}{1 - j\sqrt{k_1(1-r_1)}T_2\exp(j\beta L_1)}, \\
T &= T_1.
\end{aligned}\right\} \tag{8}$$

In Eq. (8), T_n, T_{n-1}, ... , T_2 and T_1 are the amplitude transmission coefficients of the nth, the $(n-1)$th, ... , the second, and the first rings, respectively. The overall amplitude transmission coefficient T is equal to T_1.

2.3. Microfiber multi-knot resonator with a serial structure

Analogously, we firstly study the case of microfiber double-knot resonator with a serial structure. As shown in Fig. 3, it is comprised of two rings labeled 1 and 2. The overall amplitude transmission coefficient:

$$T = T_1 T_2 \exp(j\beta L_{1,2}), \tag{9}$$

where

$$T_1 = \frac{E_4}{E_1} = \frac{j\sqrt{k_1(1-r_1)} + (1-r_1)\exp(j\beta L_1)}{1 - j\sqrt{k_1(1-r_1)}\exp(j\beta L_1)}, \tag{10}$$

173

$$T_2 = \frac{E_8}{E_5} = \frac{j\sqrt{k_2(1-r_2)}+(1-r_2)\exp(j\beta L_2)}{1-j\sqrt{k_2(1-r_2)}\exp(j\beta L_2)}. \tag{11}$$

Note that T_1 and T_2 are the amplitude transmission coefficients of the first MKR and the second one, respectively, $L_{1,2}$ is the distance between them.

It is easy to extend to the case of microfiber multi-knot resonator with a serial structure, and in this case the amplitude transmission coefficient can be expressed by

$$T = \prod_{i=1}^{n}\left[\frac{j\sqrt{k_i(1-r_i)}+(1-r_i)\exp(j\beta L_i)}{1-j\sqrt{k_i(1-r_i)}\exp(j\beta L_i)}\exp\left(j\beta L_{(i-1),i}\right)\right]. \tag{12}$$

In Eq. (12), $L_{(i-1),i}$ is the distance between the $(i-1)$th and the ith rings and $L_{0,1}=0$.

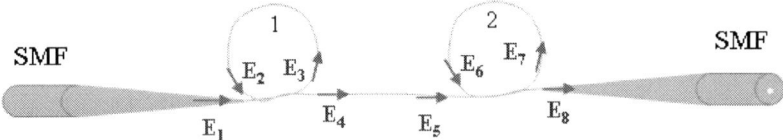

Figure 3. Microfiber double-knot resonator with a serial structure.

2.4. Slow-light time-delay characteristics of microfiber multi-knot resonator with different structures
Assuming the input light field $E_1=1$, inserting Eq. (8) and Eq. (12) into Eq. (3), respectively, and then using Eq. (4), we can obtain the group time delay features of the microfiber multi-knot resonator with a parallel structure and the microfiber multi-knot resonator with a serial structure.

Figure 4(a) compares the group time delays in the single-ring MKR, the microfiber double-knot resonator with a parallel structure, and the microfiber three-knot resonator with a parallel structure. To see clearly, Fig. 4(a) only indicates the result around the wavelength of 1550 nm. The inset of Fig. 4(a) is the zoom-in spectrum of the group time delay of the single-ring MKR. It can be seen from Fig. 4(a) that the group time delay in the microfiber three-knot resonator with a parallel structure increases drastically, but at the same time the 3-dB bandwidth of it becomes narrower than that of the microfiber double-knot resonator with a parallel structure.

Figure 4(b) compares zoom-in spectrum of the group time delay in the single-ring MKR, the microfiber double-knot resonator with a serial structure, and the microfiber three-knot resonator with a serial structure at the wavelength around 1550 nm. The group time delay in the microfiber three-knot resonator with a serial structure does not change a lot, but the 3-dB bandwidth strongly increases.

To summarize, by designing the microfiber multi-knot resonator with a parallel structure, one can obtain large slow-light time delay with a narrow bandwidth; while one can obtain the slow-light time delay with a wide bandwidth by designing the microfiber multi-knot resonator with a serial structure.

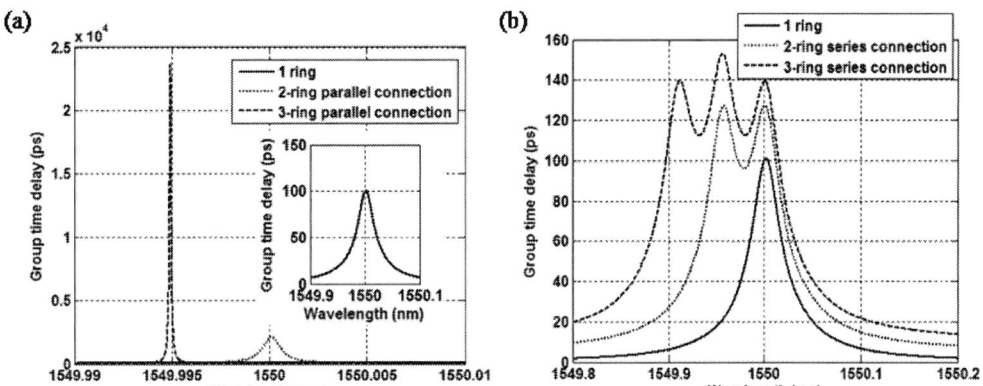

Figure 4. (a) The spectral dependence of the group time delay in the microfiber multi-knot resonator with a parallel structure. The solid line corresponds to the single-ring MKR (k=0.8, r=0.1, d=1 μm, L=904 μm, n_{eff}=n_1 =1.45 and n_2=1); the dotted line corresponds to the microfiber double-knot resonator with a parallel structure (k_1= k_2=0.8, r_1=r_2=0.1, d=1 μm, L_1=1398 μm, L_2=699 μm, n_{eff}=n_1=1.45 and n_2=1); and the dash line corresponds to the microfiber three-knot resonator with a parallel structure (k_1=k_2=k_3=0.8, r_1=r_2=r_3=0.1, d=1 μm, L_1=1584 μm, L_2=792 μm, L_3=396 μm, n_{eff}=n_1=1.45 and n_2=1). (b) Spectral dependence of the group time delay in the microfiber multi-knot resonator with a serial structure. The solid line corresponds to the single-ring MKR (k=0.8, r=0.1, d=1 μm, L=904 μm, n_{eff}=n_1 =1.45 and n_2=1); the dotted line corresponds to the microfiber double-knot resonator with a serial structure (k_1=k_2=0.8, r_1=r_2=0.1, d=1 μm, L_1=904 μm, L_2=905 μm, $L_{1,2}$=1000 μm, n_{eff} =n_1=1.45 and n_2=1); the dash line corresponds to the microfiber three-knot resonator with a serial structure (k_1=k_2=k_3=0.8, r_1=r_2=r_3=0.1, d=1 μm, L_1=904 μm, L_2=905 μm, L_3=906 μm, $L_{1,2}$=1000 μm, $L_{2,3}$=1000 μm, n_{eff}=n_1=1.45 and n_2=1).

3. Design and fabrication of tapered microfiber waveguide with good optical and mechanical performance

3.1. Theoretical analysis and simulation

In this section, we present the design and fabrication of tapered microfiber waveguide with good optical and mechanical performance. The tapered microfiber waveguide is fabricated by stretching a heated conventional SMF to form a structure with two transition regions and a uniform waist, as shown in Fig. 5. The diameter of the transition region decreases from the size of SMF down to the waist diameter. When the light propagates through the transition region, the light field distribution varies as the core and cladding diameters change along the fiber. Depending on the change rate of diameter, the energy transfers from the fundamental mode to a closest few higher order modes. Moreover, the number of the high order modes determines the loss. For the transition region with a slowly varying diameter, there always companies a less number of high order modes and thus a low loss of light. Therefore, one can minimize the microfiber loss by optimizing its profile parameters.

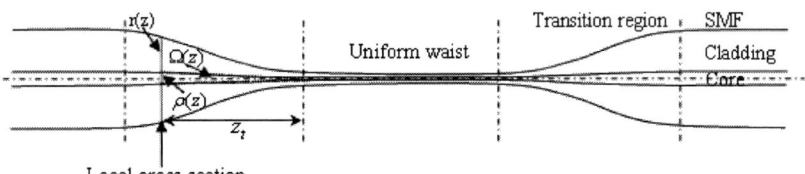

Figure 5. Typical profile of a tapered microfiber.

Based on the adiabaticity criteria [31], we use the Finite-difference Beam Propagation Method (FD-BPM) to simulate the light propagating in the tapered microfiber with different profile parameters. Firstly, the relationship between the light transmission of the tapered microfiber and the transition region length is simulated and shown in Fig. 6, one can see that when the uniform waist diameter is greater than 0.6 μm and the transition region length is longer than 5 mm, the loss becomes very small and can be neglected. Secondly, in order to analyze the dependence of the tapered microfiber optical loss on the waist diameter, in Fig. 7(a) we simulate the normalized transmission power of the tapered microfiber as a function of the uniform waist diameter. Finally, in the practical applications of the evanescent-field-based optical sensing, it is important to know the evanescent field distribution around the tapered microfiber. We therefore investigate the variation of the fractional energy of the evanescent field with the waist diameter of the tapered microfiber. The simulation result is plotted in Fig. 7(b), where the light wavelength is selected at 1550 nm as an example.

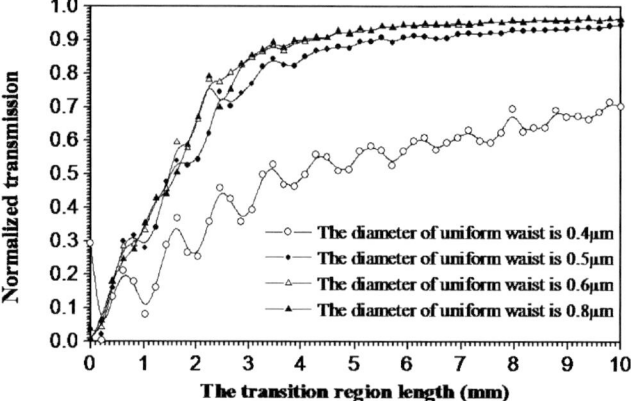

Figure 6. Normalized transmission power depends on the transition region length of tapered microfiber. The open circle line, the solid circle line, the open up-triangle line and the solid up-triangle line represent four tapered microfibers with uniform waist diameters of 0.4 μm, 0.5 μm, 0.6 μm and 0.8 μm, respectively.

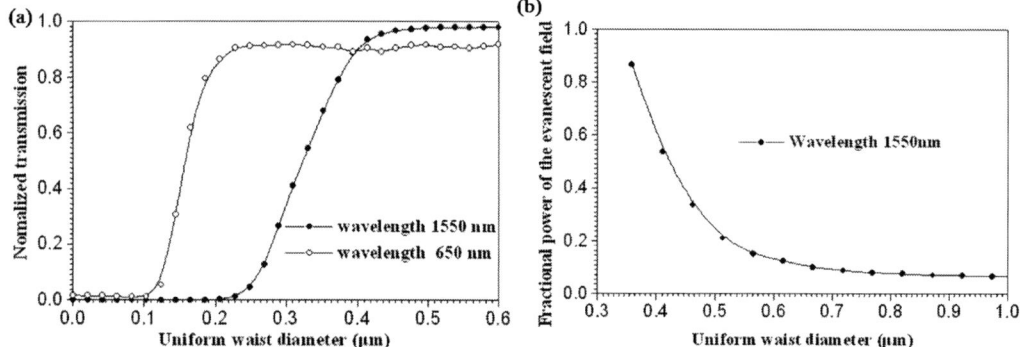

Figure 7. (a) Normalized transmission power of the tapered microfiber as a function of the uniform waist diameter. The open circle line and the solid circle one correspond to the incident light wavelengths of 650 nm and 1550 nm, respectively. (b) Dependence of the fractional power of the evanescent field on the tapered microfiber waist diameter for the incident light wavelength of 1550 nm.

Based on the analysis and simulations mentioned above, it can be concluded from Figs. 6 and 7 that the tapered microfiber with low loss, strong evanescent field and relatively shorter transition region should have two longer than 5-mm-length transition regions and a uniform waist whose diameter is larger than 600 nm and less than 1 μm. Additionally, it should be noted that the taper shape of the transition region also affects the microfiber properties. In our simulation process, we assume that the taper shape of the transition region is a decaying-exponential shape, which has been demonstrated in some experiments [32, 33].

3.2. Fabrication and performance of the tapered microfiber

The schematic experimental setup for the tapered microfiber fabrication is shown in Fig. 8. It mainly comprises of the microfiber heater (MHI FIBHEAT200, US), two translation stages with high precision stepper motors (FL110TA600, China), two fiber holders on the translation stages and the computer control system.

AOM2015

Journal of Physics: Conference Series **680** (2016) 012032

IOP Publishing

doi:10.1088/1742-6596/680/1/012032

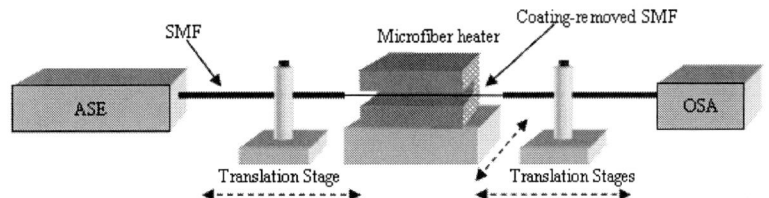

Figure 8. Schematic experimental setup for tapered microfiber fabrication.

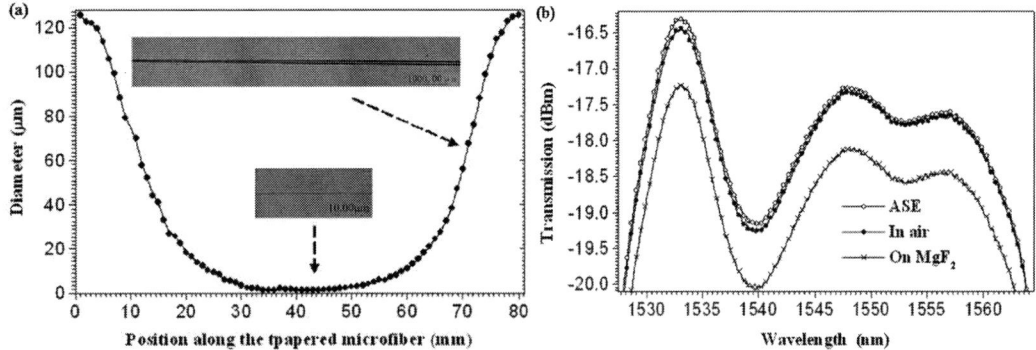

Figure 9. (a) Profile of the tapered microfiber we fabricated. Two inserted microscope images are a partial transition region and a uniform waist of the tapered microfiber, respectively. (b) The transmission spectra of the ASE light when it passes through the un-tapered SMF (the open circle line), the tapered microfiber in air (the solid circle line) and on the MgF_2 substrate (the cross line), respectively.

The profile of the tapered microfiber we fabricated is shown in Fig. 9(a). The total length of the tapered microfiber is 80 mm. The length of the transition region with an approximate decaying-exponential shape is about 30 mm and the length of the uniform waist with a diameter of 1 μm is about 20 mm. Two microscope images inserted in Fig. 9(a) show the partial transition region and the uniform waist, respectively. Note that the tapered microfiber has low surface roughness. In order to survey the optical performance of the tapered microfiber during the fabrication, ASE and OSA are connected with the two ends of the fiber.

The optical spectra are shown in Fig. 9(b). The open circle line, the solid circle line and the cross line represent the transmission spectra of the ASE light when it passes through the un-tapered SMF, the tapered microfiber in air and on the MgF_2 substrate, respectively. The loss at 1550 nm is about 0.05 dB in air. It is seen that the loss of the tapered microfiber is very low in air, while it increases up to 0.8 dB when the tapered microfiber is moved onto the MgF_2 substrate. The reason of the loss increasing is that a small quantity of evanescent filed energy is diverted to the MgF_2 surface and is lost when the microfiber is moved on the MgF_2 substrate.

4. A Simple, Polymer-Microfiber-Assisted Approach to Fabricating the Silica Microfiber Knot Resonator

4.1. Basic principle of fabrication

The schematically fabrication process is illustrated in Fig. 10, which can be divided into four steps. (1) Using the high-temperature tapered-drawing technique，one fabricates a silica microfiber with a diameter of micrometer scale from the standard single mode fiber (SMF). As a result, there is a tapered-drawing region which connects the silica microfiber to the SMF at each end of the silica microfiber. Cutting one end of the silica microfiber makes it become a freestanding end. As shown in Fig. 10(a), a polymer microfiber with a diameter of tens micrometers, drawn from solvent polymers [34], is tailored to a suitable length. Then let's manually wind a knot ring with a diameter of several millimeters from the above polymer microfiber and adhere one end of the knot ring to the freestanding end of the silica microfiber. (2) The polymer microfiber knot ring is driven to the silica microfiber with the assistance of a tapered-drawing fiber probe until the polymer microfiber is completely drawn out of the knot ring. As shown in Fig. 10(b), in this case, the knot ring is only composed of the silica microfiber. (3) As shown in Fig. 10(c), continuously drawing the polymer microfiber thus finely tunes the diameter of the

177

knot ring under an optical microscope. (4) Departing the polymer microfiber from the freestanding end of the silica microfiber and adhering the other silica microfiber produced in the step (1) to it, as shown in Fig. 10(d), one finally fabricates a silica MKR. Note that the zoom-in image in Fig. 10(d) is the intertwisted overlap at the contact area.

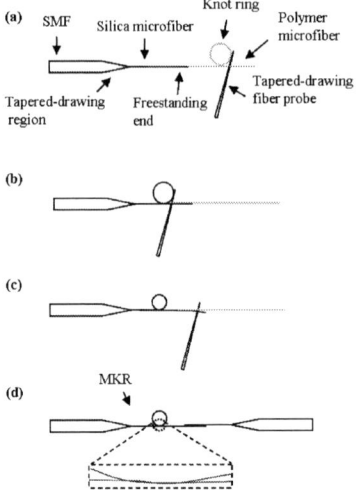

Figure 10. (a)-(d) Schematic of the fabrication process.

4.2. Experimental results and discussions

Experimentally, we fabricated a MKR with the above mentioned technique. Figure 11(a) shows an optical microscope image of the knot resonator, in which the polymer microfiber ring has being driven to the freestanding end of the silica microfiber. As one can see in Fig. 11(a), the bold line is the polymer microfiber with a diameter of about 12 μm, while the slim line is the silica microfiber with a diameter of about 1 μm. Figure 11(b) shows a zoom-in optical microscope image of the knot region in Fig. 11(a). It is easy to see that the polymer microfiber and the silica microfiber are adhered together firmly. This is attributed to the strong van der Waals and electrostatic forces between them. The final knot resonator consisted of the silica microfiber is shown in Fig. 11(c). The diameter of the fabricated MKR is about 409 μm. Figure 11(d) shows a zoom-in optical microscope image of the knot region in Fig. 11(c). To see the spectral property of this MKR, a broadband amplified spontaneous emission (ASE) light was launched into the MKR and the spectrum from the output port was collected. As an example, Fig. 11(e) shows the transmission spectrum in the wavelength range of 1548-1554 nm, where an extinction ratio of about 5 dB is obtained.

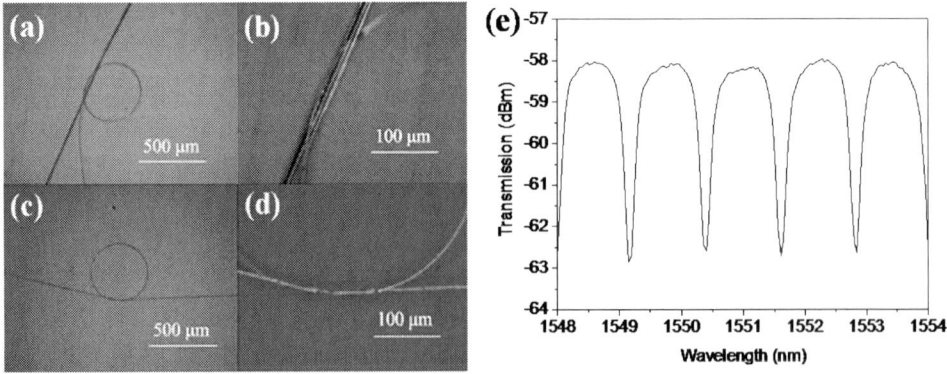

Figure 11. Optical microscope images of (a) the MKR located in the adhered region between the polymer microfiber and the silica microfiber on an MgF$_2$ substrate; (b) the zoom-in knot region in (a); (c) the final fabricated MKR; (d) the zoom-in knot region in (c); (e) Transmission spectrum of the fabricated MKR.

Using this method, we can also fabricate microfiber multi-knot resonator with different structures readily. Figure 12(a) shows an optical microscope image of the microfiber double-knot resonator with a serial structure. The diameter of the silica microfiber is about 2 μm, and the diameters of the small ring and the big ring are about 680 μm and 723 μm, respectively. A corresponding transmission spectrum is given in Fig. 12(b). The whole transmission spectrum is the overlapped consequence of the two transmission spectra produced by the two knot rings. Figure 12(c) shows an optical microscope image of the microfiber double-knot resonator with a parallel structure. The diameter of the silica microfiber is about 2 μm and the diameters of the small ring and the big ring are about 788 μm and 1605 μm, respectively. A corresponding transmission spectrum is given in Fig. 12(d). Note that the whole transmission spectrum isn't simply an overlapped consequence of the two transmission spectra produced by the two knot rings.

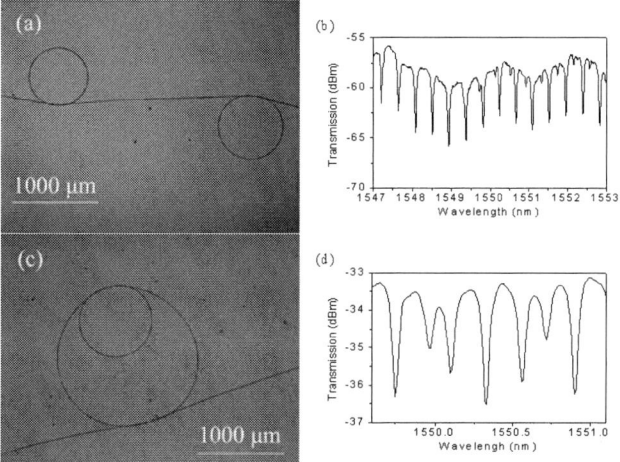

Figure 12. (a) An optical microscope image of the microfiber double-knot resonator with a serial structure. (b) Corresponding transmission spectrum of (a). (c) An optical microscope image of the microfiber double-knot resonator with a parallel structure. (d) Corresponding transmission spectrum of (c).

5. Wide-bandwidth and zero-dispersion slow light in MKRs with a two-ring parallel connection structure based on an analogue of electromagnetically induced transparency

In order to study the slow-light characteristics at the windows of the EIT-like in the microfiber double-knot resonator with a parallel structure. Firstly, we theoretically simulate the transparency window of the EIT-like [35] with the variation of the coefficients k_2 and k_1. Assuming the input light field $E_1=1$ and using Eq. (7) one can calculate the dependence of the transmission coefficient of the microfiber double-knot resonator with a parallel structure on wavelengths for varying coupling coefficient k_2. Numerical results are shown in Fig. 13(a), where, as an example, we set the parameters to $k_1=0.8$, $r_1=r_2=0.1$, $d=2$ μm, $D_1=1600$ μm, $D_2=800$ μm, $n_{eff}=n_1=1.45$ and $n_2=1$. It is easy to see from Fig. 13(a) that, when the coupling coefficient $k_2=0$ (meaning that there isn't coupling action at the knot region of MKR 2, the light field E_5 converts to the light field E_7 completely), the whole microfiber double-knot resonator with a parallel structure is equivalent to a single-ring MKR with a circumference of L_1+L_2. According to the definition of the free spectral range (FSR) of the resonator $\Delta\lambda=\lambda^2/(n_{eff} \cdot L)$, the longer the circumference of the resonator is, the narrower the FSR is. So when $k_2=0$, the FSR shown in the Fig. 13(a) is narrowest. However, when the coupling coefficient k_2 isn't equal to zero, a narrow transmission peak, the so-called EIT-like window, is produced between two wide transmission peaks due to the overlap of transmission spectra. Moreover, as the coupling coefficient k_2 increases, the induced transparency window becomes narrower and narrower. When the coupling coefficient $k_2=1.0$, the induced transparency window vanishes. This is due to the fact that, in this case, the light field E_5 converts to E_8 completely, except that there is 10% light energy loss (because $r_2=0.1$), the microfiber double-knot resonator with a parallel structure is equivalent to a single-ring MKR. Based on this characteristic, one can create narrow line-width lasers, filters, sensors and so on, through modulating the coupling coefficient k_2 and the circumferences of the two rings.

When we vary the coupling coefficient k_1 and keep the other parameters $k_2=0.8$, $r_1=r_2=0.1$, $d=2$ μm, $D_1=1600$ μm, $D_2=800$ μm, $n_{eff}=n_1=1.45$ and $n_2=1$ unchanged, we calculate the dependence of the transmission coefficient of the microfiber double-knot resonator with a parallel structure on wavelengths. The numerical results are shown in Fig. 13(b). One can see that when the coupling coefficient $k_1=0$, the FSR of the resonator is broadest. The reason is that, in this case, there isn't coupling action at the coupling region of MKR 1, the input light field E_1 converts to E_3 completely, thus the whole microfiber double-knot resonator with a parallel structure is equivalent to a sing-ring MKR with a circumference of L_2. As the coupling coefficient k_1 increases, the extinction ratio of the output spectrum increases, but the FSR of the induced transparency window between two wide transmission peaks doesn't change. When the coupling coefficient k_1 increases to 1, the input light field E_1 converts to E_4 completely, except that there is 10% light energy loss (because $r_1=0.1$). Therefore, the output spectrum is a horizontal line and the transmission coefficient T is equal to 0.9.

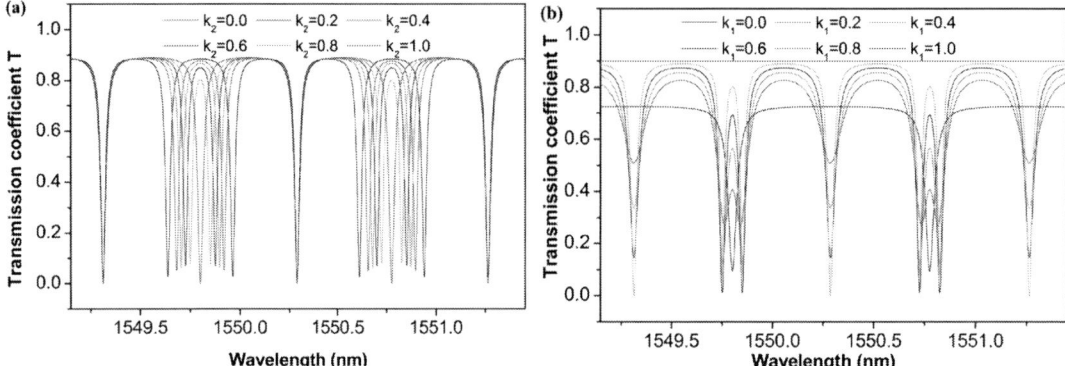

Figure 13. Dependence of the transmission coefficient of MKRs with a two-ring parallel connection structure on wavelengths for varying coupling coefficients (a) k_2 ($k_1=0.8$) and (b) k_1 ($k_2=0.8$), the other parameters are chosen to $k_1=0.8$, $r_1=r_2=0.1$, $d=2$ μm, $D_1=1600$ μm, $D_2=800$ μm, $n_{eff}=n_1=1.45$ and $n_2=1$.

Based on the analysis mentioned above, one can modulate the bandwidth and the extinction ratio of the induced transparency window by changing the coupling coefficients k_1 and k_2 simultaneously. On the other hand, as we know that group time delay is supposed to be observed at the resonance of the transmission spectrum of the resonator [13]. So, in order to study the characteristic of the group time delay, we simulate the transmission coefficient, the phase, the group time delay and the GDD with the variation of the coupling coefficient k_2 at the induced transparency window. The corresponding simulated results are shown in Fig. 14.

Note that in Fig. 14 the parameters of the theoretical model are chose to $r_1=r_2=0.1$, $d=2$ μm, $D_1=1600$ μm, $D_2=800$ μm, $n_{eff}=n_1=1.45$ and $n_2=1$ which are identical with the parameters given in Fig. 13. When the coupling coefficients $k_1=0.02$, and k_2 increases from 0.1 to 0.7, the induced transparency window firstly becomes lower gradually and then disappears, which can be easily seen in Fig. 14(a). Especially, when the coupling coefficient k_2 is equal to 0.5, the transmission coefficient T reduces to 0.48 with a flat wavelength bandwidth of about 82.7 pm. Accordingly, as shown in Fig. 14(c), the group time delay at the corresponding induced transparency window increases gradually with the increasing of the coupling coefficient k_2 from 0.1 to 0.7. And it reaches to about 72.4 ps with a flat wavelength bandwidth of about 82.7 pm, the corresponding FWHM is about 228 pm, when the coupling coefficient k_2 is equal to 0.5. The reason for this phenomenon is that the phase of the transmission light field is modulated effectively at the resonance by resonator, which makes $d\phi/d\omega$ increase rapidly. It can be seen from Fig. 14(b) that, with the increasing of the coupling coefficient k_2, the absolute value of the slope rate is increased correspondingly at the resonant wavelength of around 1549.8 nm. Therefore, from Eq. (4), one can see that the group time delay increases greatly. When the coupling coefficient k_2 is equal to 0.5, the slope rate of the phase is almost a straight line at a small wavelength range of around 1549.8 nm. Thus, the group time delay is a constant at this wavelength range, which can be seen from Fig. 14(c). On the other hand, using Eq. (5), one can calculate the GDD with respect to the variation of wavelengths. It can be seen from Fig. 14(d) that the GDD at the same induced transparency window occurs the corresponding change with the increasing of the coupling coefficient k_2 from 0.1 to 0.7. And it becomes zero with a wavelength bandwidth of about 82.7 pm which locates at the same wavelength range of the flat group time delay, when the coupling coefficient k_2 is equal to 0.5. Thus, through modulating the suitable values of the coupling coefficients k_2 and k_1, one can obtain a wide-bandwidth and zero-dispersion group time delay at the induced transparency window in the microfiber double-knot resonator with a parallel structure. We believe that the slow light with a characteristic of the wide bandwidth and zero dispersion at the induced transparency window in the microfiber double-knot resonator with a parallel structure will have significant potential applications in optical buffers, data delay lines and optical memories, etc.

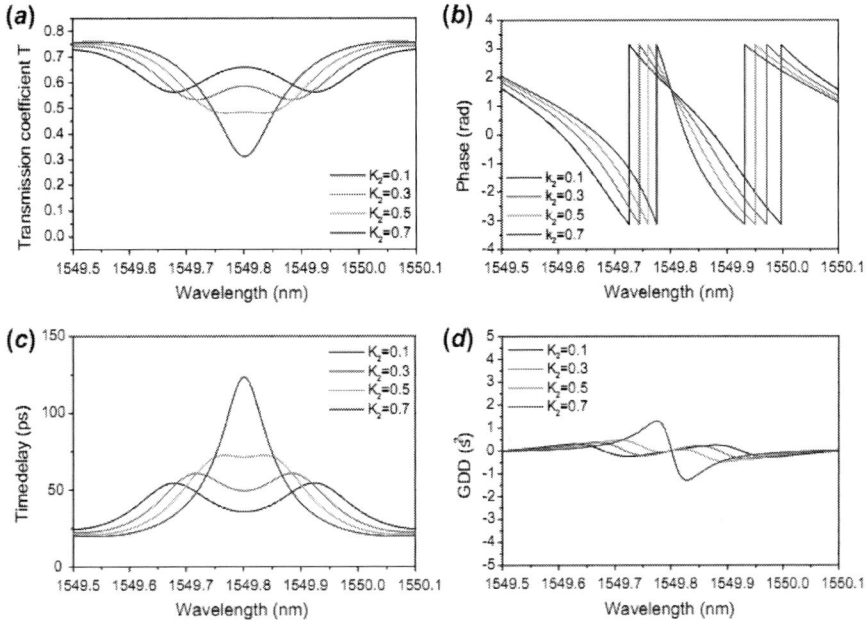

Figure 14. Dependence of the transmission coefficient (a), the phase (b), the group time delay (c), the GDD (d) at the induced transparency window in the microfiber double-knot resonator with a parallel structure on wavelengths for various coupling coefficients k_2, the other parameters are chosen to $k_1=0.02$, $r_1=r_2=0.1$, $d=2$ μm, $D_1=1600$ μm, $D_2=800$ μm, $n_{eff}=n_1=1.45$ and $n_2=1$.

6. Conclusions

In conclusion, firstly, we investigate the theoretical models of the sing-ring MKR, the microfiber multi-knot resonator with a parallel structure and the microfiber multi-knot resonator with a serial structure and numerically simulate the group time delay spectra of these MKRs with different structures. The numerical result indicates that by designing the microfiber multi-knot resonator with a parallel structure, one can obtain large slow-light time delay with a narrow bandwidth. By designing the microfiber multi-knot resonator with a serial structure, one can obtain the slow-light time delay with a wide bandwidth. Secondly, we investigate the design and fabrication of tapered microfiber waveguide with good optical and mechanical performance theoretically and experimentally. The result demonstrates that the tapered microfiber with low loss, strong evanescent field and relatively shorter transition region should have two longer than 5-mm-length transition regions and a uniform waist whose diameter is larger than 600 nm and less than 1 μm, the taper profile of the transition region should be a decaying-exponential shape. Thirdly, we present a simple, polymer-microfiber-assisted approach to fabricating the silica microfiber knot resonator. Using this technique, we have successfully fabricated several kinds of MKRs with different structures, such as those with a double-knot serial structure and with a double-knot parallel structure. Comparing with other fabrication methods, this technique is quite simple and is easy to fabricate much more complicated multi-ring MKRs. Finally, a wide-bandwidth and zero-dispersion slow light in the microfiber double-knot resonator with a parallel structure based on an analogue of EIT is demonstrated through changing the correlated parameters, such as the coupling coefficients, and the diameters of the two knot rings.

Acknowledgment

This work was supported by the National Natural Science Foundation of China under Grants 61275149, 51207159 and 61535015 as well as the Natural Science Foundation of Shaanxi province under Grant 014JM8327.

References

[1] Absil P. P., Hryniewicz J. V., Little B. E., Wilson R. A., Joneckis L. G., and Ho P.-T.,

"Compact microring notch filters," IEEE Photon. Technol. Lett. Papers 12(4), 398-400 (2000).

[2] Choi S. J., Peng Z., Yang Q., Choi S. J., and Dapkus P. D., "An eight-channel demultiplexing switch array using vertically coupled active semiconductor microdisk resonators," IEEE Photon. Technol. Lett. Papers 16(11), 2517-2519 (2004).

[3] Amarnath K., Grover R., Kanakaraju S., and Ho P. T., "Electrically pumped InGaAsP-InP microring optical amplifiers and lasers with surface passivation," IEEE Photon. Technol. Lett. Papers 17(11), 2280-2282 (2005).

[4] Jiang X. S., Tong L. M., Vienne G., Guo X., Tsao A., Yang Q., and Yang D., "Demonstration of optical microfiber knot resonators," Appl. Phys. Lett. Papers 88, pp. 223501 (2006).

[5] Tong L. M., Hu L. L., Zhang J. J., Qiu J. R., Yang Q., Lou J. Y., Shen Y. H., He J. L., and Ye Z. Z., "Photonic nanowires directly drawn from bulk glasses," Opt. Express Papers 14(1), 82-87 (2006).

[6] Li Y. H., and Tong L. M., "Mach-Zehnder interferometers assembled with optical microfibers or

[7] nanofibers," Opt. Lett. Papers 33(4), 303-305 (2008).

[8] Sumetsky M., Dulashko Y., Fini J. M., and Hale A., "Optical microfiber loop resonator," Appl. Phys. Lett. Papers 86, 161108 (2005).

[9] Xu F., and Brambilla G., "Manufacture of 3-D Microfiber Coil Resonators," IEEE Photon. Technol. Lett. Papers 19(19), 1481-1483 (2007).

[10] Jiang X. S., Chen Y., Vienne G., and Tong L. M., "All-fiber add-drop filters based on microfiber knot resonators," Opt. Lett. Papers 32(12), 1710-1712 (2007).

[11] Wu Y., Rao Y. J., Chen Y. H., and Gong Y., "Miniature fiber-optic temperature sensors based on silica/polymer microfiber knot resonators," Opt. Express Papers 17(20), 18142-18147 (2009).

[12] Jiang X. H., Song Q. H., Xu L., Fu J., and Tong L. M., "Microfiber knot dye laser based on the evanescent-wave-coupled gain," Appl. Phys. Lett. Papers 90, 233501 (2007).

[13] Vienne, G., Li, Y. H., Tong, L. M., and Grelu, P., "Observation of a nonlinear microfiber resonator," Opt. Lett. Papers 33(13), 1500-1502 (2008).

[14] Ma C. J., Ren L. Y., and Xu Y. P., "Slow-light element for tunable time delay based on optical microcoil resonator," Appl. Opt. Papers 51(26), 6295-6300 (2012).

[15] Wang T., Li X. H., Liu F. F., Long W. H., Zhang Z. Y., Tong L. M., and Su Y. H., "Enhanced fast light in microfiber ring resonator with a Sagnac loop reflector," Opt. Express Papers 18(15), 16156-16161 (2010).

[16] Xiao L. M., and Birks T. A., "High finesse microfiber knot resonators made from double-ended tapered fibers," Opt. Lett. Papers 36(7), 1098-1100 (2011).

[17] Lim K. S., Jasim A. A., and Damanhuri S. S. A., "Resonance condition of a microfiber knot resonator immersed in liquids," Appl. Opt. Papers 50(30), 5912-5916 (2011).

[18] Chen Y. H., Wu Y., and Rao Y. J., "Hybrid Mach-Zehnder interferometer and knot resonator based on silica microfibers," Opt. Commun. Papers 283, 2953-2956 (2010).

[19] Jiang X. S., Yang Q., Vienne G., Li Y. H., Tong L. M., Zhang J. J., and Hu L. L., "Demonstration of microfiber knot laser," Appl. Phys. Lett. Papers 89(14), 143513 (2006).

[20] Xu Y. P., Ren L. Y., Liang J., Ma C. J., Wang Y. L., Chen N. N., and Qu E. S., "A simple, polymer-microfiber- assisted approach to fabricating the silica microfiber knot resonator," Opt. Commun. Papers 321, 157-161 (2014).

[21] Stokes L. F., Chodorow M., and Shaw H. J., "All-single-mode fiber resonator," Opt. Lett. Papers 7(6), 288-290 (1982).

[22] Madsen C. K., and Lenz G., "Optical all-pass filters for phase response design with applications for dispersion compensation," IEEE Photon.Technol. Lett. Papers 10(7), 994-996 (1998).

[23] Paloczi G. T., Huang Y., and Yariv A., "Free-standing all-polymer microring resonator optical filter," Electron. Lett. Papers 39(23), 1650-1651 (2003).

[24] Schwelb O., "Transmission, group delay, and dispersion in single-ring optical resonators and add/drop filters-A tutorial overview," J. Light. Technol. Papers 22(5), 1380-1394 (2004).

[25] Snyder A. W., and Love J. D., [Optical Waveguide Theory], London, U.K.: Chapman & Hall (1983).

[26] Smith R. B., "Analytic solution for linearly tapered directional couplers," J. Opt. Soc. Am. Papers 66(9), 882-892 (1976).

[27] Morishita K., and Yamaguchi T., "Wavelength tunability and polarization characteristics of twisted polarization beamsplitting single-mode fiber couplers," J. Light. Technol. Papers 19(5), 732-738 (2001).

[28] Xu Y. P., Ren L. Y., Ma C. J., and Liang J., "Theoretical study on slow light in different structures of optical microfiber knot resonators (OMKRs)," Optik 125, 2856-2861 (2014).

[29] Sumetsky M., "How thin can a microfiber be and still guide light? Errata," Opt. Lett. Papers 31(24), 3577-3578 (2006).

[30] Sumetsky M., "Optical fiber microcoil resonator," Opt. Express Papers 12(10), 2303-2316 (2004).

[31] Kato T., and Kokubun Y., "Bessel-thompson filter using double-series-coupled microring resonator," J. Light. Technol. Papers 26(22), 3694-3698 (2008).

[32] Ma C. J., Ren L. Y., Xu Y. P., Wang Y. L., Liang J., and Qu E. S., "Design and fabrication of tapered microfiber waveguide with good optical and mechanical performance," J. Mod. Opt. Papers 61(8), 683-687 (2014).

[33] Tong L. M., and Sumetsky M., [Subwavelength and Nanometer Diameter Optical Fibers], Zhejiang University Press, Springer (2009).

[34] Harun S. W., Lim K. S., Tio C. K., Dimyati K., and Ahmad H., "Theoretical analysis and fabrication of tapered fiber," Optik Papers 124(6), 538–543 (2013).

[35] Harfenist S. A., Cambron S. D., Nelson E. W., Berry S. M., Isham A. W., Crain M. M., Walsh K. M., Keynton R. S., and Cohn R. W., "Direct drawing of suspended filamentary micro- and nanostructures from liquid polymers," Nano Lett. 4(10), 1931-1937 (2004).

[36] Xu Q. F., Sandhu S., Povinelli M. L., Shakya J., Fan S. H., and Lipson M., "Experimental realization of an on-chip all-optical analogue to electromagnetically induced transparency," Phys. Rev. Lett. 96, 123901 (2006).

Fluorescence enhancement with metamaterial mirrors

Jian Qin, Wei Wang, Si Luo, Xingxing Chen, Min Qiu and Qiang Li

State Key Laboratory of Modern Optical Instrumentation, College of Optical Science and Engineering, Zhejiang University, Hangzhou, 310027, China

E-mail: qiangli@zju.edu.cn

Abstract. We experimentally demonstrate the strongly enhanced photoluminescence of the fluorescent molecules on the metamaterial mirror. The matematerial mirror can optimize the reflection phase to provide a large electric field for the 20-nm-thick active layer. Compared with the smooth gold plate, the experimental result shows a nearly 45 times enhancement.

1. Introduction

Metal enhanced fluorescence is to study the metal-assisted enhanced interaction between fluorophore molecules and light. By directly placing the active layers on varied metal structures, the fluorescence intensity can be enhanced and the fluorescence lifetime can be shortened. Research in this area includes designing suitable metal nanostructures, choosing proper exciting light and wavelength-matched fluorophore molecules or quantum dots and obtaining metal materials with better qualities. Many metallic nanostructures for enhanced fluorescence have been reported, such as the bowtie structures[1], metal-insulator-metal absorber structures[2,3] and the synthesis monocrystal gold nanoplates[4].

Using metamaterial mirror [5] to design the planar optoelectronic device is a kind of simple and outstanding technology. By changing the reflection phase among zero and π in a period, metamaterial mirror can control the spatial distribution of electric field in the thin active layers. In this study, we fabricate metamaterial mirror to obtain an enhanced electric field intensity in the 20-nm-thin active layer with fluorescent molecules. Compared with the smooth gold plate, the experimental result shows a nearly 45 times enhancement.

2. Materials preparation and sample fabrication

First thermolysis process is adopted to obtain monocrystal gold nanoplates. And then metamaterial mirror is fabricated on the monocrystal gold substrates by using focused ion beam. The groove depth is 85 nm and the width is 50 nm. The period is 150 nm. Then the ATTO633 dye(1 μmol/L) and the PMMA solution with the 1:1 proportion is mixed. Finally the mixture is spin-coated on the samples and solidified by heating. The thickness of the active layer is 20 nm thickness. The SEM image of the device is provided in Figure 1.

Content from this work may be used under the terms of the Creative Commons Attribution 3.0 licence. Any further distribution of this work must maintain attribution to the author(s) and the title of the work, journal citation and DOI.

Published under licence by IOP Publishing Ltd

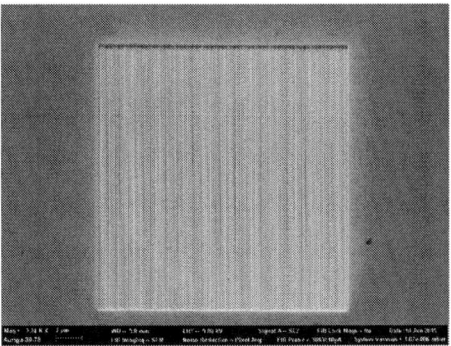

Figure 1. SEM image of the metamaterial mirrors.

3. Experiments and discussion

He-Ne laser (632nm) is used to excite the sample. The fluorescence is measured by spectroscopy via the dark-field microscope. The reflected exciting light is cut off by a 650 nm filter. The emission wavelength of fluorescence is 652 nm. From figure 2, we can see bright fluorescence of the sample under dark-field microscope.

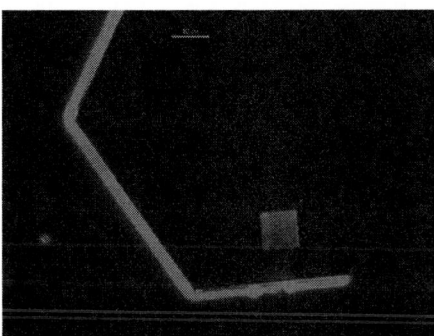

Figure 2. Fluorescence image under dark-field microscope.

The spectroscopy of the metamaterial part and the smooth mirror part are shown in Figure 3. The red line shows the fluorescence spectrum on the smooth gold film surface. And it just represents the background (BG). The black line is the fluorescence signal (SA) on the metamaterial mirror. It is obvious that the fluorescence is greatly enhanced by around 44.8 times using the metamaterial mirror.

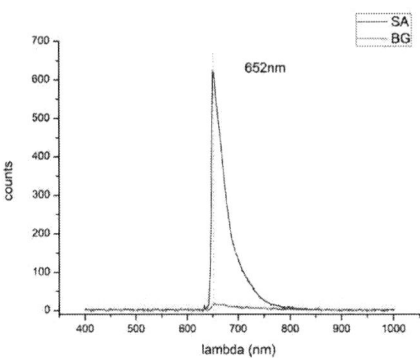

Figure 3 Experimental spectroscopy. The red line is the fluorescence on the smooth gold film and the black one is the result on the metamaterial mirror.

4. Conclusion

In summary, based on the metamaterial mirror, distribution of electric field above the metal surface can be engineered to greatly enhance the fluorescence. Compared with the smooth gold plate, the experimental result shows a nearly 45 times enhancement here. This enhanced light-mater interaction has tremendous potential in nanoscale emitter.

References

[1]   Schraml K, Spiegl M, Kammerlocher M, Bracher G, Bartl J, Campbell T, Finley J J and Kaniber M 2014 *Phys. Rev.* B **90** 035435

[2]   Hao J M, Zhou L and Qiu M 2011 *Phys. Rev.* B **83**,165107

[3]   Akselrod G M, Argyropoulos C, Hoang T B, Ciraci C, Fang C, Huang J N, Smith D R and Mikkelsen M H 2014 *Nature Photonics* **8** 835-840

[4]   Radha B and Kulkarni G U 2012 *Current science* **102**,70-77

[5]   Esfandyarpour M, Garnett E C, Cui Y, McGehee M D and Brongersma M L 2014 *Nature Nanotechnology* **9** 542-547

AOM2015   IOP Publishing
Journal of Physics: Conference Series **680** (2016) 012034   doi:10.1088/1742-6596/680/1/012034

Measuring acetone using microstructured optical fiber and Raman spectroscopy

Fenghong Chu, Jianping Wu

Shanghai University of Electric Power, Shanghai, 200090, China

chufenghong@siom.ac.cn

Abstract. A novel approach using microstructured optical fiber and Raman spectroscopy for identifying acetone is reported. This technique combines the advantage of small sampling volume of microstructured optical fiber and the specificity of Raman spectroscopy.

1. Introduction

Acetone is toxic and explosive gas. Inhale of acetone gas will do obviously harm to the mucous membrane and living in the circumstance with acetone toxic symptoms will appear [1].Besides, Acetone is an exhaled volatile organic compound that has been shown to act as a biomarker for metabolic conditions in the bloodstream [2,3]. Accordingly, persistent efforts have been directed so far to develop the detection of acetone by many analytical methods, such as gas chromatography [4], electrochemistry detection [5], fluorescence quenching method [6,7], mass spectrometry [8], etc. However, disadvantages such as poor selectivity, high operating temperature, long response times [5, 9] and sample contamination [8] are associated with these methods.

A novel approach for identifying acetone is reported, using microstructured optical fiber (MOF) and Raman spectroscopy in this paper. MOF with small core dimensions result in a significant overlap of the guided light with any liquid or gas that is loaded into the holes in the fiber, which in turn creates a response signal along the entire length of the fiber that is subsequently waveguided by the core to a detector [10], what's more enable the development of small sample volume, flexible and cost-effective sensing architectures [11]. Raman spectroscopy is a powerful analytical technique for detecting a wide range of chemicals in any state (solid, liquid, or gas phase) and is widely used in laboratories for chemical identification [12]. In this work, MOFs were used as an active dip sensor platform for Raman sensing of acetone in water solutions using the solvent's Raman signature as an internal calibration standard.

2. Experimental section

The experimental setup used in these experiments is shown in Fig.1 which is the same as Ref. [13].

Content from this work may be used under the terms of the Creative Commons Attribution 3.0 licence. Any further distribution of this work must maintain attribution to the author(s) and the title of the work, journal citation and DOI.
Published under licence by IOP Publishing Ltd

AOM2015   IOP Publishing
Journal of Physics: Conference Series **680** (2016) 012034   doi:10.1088/1742-6596/680/1/012034

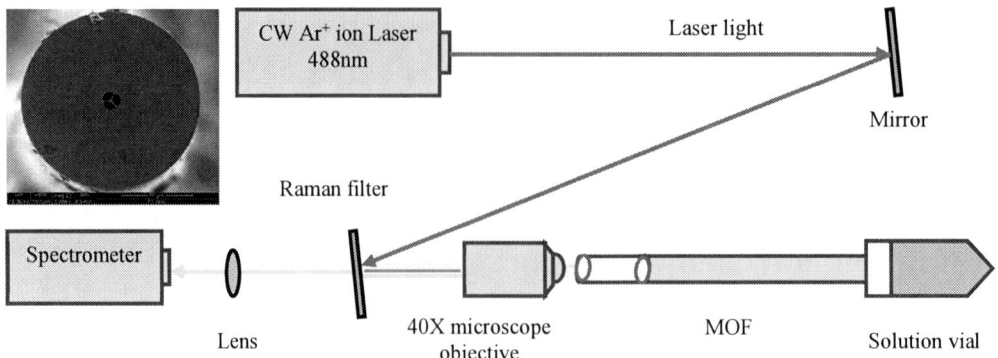

Figure 1. Schematic of experimental setup

Continuous wave light from a 488nm Ar+ laser was reflected off a long-pass Raman filter and coupled into the silica MOF (made by Institute for Photonics and Advanced Sensing, The University of Adelaide, shown in the insert of figure 1) using a 40x microscope objective. The length of MOF is 16 cm resulting in a total sampling volume of 60nL (the diameter of the air void is about 6.3μm). The other end of the fiber was dipped into the sample, capillary forces drew the acetone solution into the air voids along the length of the fiber. Backscattered Raman signal from the fiber was collected through the long pass filter and recorded by a fiber-coupled cooled-CCD spectrometer.

Acetone solutions were prepared by mixing pure acetone with milliQ water. Figure 2 is the Raman spectrum of MOF filled with 50%(volume ratio) acetone in water. The resulting spectrum for acetone and water are easily distinguishable. The strong Raman peak at 2856cm-1 is the C-H stretching peak of acetone which can be used to determine the concentration of acetone by its' relative intensity. The broad Raman peak centered at 3371cm-1 is H-O stretching peak of water which can be used as an internal calibration signal.

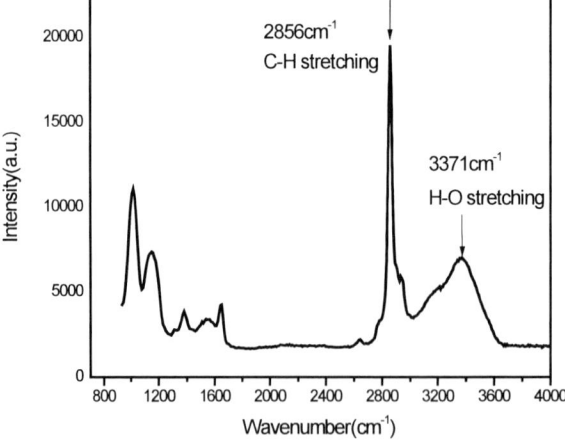

Figure 2. Raman spectroscopy of MOF filled with 50% acetone

Raman spectroscopy of MOF with different filling length (50% acetone) can be studied, as shown in figure 3(a). Both signatures increase in intensity as the fiber fills up with acetone solution by capillary force action. As an indication of the total intensity of the Raman scattering for each molecule, the area under the spectral curve is integrated and the time evolution of Raman intensity of acetone and water is

188

shown in figure 3 (b). The integrated intensities of the acetone and water Raman peaks increase in parallel as the fiber fills up.

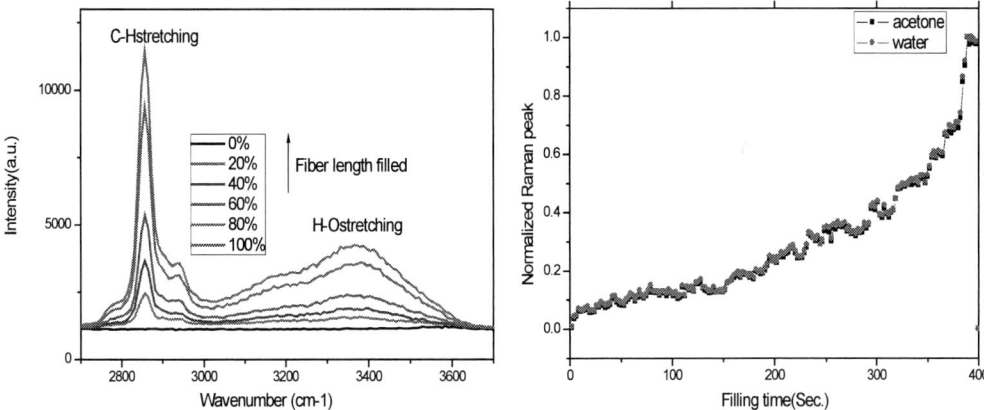

Figure 3. (a) Raman spectroscopy of MOF with different filling length (50%acetone) (b) Curves

of Raman peaks of acetone and water with filling time

The ratio of the Raman peak intensities of acetone and water is calculated throughout the measurement and the value is plotted in figure 4 against the known sample concentration. The observed linear relationship between the Raman intensities ratio and the acetone concentration allows the use of this internal calibration technique to measure the concentration of acetone in water that does not depend on input laser power fluctuations, coupling instabilities and other changes in the fiber environment [13].When the concentration of acetone is 1%, the Raman peak of acetone at 2856cm-1 is nearly invisible. However, the detection limit can be lower down by Surface Enhanced Raman Spectrum(SERS)which is a useful technique resulting in strongly increased Raman signals from molecules which have been attached to nanometer sized metallic structures [14].

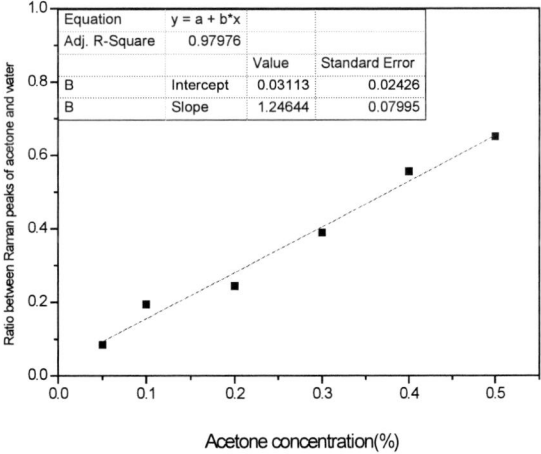

Figure 4. The relationship between acetone concentration and Raman ratio

3. Conclusions

In this paper acetone was detected by using MOF and Raman spectroscopy. This is based on an unmodified MOF as a Raman sensing platform, making use of the relatively large power fractions of excitation light available in this geometry to interact with analyte molecules along long lengths of the fiber. By using the Raman signature of water as an internal calibration standard, quantification of the acetone concentration is possible.

Acknowledgments

The authors acknowledge Institute for Photonics and Advanced Sensing, The University of Adelaide for providing the fibers, Georgios Tisminis for fruitful discussions and National Natural Science Foundation of China (Grant No. 61205081), Innovation Program of Shanghai Municipal Education Commission (15ZZ086), Opening Project from Key Laboratory of Specialty Fiber Optics and Optical Access Networks of Shanghai University for financial support.

References

[1]   Xiao Chi, Changbai Liu, Li Liu, "Tungsten trioxide nanotubes with high sensitive and selective properties to acetone," Sensors and Actuators B, 194, 33-37(2014).

[2]   Manolis, "The diagnostic potential of breath analysis," Clin. Chem,29, 5-15(1983).

[3]   AdamD.Worrall, JonathanA.Bernstein, AnastasiosP.Angel opoulos, "Portable method of measuring gaseous acetone concentrations," Talanta, 112,26-30(2013).

[4]   Dong L ing, Shen X izhong, Deng Chunhu, "Development of gas chromatography mass spectrometry following headspace single drop microextraction and simultaneous derivatizat ion for fast determination of the diabetes biomarker acetone in human blood samples," Analytica Chimica Acta,569,91- 96 (2006).

[5]   Jichao Shi, Gujin Hu, Yan Sun, "WO3 nanocrystals: Synthesis and application in highly sensitive detection of acetone," Sensors and Actuators B, 156,820- 824(2011).

[6]   Vijay Kumar Sharma, D. Mohanb, P.D. Sahare "Fluorescence quenching of 3-methyl 7-hydroxyl Coumarin in presence of acetone," Spectrochimica Acta Part A,66,111-113(2007).

[7]   FY Yi, W Yang, ZM Sun, "Highly selective acetone fluorescent sensors based on microporous Cd(II) metal–organic frameworks," Journal of Materials Chemistry, 22,23201-23209 (2012).

[8]   A. Amann, G. Poupart, S. Telser, M. Ledochowski, A. Schmidt, and S. Mechtcheriakov, "Applications of breath gas analysis in medicine,"Int. J. Mass Spec., 239, (2004).

[9]   N. Makisimovich, V. Vorotyntsev, N. Nikitina, O. Kaskevich, P. Karabun, and F. Martynenko, "Adsorption Semiconductor Sensor for Diabetic Ketoacidosis Diagnosis,"Sensors and Actuators, B: Chemical,36, 419-421(1996).

[10]   Monro, T.M. Warren-Smith, S.,Schartner, E.P.,François, A,Heng, S., Ebendorff-Heidepriem, H., Afshar, S., "Sensing with suspended-core optical fibers," Opt. Fiber Technol, 16, 343-356(2010).

[11]   Stephen C. Warren-Smith, Shahraam Afshar V., and Tanya M. Monro, "Fluorescence-based sensing with optical nanowires: a generalized model and experimental validation.," Opt. Express , 18,9474-9485(2010)

[12]   Anupam K. Misra, Shiv K. Sharma, Tayro E. Acosta, David E. Bates, "Detection of Chemicals with Standoff Raman Spectroscopy," Spectroscopy, Special Issues, Apr 1, (2011)

[13]   Tsiminis,Georgios, Chu, Fenghong,Warren-Smith, Stephen C., Spooner, Nigel A., Monro, Tanya M, ".Identification and quantification of explosives in nanolitre solution volumes by Raman spectroscopy in suspended core optical fibers," Sensors, 13,13163-13177(2013)

[14]   G. Upender, R. Satyavathi, B. Raju, K. Shadak Alee, D. Narayana Rao, C. Bansal, "Silver nanocluster films as novel SERS substrates for ultrasensitive detection of molecules," Chemical Physics Letters, 511, 309-314(2011)

Direct phase extraction of self-mixing displacement measurement using Hilbert transform

Yufeng Tao, Ming Wang*, Dongmei Guo and Jiahuan Zhang

Department of Physics Science and Technology, Nanjing Normal University, Jiangsu Key Laboratory on Opto-electronic Technology, Nanjing, 210023, P. R. China

wangming@njnu.edu.cn

Abstract. Signals of a self-mixing interferometer established on a semiconductor laser diode are analyzed. Phase is extracted out for decoding measurement which contained in self-mixing fringes. The semiconductor laser diode works as light source and receiver without modulation. By combining Hilbert transform with phase condition of self-mixing interference, micron-displacement is reconstructed by phase information at week or even moderate feedback level. Theoretical analysis and simulation results are presented before verification of experimental measurement. Practical feedback level is estimated by a data fitting technique with a programmable high-resolution PZT. Consistence of the results promises that direct phase extraction on self-mixing interferometer is available for micron-displacement measurement with a nanometer accuracy.

1. Introduction

There is a continuously rising requirement of displacement monitoring with a high accuracy as well as a wide dynamic range in semiconductor industries or biotechnology. In recent years, self-mixing interference (SMI) has been paid to considerable attention due to compact structure and easy demodulation. Fundamental theoretical three model of SMI had been constructed in Refs [1,2,3,4] which resulted in potential applications of measuring micron displacement or vibration as Refs [5,6,7]. Among piles of lasers employed in SMI, semiconductor laser diode which is equipped with an in-built photo diode gradually becomes an ideal light source and receiver in SMI interferometer without high cost. Consequently, a variety of demodulation methods have emerged as Refs[8,9,10] reported, for instance, triangle frequency modulation is introduced into SMI system by Q.Wu and his team, an accuracy more than 98% is achieved by adjusting a sampling frequency. D.Guo of our team introduces double modulations into SMI for obtaining a more competitive accuracy of $\lambda/65$.

However, technologies by injecting current modulation or introducing phase shift are imperfect. The electronics or modulators make SMI interferometer complicated and expensive, the feedback level is limited within very weak level to neglect multi-reflection effect, besides, frequency of modulation frequency will limit the bandwidth of measured displacement, all of these problems need desirable solutions. In this paper, the simplest SMI configuration is analyzed by a novel phase extraction method based on Hilbert transform, feature of displacement is captured by extracting the phase of the interference signal. Semiconductor laser driven by constant current avoids unwanted intensity fluctuation or limitation of bandwidth, which allows a larger range of feedback level comparing with previous SMI interferometer.

Content from this work may be used under the terms of the Creative Commons Attribution 3.0 licence. Any further distribution of this work must maintain attribution to the author(s) and the title of the work, journal citation and DOI.

Published under licence by IOP Publishing Ltd

Although a lot of algorithm methods have been employed in SMI for demodulation, such as Fourier transform method, beat frequency method and modified fringe counting and so on. But Fourier transform method is based on current modulation, beat frequency method is performed requiring additional reference optical path, accuracy of modified fringe counting is convenient but not high as phase demodulation technology.

The proposed phase extraction without modulation is appropriate for existing SMI system because of easy implementation and immunity to current fluctuation. Phase information of weak, moderate and high feedback level self-mixing signal is analyzed at first, then, measurement principle is demonstrated with numerical simulations before introduction of the experimental setup. In last section comparison between the SMI and Poly Tec 5000 vibration meter is given for drawing the conclusion that direct phase extraction method achieves a nanometer accuracy at weak or moderate feedback level.

2. Measurement Principle

2.1. Modulated phase obtained by Hilbert transform

The basic theory of self-mixing interference has been deeply studied by previous authors, behavior of semiconductor laser is described using a three-mirror model with optical feedback. Implementing Hilbert transform on three-mirror model is an effective method to reflect modulated phase, its process is presented as follows.

Assuming reflectivity of three mirrors are r_1, r_2 and r_3, travel time of light in resonant and external cavity are τ_1 and τ, which are expressed:

$$\tau_1 = \frac{2n_0 l}{c}, \tau = \frac{2L}{c} \tag{1}$$

where l denotes length of the resonant cavity, L denotes length of the external cavity. The whole phase condition of self-mixing interference is written:

$$\varphi_0(t) - \varphi_F(t) = C\sin(\varphi_F(t) - \arctan(\alpha))$$

$$C = (1 - r_2^2)r_3 \frac{\tau}{\tau_1}\sqrt{1 + \alpha^2}/r_2$$

$$P = P_0[1 + m\cos(\varphi_F(t))]$$

$$\alpha = \frac{\chi}{\rho} \tag{2}$$

where C stands for feedback level, α is line-width enhancement factor depending on change rate of refractive index χ and stimulated emission gain index ρ of laser , m is the undulation coefficient determined by reflectivity of the target mirror. P_0 and P are optical power without and with feedback, correspondingly, $\varphi_0(t)$ and $\varphi_F(t)$ are time-dependent phase of the external cavity without and with feedback respectively. The optical power is often observed with various levels of fringe inclination due to that SMI interference is operated at different feedback level.

For the bias part and amplitude of SMI signal can be easily and precisely measured by existing tool in software, the optical power expressed in equation (2) usually should be normalized as below before Hilbert transform:

$$\cos(\varphi_F(t)) = (P/P_0 - 1)/m \tag{3}$$

To extract modulated phase information $\varphi_F(t)$, Hilbert transform is performed on $\cos(\varphi_F(t))$ firstly, mathematical equation of which is expressed:

$$H[\cos(\varphi_F(t))] = \frac{1}{\pi}\int_{-\infty}^{\infty}\frac{\cos(\varphi_F(t))}{t - \tau}d\tau \tag{4}$$

Hilbert transform of normalized SMI signal is consisted of a real part and an imaginary part. The imaginary part is introduced with an ideal $90°$ phase shift, which is equal to $\sin(\varphi_F(t))$, so, $\varphi_F(t)$ is extracted by arc-tangent calculation:

$$\varphi_F(t) = arctangent(\frac{Imag[H(cos(\varphi_F(t))]}{cos(\varphi_F(t))}) \tag{5}$$

Unwrapping result of equation (6) is a continuous modulated phase information curve.

2.2. Correction by phase condition.

As analyzed in front selection, weak feedback level decreases the possibility of mode-hoping dramatically, laser frequency is independent to the optical feedback, the $\varphi_F(t)$ is approximate to $\varphi_0(t)$. But higher feedback level, larger the difference between $\varphi_F(t)$ and $\varphi_0(t)$ is. It is undeniable that the modulated phase is not linear to measured displacement, hence, a correction is added into its data process.

Combining phase condition of equation (2), the $\varphi_0(t)$ can be expressed as:

$$\varphi_0(t) = \varphi_F(t) + Csin(\varphi_F(t) - arctan(\alpha)) \tag{6}$$

where $Csin(\varphi_F(t) - arctan(\alpha))$ is called as modification factor in next paragraphs, the line-width enhancement factor is induced by below equation:

$$\alpha = \pm\sqrt{1 - \{\frac{\tau_1 r_2}{\tau} C/[(1 - r_2^2)r_3]\}^2} \tag{7}$$

When variation of external cavity length is at micron level, the variation of travel time τ is neglected. Reflectivity r_2 is an intrinsic parameter of semiconductor laser diode, reflectivity of the target mirror is a stable value as well, α is approximately determined by feedback level C conclusively. Instantaneous displacement between the laser and the object is linear to initial phase information as below:

$$\varphi_0(t) = 4\pi L(t)/\lambda_0 \tag{8}$$

where λ_0 is central wavelength of the single mode laser. The relationship is used to obtain displacement curve.

To illustrate the above principle, the number simulation on a computer is shown in below figures, which plot the phase curve, normalized SMI signal, imaginary part of Hilbert transform, and the results of equation (5). Data sampling ratio is set at 250KHz/s, data point number is 2500, the frequency of phase converted from displacement is 100Hz, and amplitude is 20rad, initial phase is zero. Feedback levels in Figure 1 and Figure 2 are 0.5 and 1.5 respectively. The line-width enhancement factor is equal to 4 at both figures.

The Figure 1 and Figure 2 indicate a good point and a weak point, Hilbert transform indeed has a desirable ability in phase extraction, however, arc-tangent calculation is insensitive to changes in direction of displacement.

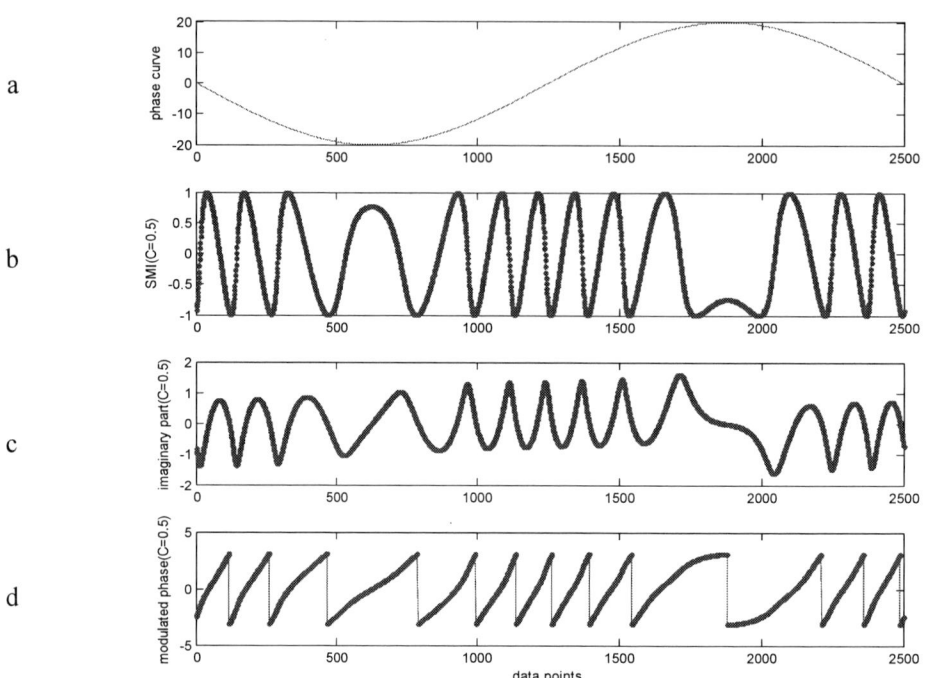

Figure 1. (a). The sinusoidal phase curve in simulation. (b). Self-mixing signal. (c). Imaginary part of Hilbert transform on picture(b). (d). The modulated phase $\varphi_F(t)$.

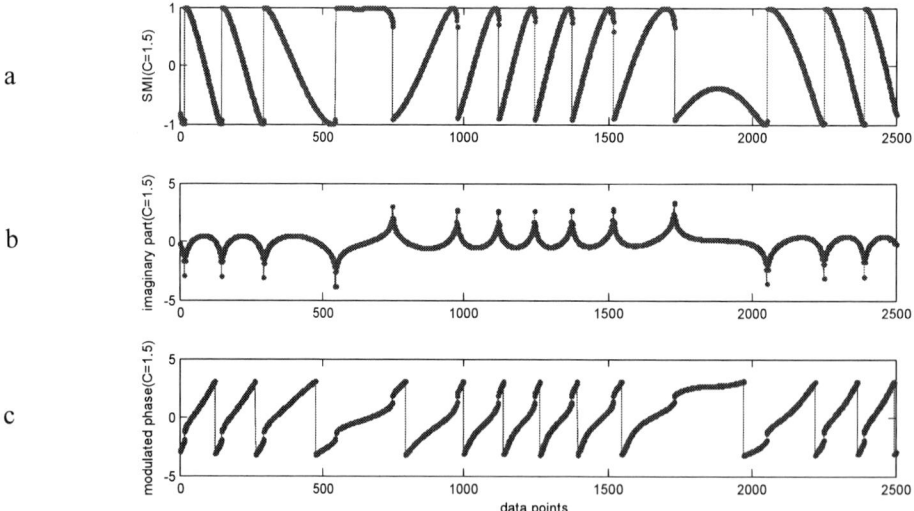

Figure 2. Self-mixing signal with feedback level is 1.5. (b). Imaginary part of Hilbert transform on picture (a). (c). The modulated phase $\varphi_F(t)$.

AOM2015   IOP Publishing
Journal of Physics: Conference Series **680** (2016) 012035   doi:10.1088/1742-6596/680/1/012035

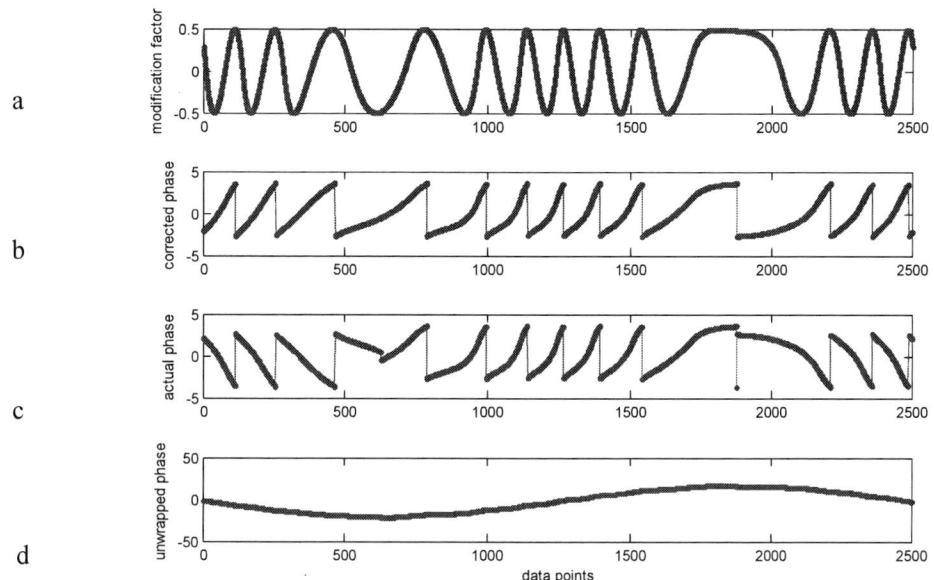

Figure 3. (a). Modification factor of Fig.1. (b). Curve of $\varphi_0(t)$.(c). Multiplication result of $\varphi_0(t)$ with sign function. (d). The unwrapped real phase $\varphi_0(t)$.

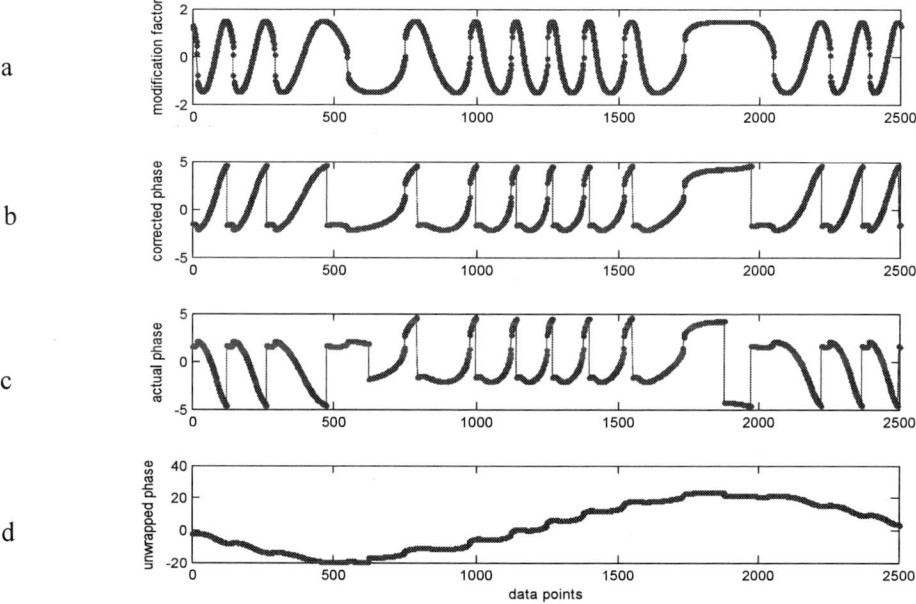

Figure 4. (a). Modification factor of Fig.2. (b). Curve of $\varphi_0(t)$.(c). Multiplication result of $\varphi_0(t)$ with sign function. (d). The unwrapped real phase $\varphi_0(t)$.

Subsequently, the equation (6) is illustrated in Figure 3 and Figure 4 with modification factor. A sign function is used to depict the phenomenon of fringe inclination in mathematics. When fringes incline to left,

195

the sign is negative, when fringes incline to right, the sign is positive. Multiplying $\varphi_0(t)$ with the sign of direction, the true phase can be reconstructed at weak and moderate feedback levels as shown in the picture(c) and (d) of Figure 3 and Figure 4. After unwrapping, actual $\varphi_0(t)$ is modified in a new expression:

$$\varphi(t) = \text{sign} \times \varphi_0(t) \pm 2n\pi \tag{9}$$

where n is an integer, $\text{sign} = \pm 1$, $\varphi(t)$ is a continuous phase curve in time domain.

2.3. Error discussion

To evaluate the accuracy, the simulative error is calculated by computing standard deviation. Peak to peak values of the reconstructed phase curve in Figure 3 are -19.91rad and 20.07rad, peak to peak values of reconstructed phase curve in Figure 4 are -19.93rad and 20.09rad. The amplitude errors are both less than 0.12rad which prove the availability of phase extraction method. Equation (10) is used to calculate the standard deviation between the reconstructed phase and the ideal phase:

$$\sigma = \sqrt{\sum_{i=1}^{1000} (\varphi_{ext}(i) - \varphi_{ideal}(i))^2 / 2499} \tag{10}$$

Assuming reconstructed phase is sampled at the instant $t_i = i/(Nf_s)$, sampling ratio is f_s Hz/s, number of data is N =2500, $\varphi_{ext}(i)$ represents the extracted phase at the i-th point, $\varphi_{ext}(i) = \varphi(t_i)$, $\varphi_{ideal}(i)$ represents an ideal phase at the i-th point. Standard deviations of two reconstructed phase curves are 0.043rad, 0.067rad.

The errors of Figure 3 and Figure 4 between reconstructed phase and initial phase curve are plotted in Figure 5.

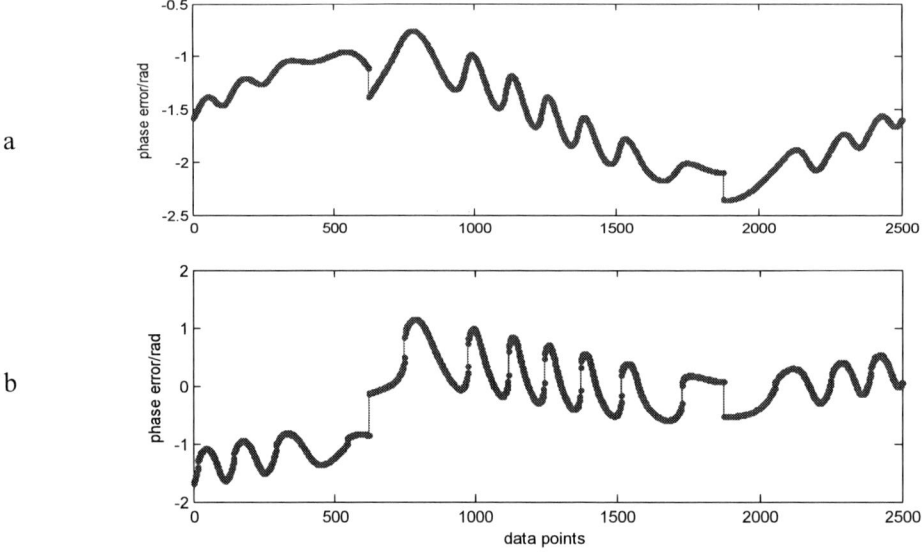

a

b

Figure 5. (a). Phase error obtained by subtraction between reconstructed phase of Figure 3 and initial phase. (b). Phase error curve of Figure 4.

Displacement $L(t)$ is calculated with equation (11):

$$L(t) = \lambda_0 \varphi_{ext}(t)/4\pi \tag{11}$$

Corresponding standard deviations of displacement are 2.47nm and 3.13nm. It is obviously seen that, proposed method is applicable at weak or moderate feedback level in displacement measurement.

3. Experimentation

3.1. Experimental setup

The experimental setup as illustrated in Figure 4 is mainly divided into 3 parts: the SMI part, the data process part and the comparison part. With the merits of long lifespan, inexpensive price and tiny size, semiconductor laser diode (Thorlabs, L650P007-650,USA) works as light source and receiver in the established experimental system. An aspheric compensating len is inserted in the optical path to focus emission of semiconductor laser into a point for alignment, 22mA constant current is output from precise supplier (ILX-Lightwave, LDX-3220,USA) for driving semiconductor. In order to control the intensity of reflected light from the target mirror, a various neutral density filter (ND) is used.

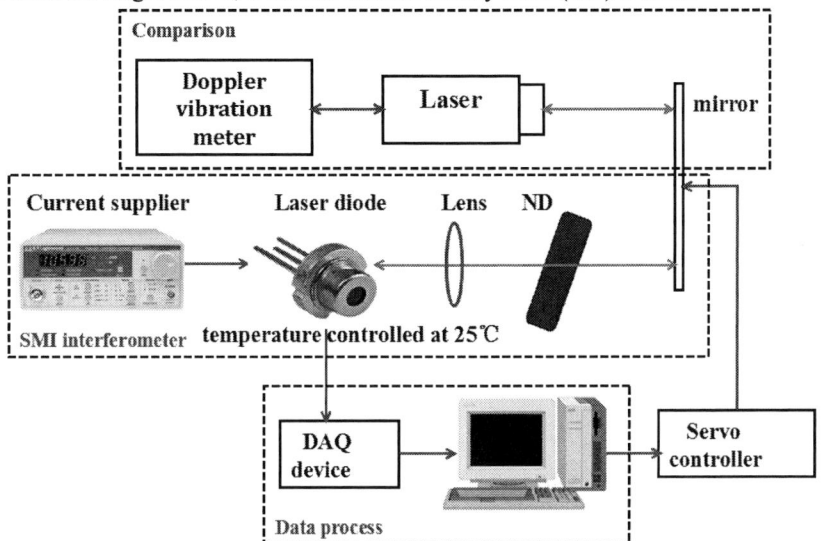

Figure 6. The experimental setup of the semiconductor self-mixing interferometer

The object mirror is fixed on a commercial PZT (P-841, Physik Instrument Co., Germany) whose displacement range is 20um and resolution is 1nm. A 16-bit A/D converter (USB-6361, National Instrument Co., USA) is used to act for data sampling and analog to digital conversion. The digital signal is filtered for removing noises and magnified in pre-process in a personal computer (PC), the SMI signal can be observed simultaneously by connecting to Tektronix oscilloscope as shown Figure 6.

3.2. Estimation of parameters in self-mixing model
All parameters appearing in simulation are well known beforehand. But before analyzing experimental data, the feedback level needed to be estimated as accurate as possible, a data fitting method in Ref [11] is adopted to estimate the value of feedback level C. With the help of programmable PZT, a group of self-mixing signals are generated in Figure 7 with displacement volts.

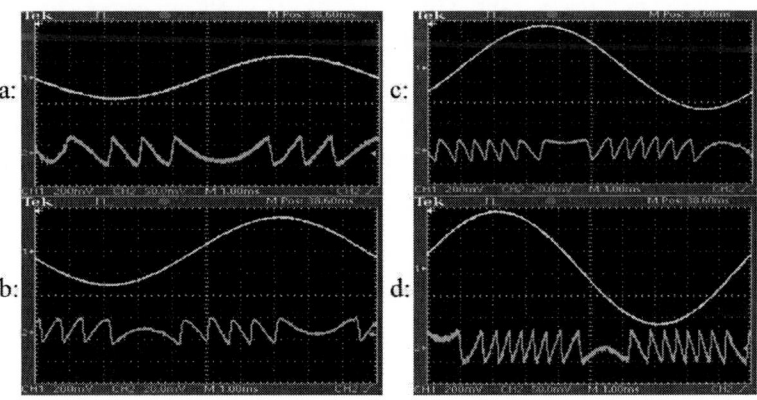

Figure7. The screen shots of digital Tektronix Oscilloscope TDS 2000C, upper curve is monitor volts of PZT, lower curve is self-mixing interference signals, displacement frequency is 100Hz, peak to peak amplitudes of displacement are (a)1.2um, (b)1.6 um, (c)2.1um and (d)2.8um.

Then, simulative self-mixing signal is generated and expressed:

$$A(C_{est}, \alpha_{est}, t_i) = \cos[\varphi_0(t_i) - C_{est}\sin((...) - \arctan(\alpha_{est}))] \quad (12)$$

where number of data point N is equal to that of observed self-mixing signals, $i \in (0,1,2....N-1)$, omitted part is $\varphi_F(t) = \varphi_0(t) - C\sin(\varphi_F(t) - \arctan(\alpha_{est}))$, C_{est} is estimated feedback level, α_{est} is estimated line-width enhancement factor, times of iteration of omitted part should exceed 10 to ensure signal veracity. Meanwhile, the direct intensity of experimental signal acquired by A/D card is removed and normalized written as:

$$B(C, \alpha, t_i) = \cos(\varphi_F(t_i)) \quad (13)$$

Set initial C_{est} and α_{est} at 0.5 and 1, then calculate average data deviation D as below:

$$D = \sum_{i=1}^{N} (A(C_{est}, \alpha_{est}, t_i) - B(C, \alpha, t_i))^2 / N \quad (14)$$

Increment of C_{est} is 0.01, equation (14) reaches its bottom value when C_{est} is most close to experimental feedback level. The estimated feedback level in Figure 6 are 0.87, 0.83 and 0.85. With the best fit C_{est}, α_{est} can be calculated on equation(7) as well.

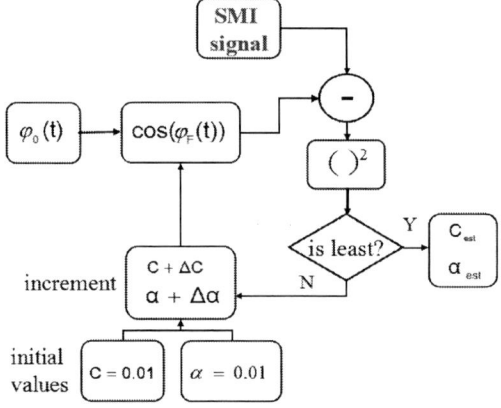

Figure 8. Process of iteration in data fitting, where SMI signal stands for observed signals.

3.3. Experimental results

One group of displacement measurement has been done to confirm the validity of the proposed method. Displacement of the target mirror was realized by driving the PZT with a frequency of 50 Hz and amplitude of 2000 nm (peak to peak) sinusoidal volts, results of SMI (red color) and Doppler vibration meter (blue color) are recorded and reconstructed error are presented, average error is 8.91nm and maximum error value is 16.7nm.

Setting PZT move in a sinusoidal wave with frequency of 100 Hz and amplitude of 3000 nm (peak to peak), another result in Figure 9 shows an average error of 18.1nm with maximum error of 29.5nm.

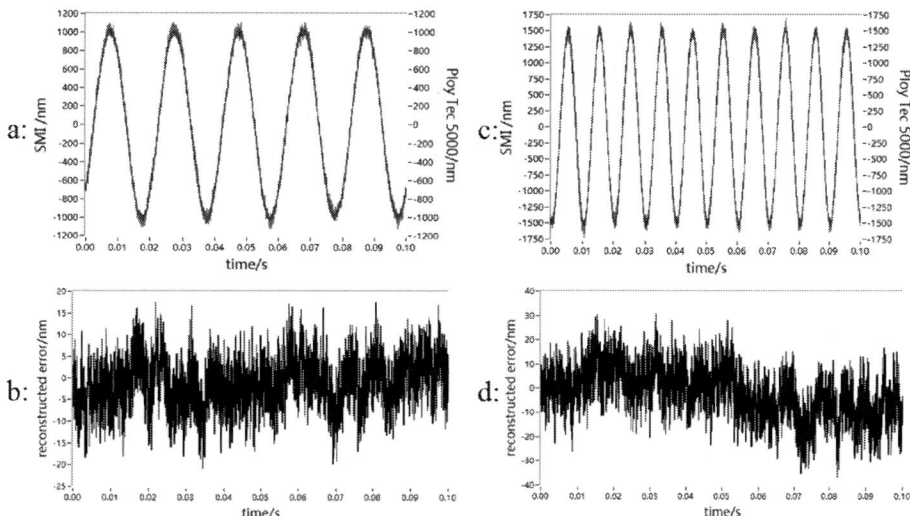

Figure 9. Experimental results: (a) The reconstructed displacement with SMI and Ploy Tec 5000 at peak-to-peak amplitude of 2000 nm and frequency of 50Hz. (b) Measurement error at amplitude of 2000 nm (p-p). (c) The reconstructed displacement at amplitude of 3000 nm (p-p) and frequency of 100Hz.(d) Measurement error at amplitude of 3000 nm (p-p).

Control program of NI DAQ device and the data process is written on Labview2014. Experimental results prove that the reconstructed displacement of the SMI is in good consistency with data from the Ploy Tec-5000 and verify that a displacement measurement accuracy of about 30 nanometers was achieved. In conclusion, the proposed method an improved phase demodulation.

4. Conclusion

The principle of direct phase extraction is presented with number simulations and experimentation. The modulated phase information is corrected by the phase condition of self-mixing and sign function denoting fringe inclination. The real phase is extracted from SMI signal without any reference signals during data process and reconstructed phase curves show an error less than 0.1rad in amplitude of 20rad. The dynamical range of the proposed method is without limitation theoretically, which is convenient and effective to implement in SMI interferometers or sensors.

To check the availability, a compact semiconductor self-mixing interferometer has been built and demonstrated as verification. No additional electronics or modulator exist in the experimental system, which is operated with a constant current at weak or moderated feedback level. The experimental signals mirror the simulative signals with laboratory noise. The parameters used in demodulation are estimated by data fitting between observed signals and simulative signals. Micron displacement measurement and comparison with Doppler vibration meter show an accuracy of less than 30nm has been achieved in measurement, which proves that the direct phase extraction is practical for monitoring micron-displacement position or displacement.

Acknowledgments

This work is supported by the National Natural Science Foundation of China (No.51405240，91123015), the Natural Science Foundation of Jiangsu Province of China (BK20140925) and the Program of Natural Science Research of Jiangsu Higher Education Institutions of China (14KJB510015).

References

[1] R. Lang, K. Lobayshi. "External optical feedback effects on semiconductor injection laser properties", IEEE.J.Quantum Elect.16(3):347-355,1980.

[2] W.M.Wang, W.J.O.Boyle, K.T.V.Grattan and A.W. Palmer. "Self-mixing interference in a diode laser: experimental observations and theoretical analysis", Applied Optics. 32:1551–8,

1993.

[3] S.Donati, G.Giuliani, and S.Merlo. "Laser Diode Feedback Interferometer for Measurement of Displacements without Ambiguity", IEEE J. Qautumn Elect.vol. 31, no. I: 113- 119, 1995.

[4] K.Petermann. "External optical feedback phenomena in semiconductor lasers ",IEEE J. Sel. Topics Quantum Electron.vol. 1, no. 2:480–489, 1995.

[5] G. Plantier, C. Bes and T. Bosch. "Behavioral Model of a Self-Mixing Laser Diode Sensor",IEEE J.Quantum Elect. vol. 41, no. 9: 1157-1167, 2005.

[6] D.Lenstra, B.Verbeek, and A.Den Boef."Coherence collapse in single mode semi-conductor lasers due to optical feedback",IEEE J. Quantum Elect.25(6):1143-1151,1989.

[7] J.Zhou, M. Wan,and D. Han. "Experiment observation of self-mixing interference in distributed feedback laser", Opt.Express. 14(12):5301-5306,2006.

[8] Q.Wu,S.Shinohara, H.Ikeda, and H.Yoshida. "Applications and Accuracy Improvement of Vibrometer Using Frequency Modulated Self-Mixing Laser Diode", IEEE Instrumentation and Measurement Technology Conference St. Paul, USA. 18-21,1998.

[9] D.Guo, M.Wang. "Self-mixing interferometry based on a double modulation technique for absolute distance measurement", Applied Optics. 46(9):1486-1491,2007.

[10] Y.Tao,M.Wang.D.Guo.X.Ni and H.Hao. "Self-mixing vibration measurement using emission frequency sinusoidal modulation", Optics Communications.340:141-150,2015.

[11] J.Xi, Y.Yu, J. F.Chicharo and T. Bosch. "Estimating Parameters of Semiconductor Lasers Based on Weak optical feedback Self-Mixing Interferometry". IEEE JOURNAL of Quantum Electonics, 41(8), 2005.

| AOM2015 | IOP Publishing |

Convex Aspherical Surface Testing Using Catadioptric Partial Compensating System

Jingxian Wang[1], Qun Hao[2], Yao Hu[#, 3], Shaopu Wang, Tengfei Li, Yuhan Tian, Lin Li

Beijing Key Lab. For Precision Optoelectronic Measurement Instrument and Technology, School of Optoelectronics, Beijing Institute of Technology, Beijing, 100081, China

[1]wangjx0506@sina.com, [2]qhao@bit.edu.cn, [3]huy08@bit.edu.cn, [#] Corresponding Author

Abstract. Aspheric optical components are the indispensable part of modern optics systems. With the constant development of aspheric optical fabrication technique, the systems with large aperture convex aspheric optical components are widely used in astronomy and space optics. Thus, the measurement of the figure error of the whole convex aspherical surface with high precision comes to be a challenge in the area of optical surface manufacture, and surface testing method is also very important. This paper presents a new partial compensating system by the combination of a refractive lens and a reflective mirror for testing convex aspherical surface. The refractive lens is used to compensate the aberration of the tested convex asphere partially. The reflective mirror is a spherical mirror which is coaxial to the refractive lens and reflecting the lights reflected by the tested convex asphere back to the convex asphere itself. With the long focal length and large aperture system we can realize a lighter and more compact system than the refractive partial compensating system because the spheric reflective mirror is more easily to realize and can bending the light conveniently.

1. Introduction

With the rapid development of astronomy, military and space technology, aspheric optical element has been widely used in optical system to correct aberrations, improve the image quality and simplify the structure of the system, so as to decrease the weight and volume of optical system and decrease the cost[1,2]. Thus, the research of high precision, simple and practical aspheric surface testing technology comes to be an urgent problem to be solved in the application of aspherical surface.

Nowadays, interference measurement is one of the most widely used method in aspheric surface testing with high accuracy[3]. The traditional null compensation testing is a small residual wavefront testing method. The normal aberrations of the aspheric surface are completely compensated by the longitudinal spherical aberration of the null compensator[2]. Thus, null compensator usually requires more lenses[4-5], which is difficult to design and manufacture[6-11]. A method for testing asphere by partial-compensator and digital moiré phase-shifting interferometry was proposed by our research group[12]. Partial compensating method can allow more residual wave aberration, so it can reduce the requirements for optical system design. In addition, the requirtment of processing of partical compensator is lower than null compensator and the requirement of system installation is lower than null testing.

Content from this work may be used under the terms of the Creative Commons Attribution 3.0 licence. Any further distribution of this work must maintain attribution to the author(s) and the title of the work, journal citation and DOI.

Published under licence by IOP Publishing Ltd

However, testing convex aspheric surface is really a trouble in the area of aspheric surface measuring because convex surface can diverge rays, which must be collected and reflected by an optical element, lens or mirror, of sufficient size. This paper presents a new partial compensating system using catadioptric partial- compensator for testing convex aspherical surface. Researches are made on measurement principle, design and optimization method of compensating system. A catadioptric partial-compensator for a specific high- order convex aspheric mirror is designed, and simulation indicates the aberration of this convex asphere is well compensated.

2. Measurement Principle and Methods

2.1. Principle of partial compensating interferometry

A novel interferometric method for measuring the surface figure error of asphere was presented by our research group. It is based on partial compensating principle and digital moiré phase-shifting technique[12].

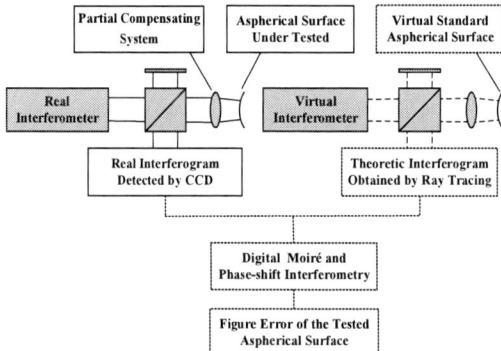

Figure 1. Measurement principle.

Figure 1 shows the measurement principle. Aspheric surface testing can be realized by an improved Twyman- Green interferometer. Partial-compensator and the tested asphere are placed into the testing arm of the real interferometer. Real interferogram including the information of the surface error of the tested asphere is detected. In the virtual interferometer, standard reference aspheric surface is produced by digital technology, and thus the fabrication errors of the standard asphere can be avoided. The nominal structure parameters of the partial- compensator are used, and the theoretical interferogram is calculated by ray tracing. The surface error of the tested asphere can be obtained by using digital moiré and phase-shifting interferometry to compare real interferogram with theoretical interferogram.

In this method, partial-compensator is not required to compensate the large wave aberration produced by the aspheric surface under test completely. Some residual wave aberration is allowed to exist, and it is limited by the spatial resolution of CCD for capture the interference fringe.

2.2. Design of catadioptric partial-compensator

As we all know, it is more difficult to test convex aspheric surface than the concave ones, because we need a converging beam produced by a lens or a mirror of sufficient size. This paper presents a new partial compensating system by the combination of a refractive lens and a reflective mirror for testing convex aspherical surface. The compensating optical path displays in Figure 2.

AOM2015

Journal of Physics: Conference Series **680** (2016) 012036

IOP Publishing

doi:10.1088/1742-6596/680/1/012036

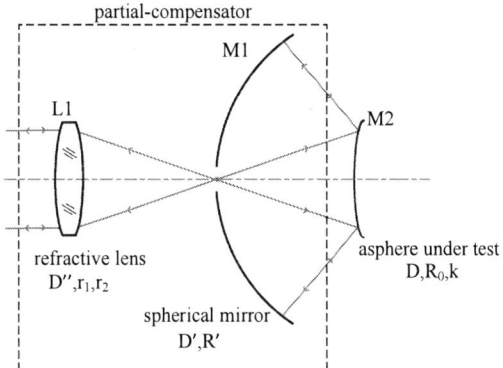

Figure 2. Compensating optical path.

L1 is a refractive lens which is used to compensate the aberration of the tested convex asphere partially. The apeture and curvature radius of L1 is D'', r_1 and r_2. M1 is a spherical reflective mirror which is coaxial to the refractive lens and reflecting the rays reflected by the tested convex asphere back to the convex asphere itself. And the apeture and curvature radius of M1 is D' and R'. The vertex of spherical mirror is located near the focal plane of the compensating lens L1. M2 is the convex aspheric surface under test with the vertex curvature radius R_0, aperture D and conicoid coefficient k. As is shown in Figure 2, a collimated beam exiting from the interferometer converged by the refractive lens firstly and then pass through the center hatch of spherical mirror and reflected by aspheric mirror directly into spherical mirror, and then returned to aspheric surface. When the beam is reflected by the aspheric mirror the second time, it transmits back into the interferometer in the original direction.

2.2.1 Calculation of initial configuration
The cross section equation of aspheric surface is described as

$$x = \frac{cy^2}{1+\sqrt{1-(1+k)c^2y^2}} + Ey^4 + Fy^6 + Gy^8 + Hy^{10} \tag{1}$$

in which, c is the vertex curvature of aspheric surface, $c=1/R_0$, and R_0 is the vertex curvature radius. k is the coefficient of conicoid surface, $k=-e^2$, and e is the eccentricity of conicoid curve. E, F, G, H are the coefficient of high-order aspheric surface.

As is shown in Figure 2, the edge ray in the system is incident to the edge of the tested aspheric mirror, and can be reflected by the sphere and transmits back to the interferometer in the original path. In this case, the parameters D' and R' of the spherical mirror can be calculated by ray tracing with some geometrical relationship and Reflection Law.

The parameters r_1 and r_2 of the refractive lens can be obtained by third-order aberration theory. As is shown in Figure 2, the collimated beam exited from the interferometer refracted by refractive lens for two times, reflected by the spherical mirror for one time and reflected by the aspheric mirror for two times, and then transmitted to the origin eventually. Thus, third-order aberration is balanced. That is to say, in order to achieve the purpose of compensation, spherical aberration generated by the small aperture compensating lens and spherical mirror with large aperture is used to counterbalance the normal aberrations of the aspheric surface. The third-order spherical aberration coefficient of aspheric surface and partial-compensator satisfy the following relations

$$2S_{IL1} + S_{IM1} + 2S_{IM2} = 0 \tag{2}$$

in which, S_{IL1}, S_{IM1}, S_{IM2} represent third-order spherical aberration coefficient of refractive lens, spherical mirror and aspheric mirror under test respectively. The expression of spherical aberration coefficient S_I is[2,13]

203

$$S_I = \Sigma hP + h^4 K \tag{3}$$

in which,

$$P = \left(\frac{u' - u}{\frac{1}{n'} - \frac{1}{n}} \right)^2 \cdot \left(\frac{u'}{n'} - \frac{u}{n} \right) \tag{4}$$

$$K = -\frac{e^2}{R_0^3} \left(n' - n \right) \tag{5}$$

h is the height of the light entering and leaveing at each mirror. n and n' are the refractive index or reflective index of light before and after the refraction or reflection. u and u' are the angle between the ray and optical axis before and after the refraction or reflection.

According to formula (2)~(5) and paraxial optical formula $n'u' - nu = \frac{h}{r}(n-1)$, the parameters r_1 and r_2 of the compensating lens can be solved.

2.2.2 Optimization of partial compensating system
According to the above analysis, the initial configuration of the compensating system are obtained. We put them into Zemax optical design software and set the curvature radius of lens and mirror and the distance between mirrors as the optimization variable. The optimization variables set in Zemax are less than those of null compensating system because the structure of partial-compensator is simpler than that of null-compensator.

The optimization target is the root-mean-square wavefront. Select the appropriate operation to control the edge ray. As the result, the edge ray refracted by refractive lens can be incident to the edge of the tested aspheric mirror. Thus, the full aperture of the aspheric surface can be well detected by controlling the edge ray.

The principle of optimization of Zemax optical design software is Damped Least Square Method. We optimize the partial compensating system in Zemax and the partial compensating system is obtained. The result of optimized design is asked to meet the requirements that the interference fringes can be detected clearly by CCD with 1024 * 1024 pixels, that is to say, the maximum slope of residual wavefront should be less than 0.45λ/pixel[14].

3. Simulation Experiment
Taking a convex aspheric mirror as an example, the partial-compensator is designed. This aspheric mirror is a high-order convex aspheric surface with the relative aperture D/R_0 equal to 1:1.5. Detailed parameters of the high-order aspheric surface are shown in Table 1.

Table 1. Parameters of the high-order convex aspheric under test

D	R_0	k	E	F	G
15.4mm	25.56mm	-1.01	3.2703958e-06	7.7205335e-10	1.6304727e-13

The design and optimization of partial-compensator is performed in the following steps:

(1) Set the system parameters

Select wavelength as 532nm. The entrance aperture of the partial compensating system is 80mm, which is the diameter of exit pupil of the interferometer.

(2) Calculate the initial configuration of partial-compensator

Set the distance between the spherical mirror and the aspheric mirror as 50mm and the refractive lens is designed with aperture D''=90mm. The parameters of spherical mirror can be obtained from the edge ray in the system by ray-tracing. As a result, the curvature radius R' is 58.244mm and the aperture D' is larger than 107.836mm. The curvature radius of refractive lens can be solved according to the third-order aberration theory mentioned above. As a result, the curvature radius r_1=-1611.514mm and r_2=-120.901mm.

(3) Design and optimize the partial-compensating system

Put the structural parameters of the partial-compensating system into Zemax. The initial configuration of partial compensating detection system are shown in Table 2.

Table 2. Initial configuration of partial-compensating system

Surf : type	Radius/mm	Thickness/mm	Glass	Semi-Diameter/mm	Conic
Refractive Lens	-1611.514	20	ZF6	45	0
	-120.901	134		45	0
Spheric Mirror	58.244	50	MIRROR	54.651	0
Even Convex Asphere	25.56		MIRROR	8.578	-1.01

Set the curvature radius, the thickness, the distance as optimization variables. Set the Default Merit Function and operating function according to Section 2.2.2 in this paper. Optimize the partial compensating system by running the tools of Automatic Optimization. The optimum design of structure parameters, optical path, residual wave aberrations and interferogram of partial compensating system are shown respectively in Table 3, Figure 3-5.

Table 3. Optimum design of structure parameters partial-compensating system

Surf : type	Radius/mm	Thickness/mm	Glass	Semi-Diameter/mm	Conic
Refractive Lens	Infinity	8.000	ZF6	45	0
	-209.900	274.635		45	0
Spheric Mirror	47.75	38.067	MIRROR	37	0
Even Convex Asphere	25.56		MIRROR	8.578	-1.01

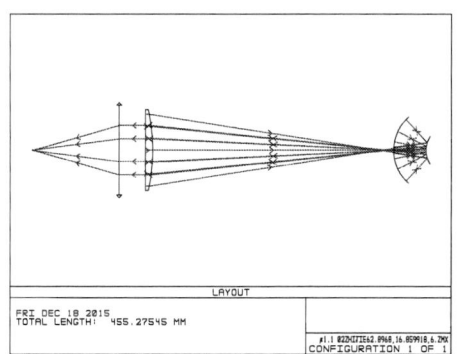

Figure 3. Optical path of partial-compensating system.

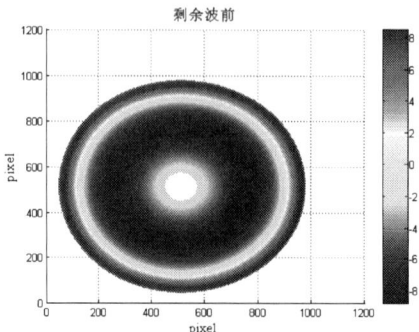

Figure 4. Residual wave aberrations of partial compensating system.

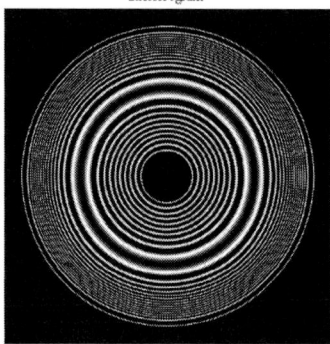

Figure 5. Simulated interferogram of partial compensating system.

If the results are acquired by CCD with 1024*1024 pixels, the maximum slope of residual wavefront is 0.2685λ/pixel, which is much less than 0.45λ/pixel, and the design requirement is well met. The result shows that the design of partial-compensating system is very satisfactory and can be applied to practical measurement.

4. Conclusions

Based on partial compensation method, a new partial compensating system using catadioptric partial-compensator for testing convex aspheric surface is discussed with a high-order convex asphere in details, whose relative aperture D/R_0 is 1:1.5. The design of catadioptric partial-compensator is the key technology in this method. Assuming the spherical aberration coefficient $\Sigma S_1=0$, the initial configuration of partial compensating system is solved based on the third-order aberration theory, and optimized by Zemax optical design software. The result shows that, on the precondition that the interference fringes are detectable, the aberration of convex aspherical surface is well compensated. Compared with null compensating system, the design of catadioptric partial-compensator is much simpler with compact structure, small volume and little optimized parameter.

Acknowledgment
This work is supported by National Natural Science Foundation of China No.51327005.

References
[1] Born M and Wolf E 1999 Principles of optics electromagnetic theory of propagation *Interference and Diffraction of Light* (Cambridge: Cambridge University Press)
[2] Pan Junhua 2004 The design, manufacture and test of the aspherical optical surface (Suzhou: Suzhou University Press)
[3] Zhu Yongjian and Pan Weiqing 2010 Measurement of aspheric surface *Laser & Optoelectronics Progress* **47** 11202

[4] John E and Greivenkamp 2005 Design and calibration of large dynamic range optical metrology systems *SPIE* **6024**

[5] Xie Yi, Chen Qiang and Wu Fan 2008 Concave aspherical surface testing with twin computer-generated holograms *Acta Optical Sinica* **28** 1313~1316

[6] Li Kexin, Yuan Liyin and Hao Peiming 2009 Design of null compensator to concave aspheric mirror with large diameter and large relative aperture *Opt. Instruments* **31** 44~48

[7] Wu Fan and Tang Jianguan 2001 Design and application of dall compensator for null testing of large aperture aspherical surface and convex lens *SPIE* **4451** 368~374

[8] Pan Junhua and Li Xinnan 2000 Compensator for even high order aspheric surfaces *SPIE* **4231** 36~38

[9] Guo Peiji, Yu Jingchi and Shun Xiafei 2002 Null lens design for small aspherical surface with large NA *Optics and Precision Engineering* **10** 518~522

[10] Chen Xu, Liu Weiqi and Kang Yusi 2010 Design and tolerance analysis of Offner compensator *Optics and Precision Engineering* **18** 88~93

[11] Shen Guiping 2007 Design of compensator for large relative apertureaspherical surface *SPIE* **6723**

[12] Liu Huilan, Hao Qun, Zhu Qiudong and Sha Dingguo 2004 Testing an aspheric surface using part-compensating lens *Transactions of Beijing Institute of Technology* **24** 625~628

[13] Yuan Xucang 1995 Optical design (Beijing: Science Publishing Company)

[14] Meng Xiaochen, Hao Qun, Zhu Qiudong and Hu Yao 2011 Influence of Interference Fringe's Spatial Frequency Measurement Accuracy in Digital Moiré Phases Shifting on the Phase Interferometry *Chinese Journal of Lasers* **38**

AOM2015   IOP Publishing
Journal of Physics: Conference Series **680** (2016) 012037   doi:10.1088/1742-6596/680/1/012037

Implement of Digital Moiré technique on DSP for alignment of partial compensation interferometer

Yuhan Tian[1], QunHao[#,2,] YaoHu[3], Shaopu Wang, Tengfei Li, Jingxian Wang

Beijing Key Lab. For Precision Optoelectronic Measurement Instrument and Technology, School of Optoelectronics, Beijing Institute of Technology, Beijing, 100081, China

[1]2120130616@bit.edu.cn, [2]qhao@bit.edu.cn, [3]huy08@bit.edu.cn, [#] Corresponding Author

Abstract. Digital Moiré technique is adopted in partial compensation interferometer (PCI) for high-precision testing of figure error of the aspheric surfaces. The figure error of the measured aspheric is obtained by a series of calculation with the real interferogram and ideal interferograms generated by computer. The dense interference fringes at the exit pupil make it difficult to align the PCI. On the contrary, digital Moiré fringes composed from real and ideal interferograms are sparse and corresponding to the figure error of the measured aspheric, making it easier to align the PCI. Generally, digital Moiré technique is processed on the computer, resulting in slow processing speed and difficult display in real time. Digital Signal Processor (DSP) can be used to implement digital Moiré technique and display digital Moiré fringes in real time with its powerful processing capacity. In this paper, digital Moiré technique is implemented on the TMS320C6455 DSP. The hardware system consists of a DSP module, a CCD camera and a monitor. Finally we experimentally obtain the digital Moiré image, and further analyze how to align the PCI theoretically.

1. Introduction

Digital Moiré technique is used in partial compensation interferometer (PCI) for the testing of aspheric surfaces[1,2].The figure error of the measured aspheric surface is obtained by a series of calculation with the real interferogram and ideal interferograms generated by computer. Usually, we monitor the interference fringes to align the PCI. However, PCI is in non-null configuration, generating dense interference fringes at the exit pupil. It's hard to align the PCI. Digital Moiré fringes composed of real and ideal interferograms are sparse and corresponding to the figure error of the measured aspheric surface, making it easier to align the PCI. Thus, real-time composition and display of the digital Moiré image is of great significance for interferometer alignment and high-precision measurement. The process of obtaining the digital Moiré image is called digital Moiré composition.

Generally, digital Moiré composition is processed on computer. It will take up the operation capacity of the CPU, resulting in slow processing speed. In order to realize display in real time, Digital Signal Processor (DSP) can be used to implement digital Moiré composition and display in high speed with its powerful processing capacity. The TMS320C6455 DSPs developed by Texas Instrument are the highest-performance fixed-point DSP generation in the C6000 DSP platform. With performance of up to 9600 million instructions per second(MIPS) at a 1.2GHz clock rate, the C6455 offers cost-

Content from this work may be used under the terms of the Creative Commons Attribution 3.0 licence. Any further distribution of this work must maintain attribution to the author(s) and the title of the work, journal citation and DOI.
Published under licence by IOP Publishing Ltd

AOM2015   IOP Publishing

Journal of Physics: Conference Series **680** (2016) 012037   doi:10.1088/1742-6596/680/1/012037

effective solutions to high-performance DSP programming[3]. It is an excellent choice for digital Moiré composition algorithm which includes large amount of multiplication and addition operations.

In this paper, we implement digital Moiré technique on TMS320C6455 DSP, experimentally obtain the digital Moiré image, and further analyze how to align the PCI according to the digital Moiré image theoretically.

2. Principle

2.1. *Principle of PCI*

A partial compensation lens of PCI partially compensates the large spherical aberration yield by the measured aspheric surface. Unlike null compensation interferometer, PCI transforms the aspheric wavefront into residual wavefront, not planar nor spherical wavefront. The magnitude of residual wavefront is limited by the spatial resolution of the camera used for capturing the interference fringes[4].

Figure1 shows the principle of PCI. We have two sets of interferometers. Set1 is the computer-simulated virtual PCI consisting of a virtual Fizeau-type interferometer, a partial compensator(PC)[5] and an aspheric surface, the parameters of which are all nominal. We can get virtual interferograms with intensity distribution I_V corresponding to residual wavefront of W_V from Set1. Set2 is the real experimental PCI, which has the same structure with Set1 except that all elements are with errors. We can get real interferograms with intensity distribution I_R.

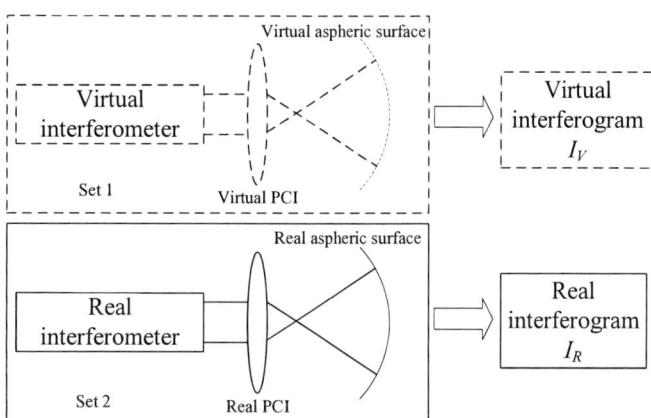

Figure 1. Principle of partial compensation interferometry

2.2. *Principle of Digital Moiré technique*

Two frames of interferograms are obtained with PCI. The real interferogram contains the information of real measured aspheric surface and the virtual interferogram generated by computer contains the information of virtual measured aspheric surface. Digital Moiré technique provides a way calculating the deviation between the standard aspheric surface and measured aspheric surface. The figure error of the measured aspheric surface can be calculated by digital phase-shifting of the Moiré image which is discussed in Refs. [6] and [7]. The principle of digital Moiré technique is as follows.

$$I_V = I_1(x,y)\{1+\cos[2\pi f_V x + \delta_V(x,y)]\} \tag{1}$$

$$I_R = I_2(x,y)\{1+\cos[2\pi f_R x + \delta_R(x,y)]\} \tag{2}$$

Equation(1) describes the ideal interferogram with intensity distribution I_V. I_1 is the direct current component of intensity distribution. $\delta_V(x,y)$ is the residual wavefront after partially compensating. Spatial carrier frequency f_V is added in simulation software for frequency filtering. Equation (2) is similar to equation (1). I_R is the light intensity distribution of the real interferogram. I_2 is the direct

209

current component and $\delta_R(x,y)$ is the real residual wavefront in the real PCI after partially compensating. f_R is real spatial frequency controlled by the tilt of the mirror.

The intensity distribution of Moiré image calculated by multiplication of I_R and I_V is shown in equation (3).

$$
\begin{aligned}
I_{MR}(x, y) &= I_R(x, y)I_V(x, y) \\
&= I_1I_2(x, y) + I_1I_2(x, y)\cos[2\pi f_R x + \delta_R(x, y)] \\
&\quad + I_1I_2(x, y)\cos[2\pi f_V x + \delta_V(x, y)] \\
&\quad + \frac{1}{2}I_1I_2(x, y)\cos[2\pi(f_R + f_V)x + \delta_R(x, y) + \delta_V(x, y)] \\
&\quad + \frac{1}{2}I_1I_2(x, y)\cos[2\pi(f_R - f_V)x + \delta_R(x, y) - \delta_V(x, y)]
\end{aligned}
\tag{3}
$$

Equation (3) shows that the light intensity distribution of digital Moiré image generates two new frequencies. $f_R + f_V$ is the sum frequency and f_R-f_V is the difference frequency. The difference between real and ideal residual wavefront $\delta_R(x,y)$-$\delta_V(x,y)$ is related to the figure error of the measured surface, so the item with difference frequency is the target of the measurement. Because sum frequency leads to dense fringes which is difficult to observe while difference frequency leads to sparse fringes, we need to use low-pass filter to remove the sum frequency and obtain the difference frequency at the same time. We get the final intensity distribution of digital Moiré image shown as in equation (4).

$$
\begin{aligned}
I_{MR}(x, y) &= I_1I_2(x, y) \\
&\quad + \frac{1}{2}I_1I_2(x, y)\cos[2\pi(f_R - f_V)x + \delta_R(x, y) - \delta_V(x, y)] \\
&= a(x, y) + b(x, y)\cos[2\pi(f_R - f_V)x + \delta_R(x, y) - \delta_V(x, y)]
\end{aligned}
\tag{4}
$$

Equation (4) shows that the light intensity distribution of digital Moiré image only contains difference frequency component, also making it an easier way to monitor and align the PCI.

3. Implement of digital Moiré technique on DSP

According to the principle of digital Moiré technique, we can get the algorithm flow chart on DSP as shown in figure 2.

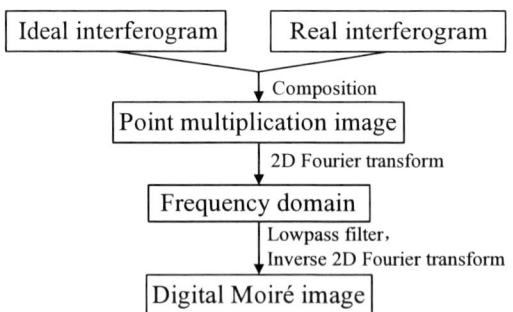

Figure 2. Flow chart of the digital Moiré technique on DSP

All the PCI sets are simulated in Zemax, including the real set2. We transmit the residual wavefront data on the exit pupil of the interferometer from Zemax to DSP to generate two interferograms: one is ideal, and the other is with figure error. The image resolution is 256 pixel × 256 pixel. We set the carrier frequency f_V=0.35λ/pixel to guarantee that there are 90 fringes in the field of view. f_R is set the same as f_V. Figure 3(a) shows the ideal interferogram and figure 3(b) shows the real interferogram with figure error. We multiplied the intensity of each corresponding pixel in these two interferograms to obtain an new image called point multiplication image shown as in figure 3(c).The light intensity distribution of point multiplication image is showed in equation (3).

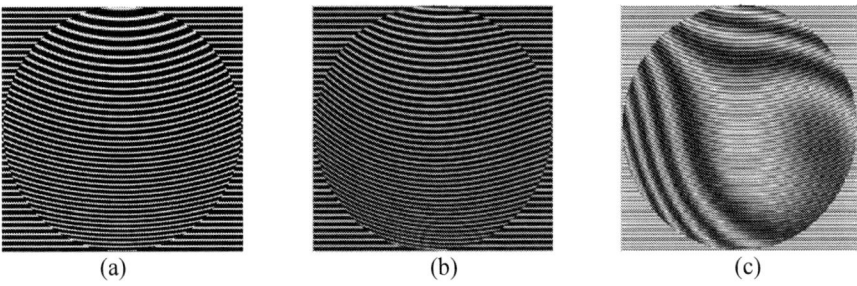

(a)   (b)   (c)

Figure 3. Two interferograms and the point multiplication image

It is required to transform the spatial domain image to frequency domain before low-pass filter. Fast Fourier Transform(FFT) is an efficient way to do this, which uses numerical multiplications and additions. DSP is very suitable for this kind of operation with its special hardware structure and CPU process capacity. We realize one-dimensional FFT algorithm on DSP according to butterfly computation[8,9], and then we extend it to two-dimensional FFT(FFT2). The logarithmic 2D frequency chart of point multiplication image is showed in figure 4. The central point is of zero-frequency so the bright spot in the center is corresponding to the difference frequency component. The two symmetric circles below and above the center are f_R and f_V fundamental frequency component. The sum frequency component is too high and goes out of the range of this chart.

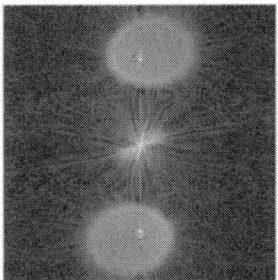

Figure 4. 2D Frequency chart of the point multiplication image

The next step is low-pass filtering, filtering out the high-frequency components. There are many kinds of low-pass filters. The Butterworth low-pass filter(BLPF)[10] is adopted, because it does not have a sharp discontinuity that establishes a clear cutoff between passed and filtered frequencies at the carrier frequency, making a good compromise between effective low-pass filtering and acceptable ringing characteristics if the order of BLPF is low. The transfer function of nth order BLPF with cutoff frequency at a distance D_0 from the origin is defined as

$$H(u,v) = \frac{1}{1+\left[D(u,v)/D_0\right]^{2n}}$$

(5)

where $D(u,v)$ is the distance from point (u,v) to the center of the frequency rectangle. It should be noted that the carrier frequency controlled by tilt should be suitable to separate the fundamental frequency and difference frequency. After low-pass filtering in frequency domain we need to transform the image from the frequency domain back to the spatial domain. Inverse Fast Fourier Transform(IFFT) is implemented and we extend it to two-dimensional IFFT(IFFT2) on DSP. Figure 5 shows the result of applying the BLPF of equation (5) with n=3 and D_0=15 and IFFT2 method.

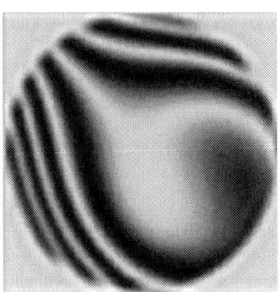

Figure 5. The digital Moiré image

4. The result of experiment and analysis of PCI alignment

4.1. The result of simulation experiment

With various figure error added to the measured surface, such as spherical, defocus, coma, astigmatism, we obtain different kinds of digital Moiré images by implementing digital Moiré technique on DSP. The digital Moiré technique efficiently transforms dense interference fringes into sparse digital Moiré fringes. Along with some surface figure errors changing, digital Moiré images also make the corresponding changes as shown in figure 6. Figure 6(a) is with spherical and defocus. Figure 6(b) is with 0° and 90° astigmatism. Figure 6(c) is with -45° and 45° astigmatism and figure 6(d) is with coma.

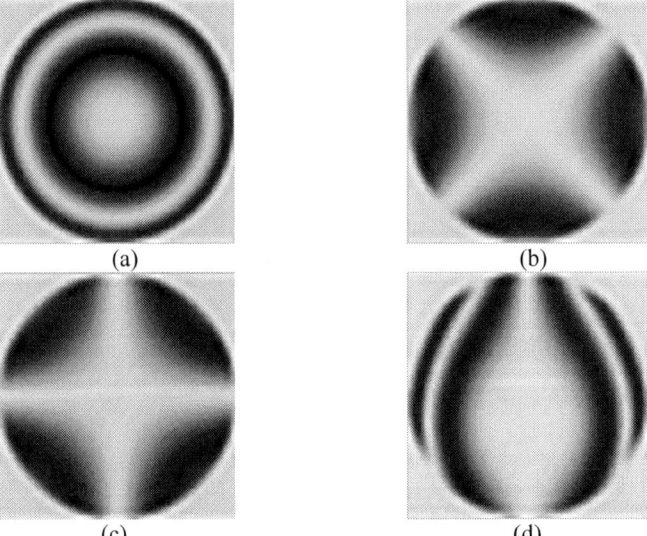

Figure.6. Digital Moiré images with corresponding surface figure error

4.2. Analysis of PCI alignment

Figure7 is the hardware system chart of a PCI system adopting digital Moiré technique for alignment. It includes a real experimental PCI, an image transmission and processing system and a monitor. The system captures real interferograms generated by PCI by the CCD camera and transmits them to the DSP first. Then DSP receives the ideal interferogram image from the PC and implements digital Moiré technique. Finally digital Moiré images are sent to the monitor for observing.

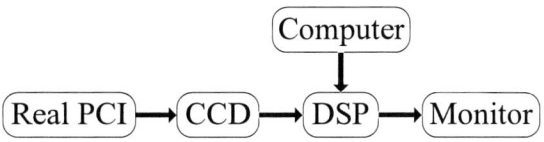

Figure.7 Hardware system chart

It is difficult to calculate the surface figure error directly from the original dense interference fringes even if the interferometer is in good alignment. However, digital Moiré fringes are sparse enough to calculate the figure error of the measured aspheric surface, and make it easier to align the PCI. In the process of alignment of the PCI, defocus and coaxiality are the key influencing factors. We simulate these two factors in Zemax and obtain the digital Moiré images by implementing digital Moiré technique on DSP. Figure 8(a) shows that defocus introduces defocus and spherical aberration and figure 8(b) shows that coaxiality introduces asymmetric aberration in the digital Moiré images.

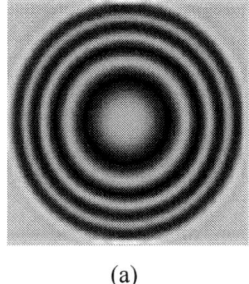 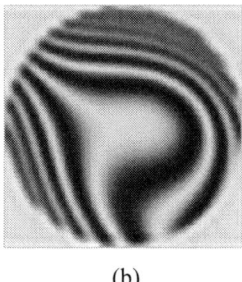

(a)   (b)

Figure.8 Digital Moiré images with defocus and coaxiality

We turn the screw in the optical mount of PCI to adjust defocus and coaxiality of the measured surface by monitoring the change of digital Moiré fringes to align the PCI. The aberration introduced by the misalignment will gradually change when defocus and misalignment are introduced to the PCI. In the process of alignment, when the digital Moiré fringes on the monitor become the sparsest which indicates that the spherical and asymmetric aberrations are the smallest, the PCI is set to the optimal measurement position.

5. Conclusions
We implement digital Moiré technique on the TMS320C6455 DSP in high speed, and finally we experimentally obtain different kinds of digital Moiré images with corresponding surface figure error, and further analyze how to align the PCI theoretically. It is of great significance for the interferometer alignment and high-precision measurement and lays the foundation for the system of real-time composition and display of the digital Moiré image.

Acknowledgment
This work is supported by National Natural Science Foundation of China No.51327005.

References
[1]   Qiudong Zhu and Qun Hao 2007 Aspheric surface test by digital Moiré method *Proc SPIE* **6723**
[2]   Liu Huilan, Hao Qun, Zhu Qiudong and Sha Dingguo 2004 Testing an aspheric surface using part-compensating lens *Transactions of Beijing Institute of Technology* **24** 625~628
[3]   R G Lyons 2005 Understanding Digital Signal Processing 2nd Edition (New Jersey: Prentice Hall)
[4]   H L Liu, Q D Zhu, Q Hao and D G Sha 2003 Design of novel part-compensating lens used in aspheric testing *Proc SPIE* **5253** 480-484.
[5]   Shen Guiping 2007 Design of compensator for large relative apertureaspherical surface *Proc*

SPIE **6723.**

[6] R Gappinger and J.Greivenkamp 2004 Iterative reverse optimization procedure for calibration of aspheric wave-front measurements on a nonnull interferometer *Appl Opt* **43** 5152-5161.

[7] Q Hao and Q D Zhu 2008 Aspheric Surface Testing Using a Partial Compensation Lens *Key Engineering Materials* **381** 263-266.

[8] Candes E, Demanet L and Ying L 2009 A Fast Butterfly Algorithm for the Computation of Fourier Integral Operators *Siam Journal on Multiscale Modeling & Simulation* **7** 1727-1750.

[9] Tanno K, Taketa T and Horiguchi S 1995 Parallel FFT algorithms using radix 4 butterfly computation on an eight-neighbor processor array *Parallel Computing* **21** 121-136.

[10] Rafael C Gonzalez and Richard E Woods 2002 Digital Image Processing (New Jersey: Prentice-Hall).

System of Thermal Micro/Nano Printing and its Application in Metallic Glass

Y. Xu[1], X. L. Hu[1], L. B. Sun[1], L. S. Wang[3], S. Q. Ding[2], J. Liu[2], J. Z. Jiang[2] and D. X. Zhang[1]

[1] State Key Laboratory of Modern Optical Instrumentation, Zhejiang University, Hangzhou, 310027, People's Republic of China
[2] International Center for New-Structured Materials, State Key Laboratory of Silicon Materials and Department of Materials Science and Engineering, Zhejiang University, Hangzhou 310027, People's Republic of China
[3] Shanghai Institute of Applied Physics, Chinese Academy of Sciences, Shanghai 201204, People's Republic of China List the author names here

Abstract. A micro/nano thermal printing system was developed in this paper. The system has the characteristics of high resolution, large imprinting areas, convenient operation and low cost. Some experiments on metallic glass (La-Co-Al) were carried out by the system. The results indicated that this system has the elegant performance and the metallic glass is one of the best materials to fabricate the microstructures.

1. Introduction:

With the development of nanotechnology, the demand for processing microstructures is great needed. Nowadays there are several methods dealing with it. Among them, LIGA [1] technology under Synchrotron Radiation Light source is an effective way to get high aspect ratio microstructure [2] but with low precision (large than 1um) and high cost-effectiveness [3]. Electron Beam Lithography (EBL) [4] or Focused-Ion-Beam (FIB) [5, 6] is another one which takes over the disadvantages of LIGA by increasing the precision [7-10] (less than hundred nanometers even tens nanometers) using electron or ion instead of light, but still with high cost as well as LIGA. A relatively cheaper method with inexpensive machine and simple process is Ultra-Violet lithography, whereas its precision is low approximately micrometer or hundred nanometers. Nevertheless, all of the methods mentioned above can't yield large area microstructure. So, it is necessary to develop new methods which can convenient manufacture micro/nano structures with large area, high precision and low cost. Here we presented a micro/nano thermal printing method to address these issues. The basic concept and controlling system of the micro/nano thermal printing were introduced, and some experiments were carried out with the thin metallic glasses films.

2. Working principle and operating methods of the system

This system was automatic controlled by computer. With inputting control parameters at first, the system can implement the fabricating procedure automatically. The working principle and operating method is shown in Figure 1.

Content from this work may be used under the terms of the Creative Commons Attribution 3.0 licence. Any further distribution of this work must maintain attribution to the author(s) and the title of the work, journal citation and DOI.
Published under licence by IOP Publishing Ltd

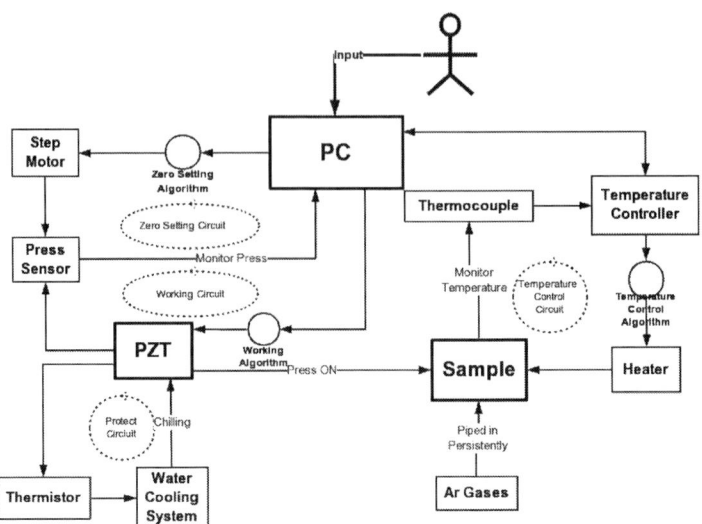

Figure 1. A schematic diagram of the system. It mainly includes: temperature control circuit, protect circuit, zero circuit, circuit and anti-oxidation air source.

This system was consisting of five parts, temperature control circuit, protect circuit, zero circuit, circuit and anti-oxidation air source, respectively. Some parameters need to be input before system working. And then the sample needs to be heated to a value as preset. During the heating process, protect circuit must be open to protect piezoelectric ceramics (PZT) to a higher temperature. Once the sample is heated to the appointed value, the zero-circuit starts to work. The zero-circuit adjusts the system to zero-position where the press on sample happens to result in sample's deformation. Next, the system begins to work with arithmetic. All the follow-up work is automatically finished by the system.

3. System Design

3.1. Mechanical design

A vertical pressure transmission structure was applied in the machinery, which exerts a pressure force on the system from top to bottom. Compared with horizontal pressure transmission structure, the sample can be avoided to be asymmetrical due to the obliquity of the pressure. The mechanical design of the system was shown in Fig. 2. It was mainly composed of 6 parts. They were stepping motor, loading block, pressure sensor, PZT, pressure head and sample stage in turns. Component 1 was the stepping motor, playing a key role in adjusting system's position to a proper altitude especially during the zero circuit. The loading block (component 2) made a difference to adjust the system's operating distance as well as helping easily adjust zero position of the pressure. Due to the different sensitivity of the pressure sensor in its measuring range, the loading block assists to adjust the pressure sensor to work in its most sensitive range.

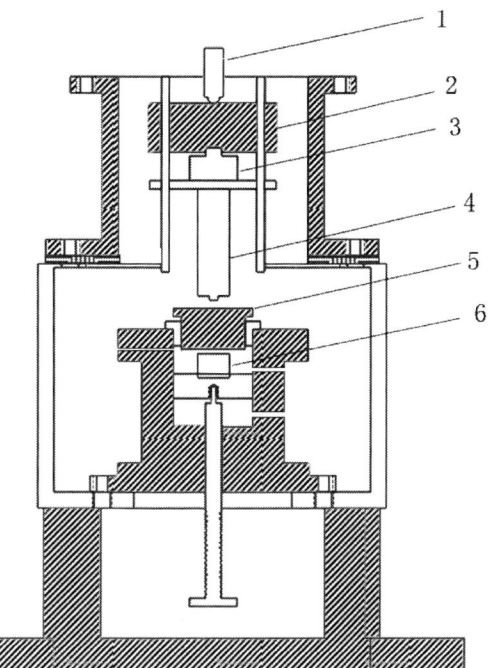

Figure 2 Pressure Transmission Structure.

Component 3 was pressure sensor, playing a part as monitor and feedback source. It monitored the press of system vertically meanwhile gave feedback to the computer across all periods especially in zero circuit and work circuit. The key part of the system was PZT component 4 in Fig.2, acting as the power source of the system by pressurizing high pressure on the sample. The pressure was operated by computer arithmetic. Component 5 was a pressure head, serving as transmitters and balancer of pressure between PZT and sample. Component 6 was sample stage.

3.2. Control system

The control part is the kernel of the system. It helps the individual section working together systematically. The control principle was shown in Fig. 3. The five parts in Fig.3 from top to bottom were corresponding to zero circuit, working circuit, protection circuit, gas protection and temperature control circuit in Fig.1, respectively. Zero circuit including stepping motor, pressure sensor and computer, was on the top of the system far away from the sample. By the way of zero algorithm, the stepping motor forwards and backwards until reach zero position.

The working circuit was blew zero circuit by sharing the pressure sensor with each other. The PZT forwards and backwards with the method of algorithm as the system running. Protected circuit situated in the head of PZT where the temperature can be reduced directly. Due to the protected circuit, the PZT can remain safe temperature, otherwise, the PZT can't work regularly. In practice, the protected circuit was continuous working until the end of the process. Gas protection was in the heat chamber to prevent the sample from oxygenizing and continuous working as well as protected circuit. Specifically, the noble gas is pumped in at low position while the air is pumped out at high position. Compared to vacuum chamber, it was more convenient and effective way with low cost.

The temperature circuit situating at the bottom of the system was to uniformly heat the sample to its softening temperature. An integral PID was used in temperature control algorithm where the current temperature, direct temperature and total power were taken into account to work out the current power.

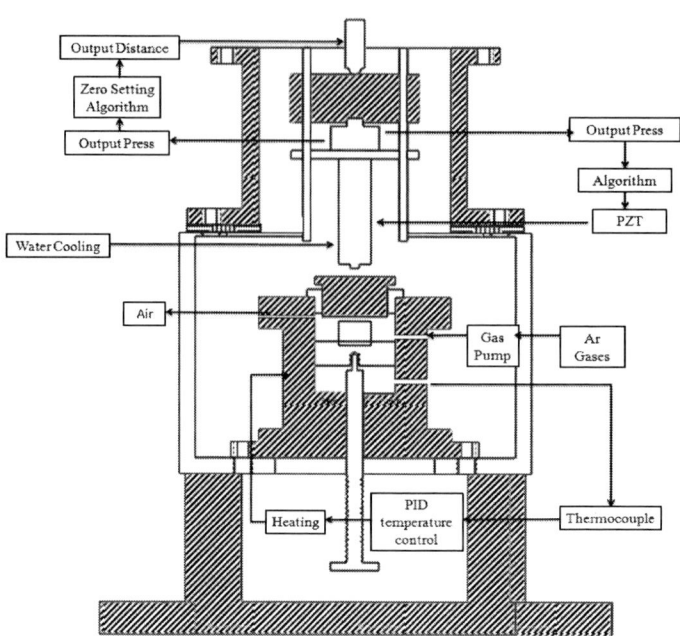

Figure 3 Control System.

2.3 software and algorithm

The system was controlled by the computer where several parameters were input in advance. Compared with traditional instrument, more parameters can be monitored to established database for future research and data analytics. Algorithms were displayed in Figure 4-1 and Figure 4-2.

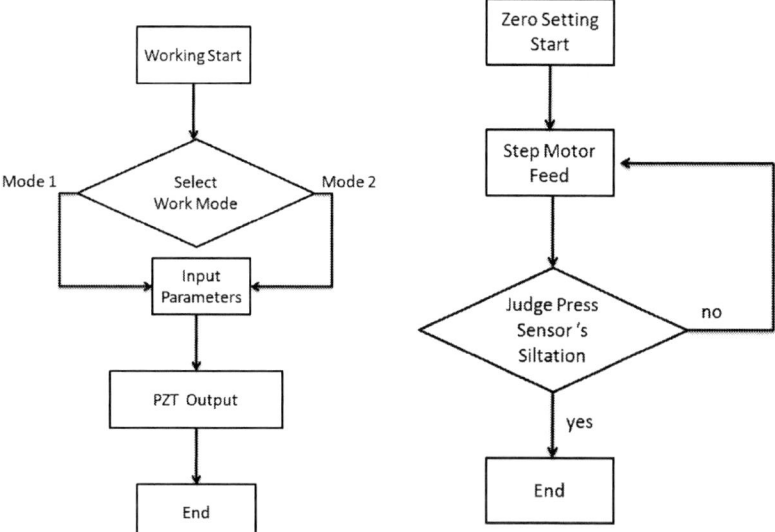

Figure 4-1 Zero Setting Algorithms.   **Figure 4-2** Working Algorithm.

Figure 4-1 is zero algorithm. Although the zero position refers to a condition that the pressure on sample happens to result in sample's deformation, it is judged by analyzing the value of pressure sensor in reality. Specifically, assuming the pressure before zero position changes slowly rather than abruptly, it is judged by pressure's first-order derivative and second-order derivative. In practice, the computer starts to monitor the pressure's first-order derivative until it is abruptly changing, and then to monitor the pressure's second-order derivative. Usually, the pressure's second-order derivative is slowly decreasing until the value turned to be a constant, i.e., the zero setting is finished.

Figure 4-2 is working algorithm. There are two work algorithms classified by displacement and pressure. One method was to control the voltage to make the PZT forwarding nearly linear. However, due to the nonlinearity and hysteresis effect of PZT(as shown in Fig. 4-3), the distance cannot be directly controlled by just linear increasing the voltage of PZT. So, linear interpolating was used in rising curve in Fig. 4-3 to get sufficient points and then filtrating all these points to select linear change points. Another way was to control pressure to put the PZT forwarding by a specific pressure curve. Specifically, if the pressure value is superior to the curve, the PZT will forward smaller, stop forwarding or backwards until the pressure nearly approaches the curve; if the pressure value is inferior, the PZT will forward larger to approaches the curve. Practically, mode two is more in common use, because pressure makes direct and large contribution to the final result.

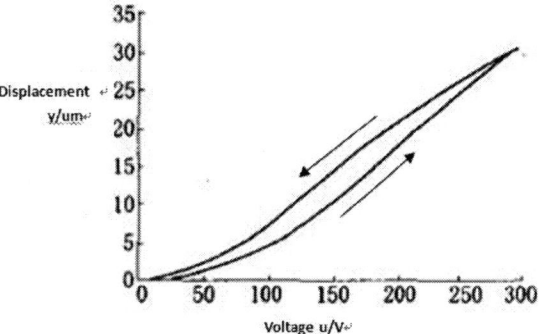

Figure 4-3 PZT hysteresis curve

4. Experiments and Applications in Metallic Glass Film

Some experiments were carried out by using this printing system with Thin-Film of metallic glasses [11, 12] (TFMGs). TFMGs are characterized by an absence of size effect, high strength and high elastic limit due to their amorphous nature. So, these materials are considered to be ideal candidates for micro electromechanical systems [13-15] and microstructures. However, till now seldom reports have been done about the microstructures in TFMGS. By this printing system, we fabricate microstructure on TFMGs with different temperature (Table 1).

Table 1 Different result under different temperature

Temperature/°C	Results
<125	No pattern or shallow pattern
125-130	Good pattern
130-135	Probabilistic good pattern or film breakup
>135	Film breakup

Firstly, some moulds were fabricated on Si surface by EBL or FIB. As shown in Fig. 5(a) and 5(c), the patterns of the moulds include grating, word, digit, and arrows. The thickness of mould is approximately 2 μm. And then TFMGs (MgZnCa) samples were fabricated with a 2μm thickness, which were placed just under the mould on the sample stage in order to be printed automatically with control parameters which were inputted in advance.   The microscope pictures shown in Fig. 5(a) and 5(c) were patterns of models, and 5(b) and 5(d) were the imprinted microstructures of TFMGs with the heated temperature among 125 to 130 °C (i.e., glass transition temperature of the TFMGs), while as the temperature out of this range, the results shown in Fig. 5(e) and Fig. 5(f) were not so good as (b) and (d).

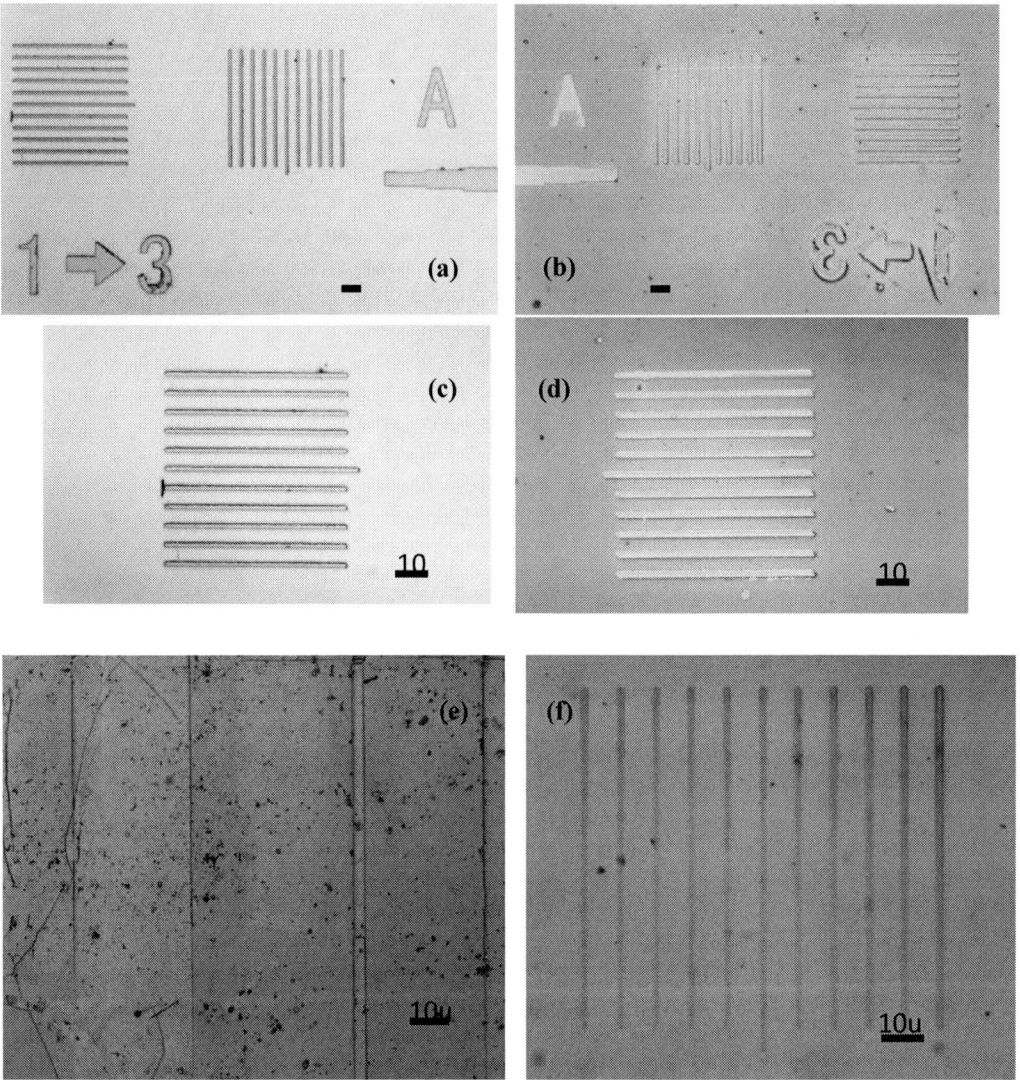

Figure 5. Microscope pictures of mould (a), (c) and imprinted microstructures of TFMGs (b), (d)

5. Conclusions

This paper presented how to develop a micro/nano thermal printing system with high precision and efficiency. Compared with traditional printing system, the system can avoid to oxidation by the gas protected circuit. Some experiments were conducted with this system. The results verified the feasibility of the system and can be used to investigate the characteristics of the TFMGs. In future, some adjustment to the system will be taken to improve the precision and stability, and more applications with this system will be explored on TFMGs.

Acknowledgements

Financial supports from National Natural Science Foundation of China (Grant No. U1432110), National Key Basic Research Program of China (2012CB825700), China Scholarship Council (Grant No. 201400260166) and the support of Soft-X Ray Interference Lithography Beamline (BL08U1B) in SSRF for sample preparation are gratefully acknowledged.

References

[1] Malek C K 2004 Saile V. Applications of LIGA technology to precision manufacturing of high-aspect-ratio micro-components and -systems a review[J]. (Microelectronics Journal, , volume 35(2):131-143(13))

[2] Ehrfeld W, Hessel V, L02we H, et al. 1999 Materials of LIGA technology[J]. (Microsystem Technologies, 5(3):105-112)

[3] Williams J D, Wang W. 2004 Microfabrication of an electromagnetic power relay using SU-8 based UV-LIGA technology[J]. (Microsystem Technologies, 10(10):699-705.)

[4] Vieu C, Carcenac F, Pépin A, et al. 2000 Electron beam lithography: resolution limits and applications[J]. (Applied Surface Science, 164:111–117.)

[5] Reyntjens S, Puers R. 2001 A review of focused ion beam applications in microsystem technology[J] (Journal of Micromechanics & Microengineering, 11(4):287-300)

[6] Matsui S, Mori K, Saigo K, et al. 1986 Lithographic approach for 100 nm fabrication by focused ion beam[J]. (Journal of Vacuum Science & Technology B, 4(4):845 - 849.)

[7] Craighead H G, Howard R E, Jackel L D, et al. 1983 10- nm linewidth electron beam lithography on GaAs[J]. (Applied Physics Letters, 42(1):38 - 40.)

[8] Mankiewich P M, Craighead H G, Harrison T R, et al. 1984 High resolution electron beam lithography on CaF2[J]. (Applied Physics Letters, 44(4):468 - 469.)

[9] Matsui S, Kojima Y, Ochiai Y, et al. 1991 High- resolution focused ion beam lithography[J]. (Applied Physics Letters, 9(10):427–430.)

[10] Altun A O, Jeong J H, Rha J J, et al. 2007 Boron nitride stamp for ultra-violet nanoimprinting lithography fabricated by focused ion beam lithography.[J]. (Nanotechnology, 18(46):5721-5723.)

[11] Jeong H W, Hata S, Shimokohbe A. 2003 Microforming of three-dimensional microstructures from thin-film metallic glass[J]. (Journal of Microelectromechanical Systems, 12(1):42 - 52.)

[12] Liu Y, Hata S, Wada K, et al. 2001 Thermal, Mechanical and Electrical Properties of Pd-Based Thin-Film Metallic Glass[J]. (Japanese Journal of Applied Physics, 40(9A):5382-5388.)

[13] Hata S, Goto J, Sato K, et al. 2000 Fabrication of Micro Structures using Thin Film Metallic Glass : Fabrication of Thin Film Metallic Glass and Micro-Forming using the Supercooled Liquid State[J]. (Journal of the Japan Society of Precision Engineering, 66(1):96-101.)

[14] Liu Y, Hata S, Wada K, et al. 2001 Thermal, Mechanical and Electrical Properties of Pd-Based Thin-Film Metallic Glass : Surfaces, Interfaces, and Films[J]. (Japanese Journal of Applied Physics.pt Regular Papers & Short Notes, 40.)

[15] Jeong H W, Hata S, Shimokohbe A. 2002 Micro-forming of thin film metallic glass by local laser heating[J]. (Proceedings of the IEEE International Conference on Micro Electro Mechanical Systems:372 - 375.)

Ultra-broad band absorber made by tungsten and aluminium

Wei Wang, Ding Zhao, Qiang Li, Min Qiu

State Key Laboratory of Modern Optical Instrumentation, College of Optical Science and Engieering, Zhejiang University, Hangzhou 310027, China

minqiu@zju.edu.cn

Abstract. A broadband absorber comprising tungsten cubic arrays, a alumina layer and a tungsten film, is numerically and experimentally investigated, which exhibits near-unity absorption of visible and near-infrared light from 400 nm to 1150 nm. Benefiting from high melting points of tungsten and alumina, this device has great application potential in solar cells and thermal emission.

1. Introduction

As a critical component, plasmonic absorbers have already attracted plenty of attention and shown great promise for devising solar cells[1], microbolometers[2], biosensors[3], and photodetectors[4,5,6,7,]. Attributing to Kirchhoff's law of thermal radiation, they can also be utilized as thermal emitters [8]. For instance, the emitter serves as a key component to radiate thermal radiation to photovoltaic (PV) cells in thermophotovoltaic (TPV) systems. To achieve high efficient thermoelectric conversion, emitter's thermal radiation spectrum should ideally coincide with the PV cell's response spectrum, which mostly ranges from visible to infrared frequency. Various broadband absorbers of exotic geometry have been proposed during recent years, such as complicated trapezoid [9], truncated spherical voids [10] and tapered pyramid structures [11]. Due to excellent properties and simple fabrication, metal/insulator/metal (MIM) architecture absorbers have become the focus of much attention. In this paper, we present a three-layer MIM plasmonic absorber, which works at the visible and near-infrared region. Broadband absorption from 400 nm to 1150 nm and high operating temperature of this device could match requirements for TPV application.

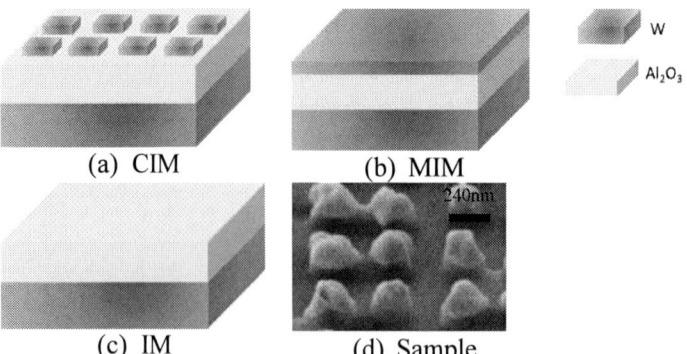

Figure 1. Geometry of CIM, MIM and IM structure. (d) top view of CIM sample.

Content from this work may be used under the terms of the Creative Commons Attribution 3.0 licence. Any further distribution of this work must maintain attribution to the author(s) and the title of the work, journal citation and DOI.

Published under licence by IOP Publishing Ltd

2. Simulation and experimental results

2.1. Simulation result

Figure 1(a) shows the proposed structure in which Al2O3 dielectric layer sandwiched by a tungsten layer and tungsten blocks. In our simulation, cubic arrays are consisting of tungsten particles with width 225nm, length 225nm, and height 225nm. Distance between each unit cell is 275nm. Height of the Al2O3 layer is 250nm, and tungsten layer below it is 140nm. Here tungsten is thick enough to prevent light from transmitting. Therefor, absorption can be defined by reflection subtracted from one.

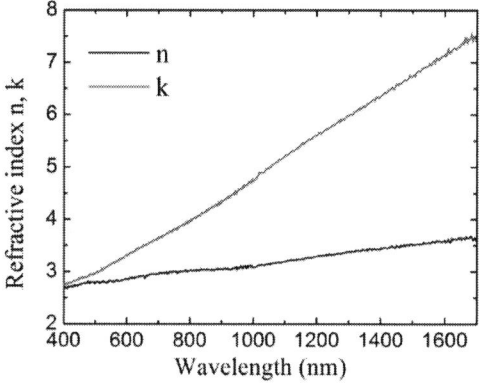

Figure 2. Refractive index of tungsten film ensured by ellipsometer.

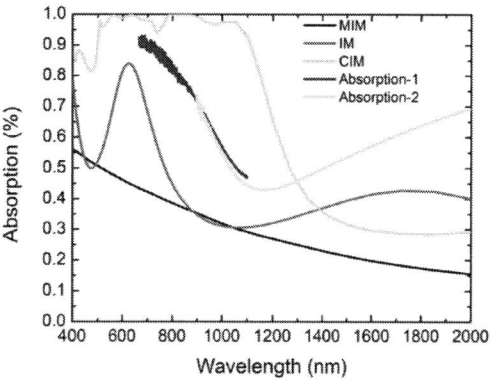

Figure 3. Comparison of absorption property for MIM, IM and CIM structure.

In this paper, tungsten film is fabricated by magnetron sputtering. Variable angle spectroscopic ellipsometry is used to extract the complex optical constants of this film. Refractive index results are shown in figure 2.

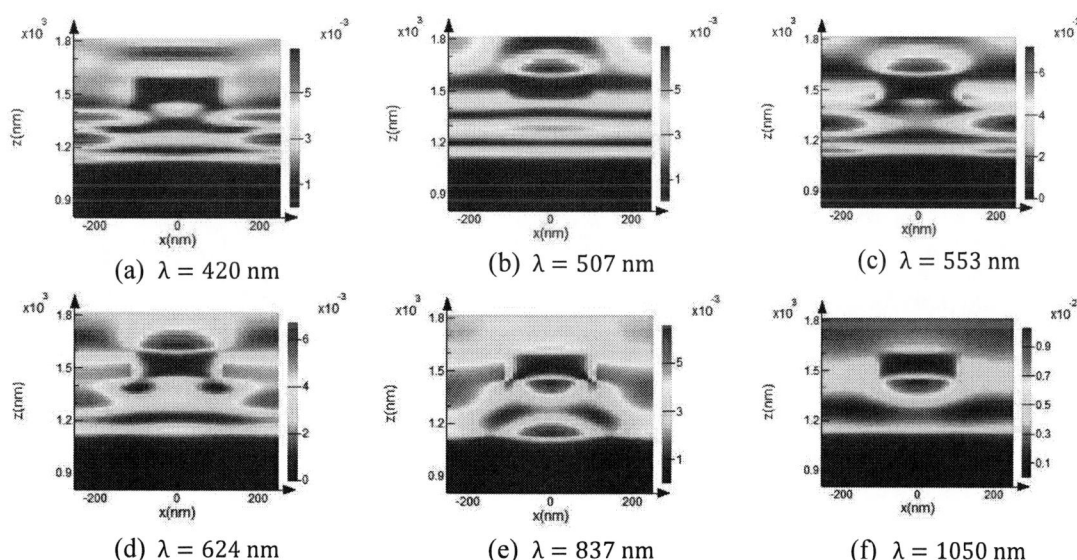

(a) λ = 420 nm (b) λ = 507 nm (c) λ = 553 nm

(d) λ = 624 nm (e) λ = 837 nm (f) λ = 1050 nm

Figure 4. Magnetic field profiles at
λ = 420 nm, 507 nm, 553 nm, 624 nm, 837 nm and 1053 nm.

During the simulation process, periodic condition is set as boundary condition. Absorption property of metal-insulator-metal (figure 1b) and insulator-metal (figure 1c) geometry are also simulated as references for cubic-insulator-metal structure (figure 1a). These structure's absorption abilities are shown in Figure 3. CIM structure shows nearly perfect absorption property and its bandwidth range from 400nm to 1150nm which covers visible and near-infrared regions. Compared with IM and MIM geometry's absorption ability, additional tungsten cubic array strongly enhanced its absorption property.

Such high and wide absorption of this device is attributed to the strongly distributed electromagnetic field. Figure 4 shows the magnetic field profiles at the spectrum where resonance appear.

2.2. Experimental result

Tungsten absorber sample is etched by Focused-ion-beam (FIB). Inset in figure 4 shows SEM image of that sample. Line named absorption-1 and absorption-2 in figure 3 both are measured by FTIR at different detector mode. As shown in figure 4, blue line (absorption-1) gradually increases to reach 90% when wavelength decrease to be 620nm.

Experimental absorption result is narrow than numerical result. This is because when FIB etch top tungsten layer, part of Al2O3 layer could be etched out together. This lead to electromagnetic wave couldn't be confined perfectly in the dielectric layer. Thus part of spectrum couldn't be absorbed by this device.

3. Conclusion

In conclusion, we have simulated a nearly perfect broadband absorber spanning from visible to near-infrared spectrum. The experimental result partially agree with theoretical predictions. There still have promotion prospect for broader absorption width which using EBL method which can circumvent the deeper etching deficiency. Furthermore tungsten and aluminium oxide owning high melting point. Such material used in this device enable this absorber would endure more thermal energy transformed from external optical stimulus in applications such as solar thermophotovoltaics (STPV).

References

[1]   N. C. Panoiu and R. M. Osgood 2007 Opt. Lett. **32** 2825
[2]   P. L. Richards 1994 J. Appl. Phys. **76** 1
[3]   B. Sepulveda, L.G. Carrascossa, D. Regarots, M. A. Otte, D. Farina, and L.M. Lechuga 2009 Proc. SPIE **7397** 1
[4]   Y. Ahn, J. Dunning, and J. Park 2005 Nano Lett. **5** 1367
[5]   Y. Gu, E.-S. Kwak, J. L. Lensch, J. E. Allen, T. W. Odom, and L. J. Lauhon 2005 Appl. Phys. Lett. **87** 43111
[6]   O. Hayden, R. Agarwal, and C. M. Lieber 2006 Nat. Mater. **5** 352
[7]   J. Rosenberg, R. V. Shenoi, T. E. Vandervelde, S. Krishna, and O. Painter 2009 Appl. Phys. Lett. **95** 161101
[8]   J. J. Greffet, R. Carminati, K. Joulain, J. P. Mulet, S. P. Mainguy, and Y. Chen 2002 Nature **416** 61
[9]   V. E. Ferry, K. Aydin, R. M. Briggs, and H. A. Atwater, 2011 Nat. Commun. **2** 517
[10]   M. Wang, CG Hu, MB Pu, C. Huang, ZY Zhao, Q. Feng, and XG Luo 2011 Opt. Express **19** 20642
[11]   E. Rephaeli, and S. Fan 2008 Appl. Phys. Lett. **92** 211107

Institute of Physics
Dirac House, Temple Back
Bristol BS1 6BE UK

ISSN: 1742-6588
ISBN 978-1-5108-2090-6

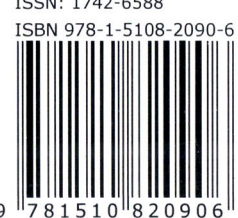

9 781510 820906